Lecture Notes in Networks and Systems

Volume 214

The series "Lecture Notes in Networks and Systems" publishes the latest developments in Networks and Systems—quickly, informally and with high quality. Original research reported in proceedings and post-proceedings represents the core of LNNS.

Volumes published in LNNS embrace all aspects and subfields of, as well as new challenges in, Networks and Systems.

The series contains proceedings and edited volumes in systems and networks, spanning the areas of Cyber-Physical Systems, Autonomous Systems, Sensor Networks, Control Systems, Energy Systems, Automotive Systems, Biological Systems, Vehicular Networking and Connected Vehicles, Aerospace Systems, Automation, Manufacturing, Smart Grids, Nonlinear Systems, Power Systems, Robotics, Social Systems, Economic Systems and other. Of particular value to both the contributors and the readership are the short publication timeframe and the world-wide distribution and exposure which enable both a wide and rapid dissemination of research output.

The series covers the theory, applications, and perspectives on the state of the art and future developments relevant to systems and networks, decision making, control, complex processes and related areas, as embedded in the fields of interdisciplinary and applied sciences, engineering, computer science, physics, economics, social, and life sciences, as well as the paradigms and methodologies behind them.

Indexed by SCOPUS, INSPEC, WTI Frankfurt eG, zbMATH, SCImago.

All books published in the series are submitted for consideration in Web of Science.

More information about this series at http://www.springer.com/series/15179

Pankaj Srivastava · S. S. Thakur ·
Georgia Irina Oros · Ali A. AlJarrah ·
Vichian Laohakosol
Editors

Mathematical, Computational Intelligence and Engineering Approaches for Tourism, Agriculture and Healthcare

Springer

Editors
Pankaj Srivastava
General Secretary
FATER Academy of India
Department of Mathematics
MNNIT Allahabad
Prayagraj, Uttar Pradesh, India

Georgia Irina Oros
Department of Mathematics and Computer Science
University of Oradea
Oradea, Romania

Vichian Laohakosol
Department of Mathematics
Kasetsart University
Bangkok, Thailand

S. S. Thakur
Jabalpur Engineering College
Jabalpur, Madhya Pradesh, India

Ali A. AlJarrah
Sur University College
Sur, Oman

ISSN 2367-3370 ISSN 2367-3389 (electronic)
Lecture Notes in Networks and Systems
ISBN 978-981-16-3806-0 ISBN 978-981-16-3807-7 (eBook)
https://doi.org/10.1007/978-981-16-3807-7

Mathematics Subject Classification: 93A30, 68T35, 62C86, 90B50, 97R40, 62P30, 97M50, 05c15

This Springer imprint is published by the registered company Springer Nature Singapore Pte Ltd.
The registered company address is: 152 Beach Road, #21-01/04 Gateway East, Singapore 189721, Singapore

Preface

With the zest to boost research and development at the international platform, the FATER Academy of India has organized the 17th International Conference on "Engineering and Mathematical Computational Intelligence" from 21–23 December 2019 at Jabalpur Engineering College, Madhya Pradesh, India. The collaborating partners involved the FATER Academy of India, Jabalpur Engineering College, Walchand Engineering College, and TEQIP-III sponsored the conference.

It gives us immense pleasure to write this message as editors of this proceeding. These proceedings include research papers provided by enthusiastic students, practicing researchers, industry personalities, and academicians. The idea behind the publication of this conference proceeding was to help and motivate researchers to carry forward their work to the next level.

The multidisciplinary forum was successfully executed and delivered the golden value of being multidisciplinary, especially involving mathematics, computation and engineering, tourism, healthcare, and agriculture. While mathematics serves as an essential foundation and takes on a natural complementary disciplinary role to support and drive the development of engineering and computational sophistication, the engineering and computational professions apply mathematics knowledge to develop ways to utilize the materials and forces of nature economically for the benefit of humanity. Thus, together, they form the backbone of a nation's economy.

Readers could be interested in some chapters discussing many aspects of computational intelligence, especially in tourism, medical diagnosis, agriculture, and decision sciences.

Studies related to tourism are presented in Chap.1 by using notions of fuzzy sets theory focused on the design and development of an intelligent tourist destination image decision information system reflecting the influence of criteria of selection of tourist destination based on destination image features. The system has an added advantage to redesign the tourist destination selection plan as per destination image criteria. Also, Chap.3 reflects research demonstrating the utility of neural network simulation to support structural equation modeling (SEM) analysis. Applying the data from the hotel industry in Aceh, Indonesia, this research confirms the mediating role of organizational learning. A study fills the gap in topics relating to spiritual and pilgrimage tourism in the "tourist experience" domain based on a stimulus-organism-response (S-O-R) framework and Chap. 25 highlights the spiritual experience as the stimulus domain, demonstrating the S-O-R paradigm's generic validity.

For the medical domain, especially heart-related issues, Chap. 8 is focused on the design and development of a soft computing diagnostic intelligent system that classifies ECG beats in different phases and enables the identification of the status of cardiac health normalities and abnormalities as per available ECG graphs showing that fuzzy tools can be very fruitful and applicable tools in medical science. Chapters 20 and 28 are also reflecting the heart issues. The agile software development methodology is used in association with fuzzy logic for evaluating technical artifacts during the development of the heart disease system, the experimentation being done on the Cleveland Heart disease data sets present in the UC Irvine repository. The simulation study is done by a finite element method (FEM) based on Ansoft's HFSS software. Performance prediction of technical artifacts in heart disease diagnosis is investigated, and computational techniques for investigation of the severity of heart diseases technically by using technical artifacts are proposed; the work is helpful in diagnosing with optimized risk factors in a low resource setting.

As far as agriculture is concerned, Chap. 9 proposes an algorithm for efficient utilization of water and for checking any wastage/leakage of it in the field in order to help the farmers for efficient and optimal utilization of water resources in any field.

For sharpening the decision-making process, Chap.15 highlight issues based on the design and development of an intelligent information system (IIS) that runs on the knowledge base of the hesitant fuzzy weighting linguistic term set (HFWLTS) for computing with words with reference to decision-making phenomena is presented. The concept of HFWLTS introduced makes an IIS with flexibility and more realistic as per human nature. Chapter 17 presents different soft set-based NPR algorithms for information systems that give optimal solutions for big data focusing on the dimensionality reduction approach of data mining using soft set theory, which can be applied to the machine learning field. Thus normal dimensionality reduction of soft set (NDRSS) algorithm is proposed for real-time data to give a better optimal solution than existing soft set algorithms.Chap.26 shares a correlational analysis of the factors influencing the transfer of research and development to entrepreneurs.

Mathematicians could be interested in some chapters discussing classical aspects of the mathematical domain. Several differential subordinations are studied by using a new differential-integral operator and a new class of univalent functions defined with it, and from this study, a condition for univalence is derived. A convolution involving a summation over the greatest common divisor of two integers, which extends the classical Dirichlet convolution, is investigated, and basic properties are derived. Applying these results to a periodic function yields a Fourier series expansion, referred to as its finite Fourier series. Several recent results and applications related to the Fourier series of the greatest common divisor function are derived. Some standard fixed point theorems in generalized fuzzy metric spaces for weakly compatible with property are proved, and the role of the "common limit range" property is foreground to find out the presence of a common fixed point in generalized fuzzy metric spaces.

Two classes of bi-univalent-analytic functions are introduced, and the initial coefficient bounds and Fekete–Szegő problem are obtained by using Chebyshev polynomials expansions. By a 1937 method of K. Mahler, it is proved by using the Thue–Siegel–Roth theorem that a certain decimal expansion is a transcendental number but is not a Liouville number. The error estimate in terms of higher modulus of smoothness is obtained for hybrid type operators defined by combining the weight of Szasz and Baskakov operators, highlighting the idea that by taking a linear combination of our hybrid operators rate of convergence is improved. Several generalized forms of continuous maps such as totally -continuous maps, totally -continuous, totally -semi-continuous maps, and totally -pre-continuous maps in an ideal generalized binary topological spaces are introduced, and various relationships of these maps are investigated by making use of some examples. General fractional calculus operators are studied, which would widely be applicable to several classes of higher transcendental functions. Nonstandard analysis of almost periodicity for the "traditional" functions and then for the operators is presented.

Related to computation and engineering topics, the notion of totally continuous function in Čech rough closure space and nano soft topological space is developed, and the relationships between these new classes of Čech rough totally continuous function and nano soft totally continuous function with the existing composition function through computing technique are studied. These results are expected to enhance the study of developments of information systems through nano soft topological spaces and will help researchers to establish real-life applications through nano topology. A chapter discusses the functioning of ideals in topological space and also applies various topological properties to ideal N-topology, proving that many of N-topological concepts hold good for the ideal N-topology too, a computer program along with its algorithm being derived from establishing the relationship between the resultant notions of weak and strong forms of local functions. All possible graphs are studied by an author, for coloring, with "almost twin" perfect outer domination numbers $2n-5$ and $2n-6$ for $n \geq 4$. The new concept of quotient supra topology is introduced, and its properties are studied. Using this quotient supra topology, a quotient supra space is constructed in terms of quotient set originating the new notion of vivid supra topology and providing its existence and uniqueness theorem. Moreover, the universal property of vivid supra topology is established, and a new approach of representation of concept hierarchy based on quotient supra space is given.

A chapter gives the classification and evaluation of the quality of service for wireless networks based on the K-nearest neighbor algorithm. The proposed algorithm may be utilized in the network design and site optimization by telecom network operators. Management of high volume clashes during clash detection is investigated.

In this book, certain chapters are focused on environmental-related issues, and readers will find them useful in further investigations. A study of nanoscale zeolitic imidazole framework-8 in the removal of humic acid from wastewater by adsorption is presented. A circular ring slot patch antenna for hyperthermia applications is presented at 2.45 GHz.

The development of dust repellent coating for a solar panel is analyzed, and an evaluation of energy efficiency is done. Also, a study of nanoscale zeolitic imidazole framework-8 in the removal of humic acid from wastewater by adsorption can be read.

On behalf of the editorial board, we extend our warmest thanks to the authors for their interest, enthusiasm, and timely submission of full-length research papers and reviewers for their quick responses with valuable suggestions to improve the articles. We anticipate that the proceedings would be of immense value and be definitely useful to researchers in their further research. This collection will also offer a window for new perspectives and directions in the respective horizons.

We express our sincere thanks to all those who made the conference and these proceedings possible with wholehearted gratitude and dedication.

Prayagraj, India — Pankaj Srivastava
Jabalpur, India — S. S. Thakur
Oradea, Romania — Georgia Irina Oros
Sur, Oman — Ali A. AlJarrah
Bangkok, Thailand — Vichian Laohakosol

Contents

About the Editor

Pankaj Srivastava is Professor and Former Head, Department of Mathematics, Motilal Nehru National Institute of Technology Allahabad, Prayagraj, India. With over 32 years of teaching and research experience, his research areas include design and development of soft computing intelligent systems with reference to medical diagnosis and tourism, fuzzy analytics, Ramanujan's computational systems, tensor analysis and market research. He has published over 90 research papers in respected international journals and conference proceedings. Currently, he is also serving as General Secretary at the FATER Academy of India (FAI), since August 2016.

S. S. Thakur is Principal of Jabalpur Engineering College, Jabalpur, Madhya Pradesh, India. With over 39 years of teaching and research experience, his areas of research include fuzzy theory and intuitionistic fuzzy theory on which he has published more than 100 research papers in renowned journals.

Georgia Irina Oros is Associate Professor at the Faculty of Informatics and Sciences, Department of Mathematics and Computer Science, University of Oradea, Romania, since 2013. Her research interests include complex analysis, geometric function theory and fuzzy differential subordination. She has published over 100 research papers in respected international journals and conference proceedings.

Ali A. AlJarrah is Dean of Sur University College, Sur, Oman, since September 2015. He received his Ph.D. in Mathematics from Clarkson University, New York, USA. With over 30 years of teaching and research experience, his research areas include special functions, integral operators, vibration theory and numerical analysis. He has published over 40 research articles in respected journals and conference proceedings.

Vichian Laohakosol was Professor of Mathematics at Kasetsart University, Jatujak District Bangkok, Thailand. He was Head of the Department of Mathematics at Kasetsart University, during 1995–1999. With over 37 years of teaching and research experience, his major fields of research interest are number theory and classical analysis. Currently, he has published 17 research articles since 2019 in the international journals of repute.

Spectrum of Intelligent Tourist Destination Image Decision Information System

Pankaj Srivastava, Saurabh Srivastav, and Yannakorn Toprayoon

1 Introduction

Tourism includes the activities of persons traveling and staying at places outside their usual environment for not more than one consecutive year for leisure, business, and other purposes. In recent years, it has been seen that the tourism industry in India took an exponential rise. Nowadays, different types of tourism are in fashion like Heritage tourism, Eco-Tourism, Pilgrimage Tourism, Medical Tourism, Business Tourism, Leisure Tourism, Cultural Tourism, etc. The motivation and the reasons behind visiting a place may be different for different people; an escape from routine life for relaxation may also be one of the main reasons to make a trip. The current practice for the selection of tourist destinations in India depends mainly on an individual's interest. Mostly, Indian people plan their trips to different destinations as per their attention, and their decisions for trips depend upon the climate conditions; people visit hill stations during summer and southern parts during cold weather for a little escape from routine life. Some of them have an interest in religious and spiritual views as spirituality and mental peace are the only things that attract them to travel for pilgrimage tourism and spiritual tourism. It is observed that many people visit some particular type of places frequently because these places are their favorite destinations, and the overall experience at these places is very satisfactory.

India is counted as one of the most preferred destinations for both foreign and domestic travelers. Tourism allows the tourist visiting India to understand and experience the cultural diversity of India. The Indian tourism industry is outperformed in

P. Srivastava · S. Srivastav (✉)
Department of Mathematics, Motilal Nehru National Institute of Technology, Allahabad, Prayagraj, India
e-mail: rma1604@mnnit.ac.in

Y. Toprayoon
Research and Development Center, Rajapark Institute, Bangkok, Thailand

P. Srivastava et al. (eds.), *Mathematical, Computational Intelligence and Engineering Approaches for Tourism, Agriculture and Healthcare*, Lecture Notes in Networks and Systems 214, https://doi.org/10.1007/978-981-16-3807-7_1

the global tourism industry for growth in the volume of foreign visitors as well as revenue.

There are many criteria behind the decision of selection of a tourist destination such as brand awareness, brand attitude, spirituality, romantic, perceived quality, and adventure, which are of utmost importance in the tourists' decision-making process of selection of a tourist destination and tangled with the image that people have about a particular tourist destination.

2 Literature Review

In the literature of tourism, there are many studies of destination choice concerning destination image and destination branding, and both have their separate streams of research. In 2011, Qu et al. [1] stated the collaboration of these two concepts as overall destination image and consumer behavior to revisit at the destination. Konecnik and Gartner [2] proposed a model which interrelates the four components: destination image, brand awareness, perceived quality, and tourist loyalty.

Dias and Cardoso [3] suggested that the experience of the tourist depends on impalpable attributes, and the quality aspect of the destination depends upon perceived quality. These experiences, either good or bad, remains in tourists' mind as a memory. Rajesh [4] observed components, attributes, and factors inducing the destination image,and examined the tourist satisfaction and determinants of destination loyalty. The terms such as spirituality, adventure, romantic, tourist loyalty, and tourists' satisfaction are very responsible factors for creating a positive or negative destination image for a tourist place. Hunt [5], Gunn [6], and Mayo [7] introduced the concept of destination image in the tourism field. And since then, destination image became the most popular topic for researchers in tourism studies. It is found that experiences of previous visits and destination image are very essential terms for destination choice for a future visit to the destination or revisits to the same destination (e.g., which destinations are to be selected in the next vacation or the plan of next trip) (McFarland and Glisky [8]).

It has been observed by academicians (Crompton [9], Rajesh [4], and Gunn [6]) that the tourism market is profoundly affected by tourists' perceptions, branding of the tourist destinations, and overall tourist satisfaction. It is already discussed above that the Destination image is composed of all types of impressions and experiences. It can be easily observed that each factor mentioned above is a part of the destination image. Although tourism has vast research areas that involve terms such as spirituality, adventure, romantic, tourist loyalty, and tourist satisfaction, all these terms have independent lines of research. It seems evident that all these criteria can be clubbed in one common framework. The selection of a tourist destination has always been a task of dilemma (Thai and Yuksel [10]) and depends on the different types of criteria and the travel experiences. Previously, Ao [11] and Goksu and Kaya [12] used fuzzy methods and rules for extracting the main criteria influencing the tourism industry and for predictions in the tourism industry as the choice of tourist destinations reflects the intangibility and vagueness; that's why in the present work, we are proposing a soft computing technique for the best selection of destination based on specific criteria.

3 Material and Methods

In order to develop the system, we use essential soft computing tools such as Fuzzy sets and Fuzzy TOPSIS [13] related tools and customize them to generate the desired intelligent system, and the working of an intelligent system is investigated with reference to the generated fuzzy data via questionnaire, under the positive ideal and negative ideal solutions and alternatives for authentication of outputs.

Definitions

Fuzzy Sets

A fuzzy set $\widetilde{A}$ in a universe of discourse X is defined by a membership function $\widetilde{\mu}_A(x)$ such that

$$\tilde{\mu}_A(x) \to [0\ 1] \quad , \forall x \in X$$

Fuzzy Numbers

A fuzzy number is a fuzzy subset in the universe of discourse, X that is both convex and normal. The triangular fuzzy number (TFN) is presented by a triplet $\widetilde{A} = (a_1, a_2, a_3)$ and its membership function is given by (Fig. 1)

$$\tilde{\mu}_A(x) = \begin{cases} \frac{x-a_1}{a_2-a_1}, & a_1 \leq x \leq a_2 \\ \frac{a_3-x}{a_3-a_2}, & a_2 \leq x \leq a_3 \\ 0 & else\ where \end{cases}$$

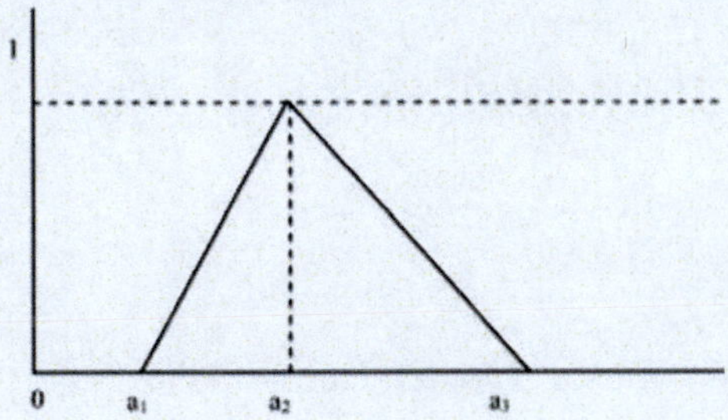

Fig. 1 Triangular fuzzy number

Fig. 2 Linguistic variables as triangular fuzzy number

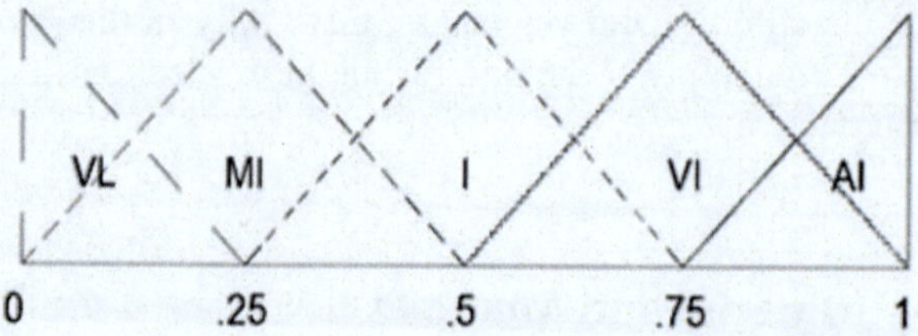

Normalized Fuzzy Sets

A fuzzy set $\widetilde{A}$ of the universe of discourse X is said to be a normalized fuzzy set if $\exists\ an\ x \in X$ such that

$$\tilde{\mu}_A(x) = 1$$

Distance Between Two Fuzzy Numbers

Distance between two fuzzy numbers $\widetilde{A} = (a_1, a_2, a_3)\ and\ \widetilde{B} = (b_1, b_2, b_3)$ is given by

$$\mathrm{d}(\widetilde{A}, \widetilde{B}) = \sqrt{\frac{1}{3}\left[(a_1 - b_1)^2 + (a_2 - b_2)^2 + (a_3 - b_3)^2\right]}$$

Linguistic Variables

A variable whose values are linguistic terms. The linguistic values can also be expressed as fuzzy numbers.

For example, "weight" is a linguistic variable; its values are very low, moderately important, important, very important, absolutely important, etc. (Fig. 2).

Linguistic terms and fuzzy rating scales of different criteria are listed in the following tables.

Algorithm of the proposed Intelligent System is as follows

Input: Number of alternatives (tourist destinations) (m)

(1) A_i, $i = 1, 2, 3, \ldots,$ m
(2) $i \leftarrow 1$

Input: Number of criteria (n)

(1) C_j, $j = 1, 2, 3, \ldots,$ n
(2) $j \leftarrow 1$

Input: Weights of criteria (n)

(1) w_j, $j = 1, 2, 3, \ldots, n$
(2) $j \leftarrow 1$

Output: Order of preference of tourist destinations

Step 1: Create decision matrix-

Input: number of decision-makers K

$$k \leftarrow 1$$

(1) **While (i ≤ K) do**
$\widetilde{D}^k = [\tilde{x}^k{}_{ij}], \quad i = 1, 2, 3, \ldots, m, \; j = 1, 2, 3, \ldots n$
end While
$\tilde{x}^i{}_{jl}$, Fuzzy ratings of linguistic terms given by kth decision-maker for jth alternative with respect to lth criterion

(2) Aggregate decision matrix
While (k≤ K) do
$\tilde{x}_{ij} \leftarrow (a_{ij}, b_{ij}, c_{ij}),$
$a_{ij} \leftarrow \min_k \{a_{ij}{}^k\},$
$b_{ij} \leftarrow \operatorname{mean}_k \{b_{ij}{}^k\}, c_{ij} \leftarrow \max_k \{c_{ij}{}^k\}$
end While
$\widetilde{D} = [\tilde{x}_{jl}], \quad j = 1, 2, 3, \ldots, m, l = 1, 2, 3, \ldots n$

(3) Weight assigning
While (k≤ K) do
$\tilde{w}_j{}^k \leftarrow (w_{j1}{}^k, w_{j2}{}^k, w_{j3}{}^k), j = 1, 2, 3, \ldots, n$
$w_{j1} \leftarrow \min_k \{w_{j1}{}^k\}, w_{j2} \leftarrow \operatorname{mean}_k \{w_{j2}{}^k\}, w_{j3} \leftarrow \max_k \{w_{j3}{}^k\}$
end While
$\widetilde{W} \leftarrow [\tilde{w}_j], \quad , j = 1, 2, 3, \ldots n$

Step 2: Normalization

While (k ≤ K) do
$\tilde{w}_j{}^k \leftarrow (w_{j1}{}^k, w_{j2}{}^k, w_{j3}{}^k), \quad j = 1, 2, 3, \ldots n$
$w_{j1} \leftarrow \min_k \{w_{j1}{}^k\}, \quad w_{j2} \leftarrow \operatorname{mean}_k \{w_{j2}{}^k\}, \quad w_{j3} \leftarrow \max_k \{w_{j3}{}^k\}$
end While
$\tilde{r}_{ij} \leftarrow \left(\frac{a_{ij}}{c_j^+}, \frac{a_{ij}}{c_j^+}, \frac{a_{ij}}{c_j^+}\right), c_j^+ \leftarrow \max_i (c_{ij})$, for benefit criteria
$\tilde{r}_{ij} \leftarrow \left(\frac{a_j^-}{c_{ij}}, \frac{a_j^-}{b_{ij}}, \frac{a_j^-}{a_{ij}}\right), a_j^- \leftarrow \min_i (a_{ij})$, for cost criteria
$\widetilde{R} \leftarrow [\tilde{r}_{ij}]_{m \times n}$

Step 3: Weighted normalized decision matrix

$[\tilde{v}_{ij}] \leftarrow [\tilde{r}_{ij} * \tilde{w}_j]_{m \times n}$, where $i = 1, 2, 3, \ldots, m$ and $j = 1, 2, 3, n$

$\widetilde{V} \leftarrow [\tilde{v}_{ij}]$

Step 4: Calculation of FPIS and FNIS

While (i ≤ m) do

$$\tilde{v}_j^+ \leftarrow \max_i (v_{ij3})$$

end While

$$A^+ \leftarrow \tilde{v}_j^+, j = 1, 2, 3, \ldots, n$$

While (i ≤ m) do

$$\tilde{v}_j^- \leftarrow \min_{\mathrm{i}} (v_{ij1})$$

end While

$$A^- \leftarrow \tilde{v}_j^-, j = 1, 2, 3, \ldots, n$$

Step 5: Distances of FPIS and FNIS

While (i ≤ m) do

$$d_i^+ \leftarrow \sum_{j=1}^{n} d_v\left(\tilde{v}_{ij}, \tilde{v}_j^+\right)$$

end While
While (i ≤ m) do

$$d_i^- \leftarrow \sum_{j=1}^{n} d_v\left(\tilde{v}_{ij}, \tilde{v}_j^-\right)$$

end While

Step 5: Calculate closeness coefficient

While (i ≤ m) do

$$CCi \leftarrow \frac{d_i^-}{d_i^- + d_i^+}$$

end While

Step 6: rank the alternatives as per closeness coefficient

A flowchart of the proposed Intelligent Information System functioning is given in (Fig. 3).

4 Methodology

A. **Selection of tourist destinations:**

Since our working domain is Varanasi, India, therefore tourist destinations ratings are as per our criterion of decision-making.

B. **Identification of criteria:**

The identification and selection criteria are based on literature review, reviews of tourists, and experts' suggestions.

C. **Fuzzy TOPSIS:**

The Fuzzy TOPSIS works on the concept of distance to positive ideal solution and negative ideal solution; the basic procedure of Fuzzy TOPSIS is given as follows:

Step 1:

Suppose there are k decision-makers. If the fuzzy rating and weight of importance of kth decision-maker about ith alternative concerning jth criteria are $\tilde{x}_{ij}{}^k = \left(a_{ij}{}^k, b_{ij}{}^k, c_{ij}{}^k\right)$ and $\widetilde{w}_j{}^k = \left(w_{j1}{}^k, w_{j2}{}^k, w_{j3}{}^k\right)$ respectively, where $i = 1, 2, 3, \ldots, m$ and $j = 1, 2, 3, \ldots, n$

Step 2:

Aggregated fuzzy rating of ith alternative concerning jth criteria is given by $\tilde{x}_{ij} = \left(a_{ij}, b_{ij}, c_{ij}\right)$ such that

$$a_{ij} = \min_k\left\{a_{ij}^k\right\}, b_{ij} = \underset{k}{\text{mean}}\{b_{ij}^k, c_{ij} = \max_k\left\{c_{ij}^k\right\}$$

And the aggregated fuzzy weights of importance are $\tilde{w}_j = \left(w_{j1}, w_{j2}, w_{j3}\right)$

$$w_{j1} = \min_k\left\{w_{j1}{}^k\right\}, \quad w_{j2} = \underset{k}{\text{mean}}\left\{w_{j2}{}^k\right\}, \quad w_{j3} = \max_k\left\{w_{j3}{}^k\right\}$$

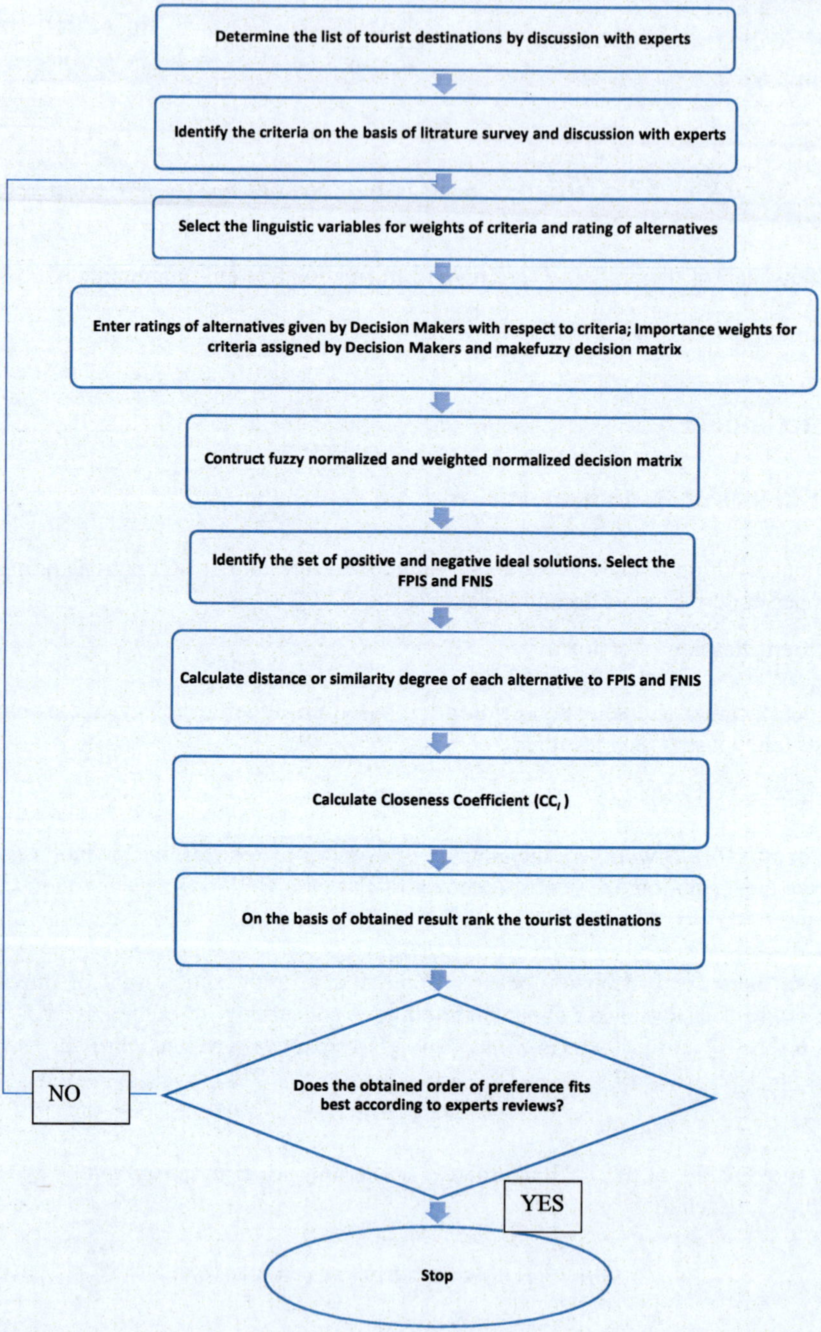

Fig. 3 Flowchart for the process

Step 3:

The fuzzy decision matrix of aggregate values is expressed as follows:

$$\widetilde{D} = \begin{matrix} \\ A_1 \\ A_2 \\ \ldots \\ A_m \end{matrix} \begin{matrix} C_1\ C_2 \ldots C_n \\ \begin{bmatrix} \tilde{x}_{11}\ \tilde{x}_{12} \ldots \tilde{x}_{1n} \\ \tilde{x}_{21}\ \tilde{x}_{22} \ldots \tilde{x}_{2n} \\ \ldots\ \ldots\ \ldots \\ \tilde{x}_{m1}\ \tilde{x}_{m2} \ldots \tilde{x}_{mn} \end{bmatrix} \end{matrix}$$

And the aggregate fuzzy importance weights are

$$\widetilde{W} = [\, \tilde{w}_1\ \tilde{w}_2\ \tilde{w}_3 \ldots \tilde{w}_n \,]$$

where each $\tilde{x}_{ij}, \tilde{w}_j \forall\ i = 1, 2, 3, \ldots, m$ and $j = 1, 2, 3, \ldots, n$, are linguistic variables and which can be interpreted by triangular fuzzy numbers $\tilde{x}_{ij} = (a_{ij}, b_{ij}, c_{ij})$ and $\tilde{w}_j = (w_{j1}, w_{j2}, w_{j3})$.

Step 4:

The normalized fuzzy decision matrix is given by
$\widetilde{R} = [\tilde{r}_{ij}]_{m \times n}$, where $i = 1, 2, 3, \ldots, m$ and $j = 1, 2, 3, \ldots, n$
Where

$$\left.\begin{aligned} \tilde{r}_{ij} &= \left(\frac{a_{ij}}{c_j^+}, \frac{a_{ij}}{c_j^+}, \frac{a_{ij}}{c_j^+}\right)\ And \\ c_j^+ &= \max_i (c_{ij})\ (benifit\ criteria) \\ \tilde{r}_{ij} &= \left(\frac{a_j^-}{c_{ij}}, \frac{a_j^-}{b_{ij}}, \frac{a_j^-}{a_{ij}}\right)\ And \\ a_j^- &= \min_i (a_{ij})\ (cost\ criteria) \end{aligned}\right\}$$

In this normalization process, the ranges of normalized fuzzy numbers lie in the interval [0, 1].

Step 5:

The weighted normalized fuzzy matrix $\widetilde{V}$ can be evaluated by
$\widetilde{V} = [\tilde{v}_{ij}] = [\tilde{r}_{ij} * \tilde{w}_j]_{m \times n}$, where $i = 1, 2, 3, \ldots, m$ and $j = 1, 2, 3, n$

Step 6:

The fuzzy positive ideal solution (FPIS) and fuzzy negative ideal solutions (FNIS) are defined as follows:

$$A^+ = \left(\tilde{v}_1^+, \tilde{v}_2^+, \ldots \tilde{v}_n^+\right) \ \ And$$
$$\tilde{v}_j^+ = \max_i\left(v_{ij3}\right) \ i = 1, 2, 3, \ldots, m \ and \ j = 1, 2, 3, \ldots, n$$
$$A^- = \left(\tilde{v}_1^-, \tilde{v}_2^-, \ldots \tilde{v}_n^-\right) \ \ And$$
$$\tilde{v}_j^- = \min_i\left(v_{ij1}\right) \ i = 1, 2, 3, \ldots, m \ and \ j = 1, 2, 3, \ldots s, n$$

Step 7:

The distances d_i^+ *And* d_i^- of each weighted alternative from fuzzy positive ideal solution (FPIS) and fuzzy negative ideal solutions (FNIS) can be evaluated by the following equations:

$$d_i^+ = \sum_{j=1}^{n} d_v\left(\tilde{v}_{ij}, \tilde{v}_j^+\right) \quad i = 1, 2, 3, \ldots, m$$

$$d_i^- = \sum_{j=1}^{n} d_v\left(\tilde{v}_{ij}, \tilde{v}_j^-\right) \quad i = 1, 2, 3, \ldots, m$$

where $d_v\left(\tilde{a}, \tilde{b}\right)$ is the distance between two triangular fuzzy numbers $\tilde{a}$ *and* $\tilde{b}$.

Step 8:

The distance from fuzzy positive ideal solution (FPIS A^+) and fuzzy negative ideal solutions (FNIS A^-) is simultaneously represented by the closeness coefficient CCi

$$CCi = \frac{d_i^-}{d_i^- + d_i^+}, \quad i = 1, 2, 3, \ldots, m$$

The highest value of the closeness coefficient represents the best alternative, and it shows the best alternative is closest to the FPIS and farthest from FNIS.

Step 9: Rank the alternatives with the help of the closeness coefficient.

5 Performance Assessment of the Proposed Intelligent System

A. Selection of tourist destinations:

As we have incorporated the concept of destination image to design the intelligent system, therefore, we proceed first to develop the system with the following destinations for investigation.

1. A_1-Ghats Of Varanasi.
2. A_2-Kashi Vishwanath Temple.
3. A_3-Sankat Mochan Hanuman Temple.
4. A_4-Sarnath.
5. A_5-Ramnagar Fort.
6. A_6-B.H.U Shiva Temple.
7. A_7-Famous Varanasi Cuisines Shops.

B. **Identification criteria:**

The identification features are as follows; the respective table is given below:

1. C_1-Brand Awareness (How much tourists are aware of the destination).
2. C_2-Tourist Satisfaction.
3. C_3-Perceived Quality (Overall evaluation of services and products at destination).
4. C_4-Brand Attitude (Overall perception about destination).
5. C_5-Tourist Loyalty.
6. C_6-Spirituality.
7. C_7-Romantic.
8. C_8-Adventure.

The decision-makers rated the weights of each criterion, and corresponding to each criterion, they rated the alternatives based on collected data in the form of linguistic variables scale which are given in Tables 1, 2, 3, 4, 5, 6, 7, 8, and 9, and all criteria we have taken are benefit criteria. The rankings of decision-makers are given below (Tables 10 and 11).

The aggregate fuzzy weight $\tilde{w}_j = \left(w_{j1}, w_{j2}, w_{j3}\right)$ for criterion C_1 (Brand awareness) is calculated with the help of the process given in step (2)

$$w_{11} = \min_{k=4}(3, 7, 7, 7) = 3, \quad w_{12} = \frac{1}{4}\sum_{k=1}^{4}(5, 9, 9, 9) = 8,$$

$$w_{13} = \max_{k=4}(7, 9, 9, 9) = 9$$

Table 1 Linguistic terms for weights

Linguistic term	Crisp number	Fuzzy rating
Very less important (VLI)	1	(0, 1, 3)
Less important (LI)	3	(1, 3, 5)
Important (I)	5	(3, 5, 7)
Very important (VI)	7	(5, 7, 9)
Excellent (E)	9	(7, 9, 9)
Intermittent values	2, 4, 6, 8	(1, 2, 3), (3, 4, 5), (5, 6, 7), (7, 8, 9)

Table 2 Linguistic terms for brand awareness

Linguistic term	Crisp number	Fuzzy rating
Not at all aware (NA)	1	(0, 1, 3)
Not so much aware (NSMA)	3	(1, 3, 5)
Somewhat aware (SA)	5	(3, 5, 7)
Very much aware (VMA)	7	(5, 7, 9)
Extremely aware (EA)	9	(7, 9, 9)
Intermittent values	2, 4, 6, 8	(1, 2, 3), (3, 4, 5), (5, 6, 7), (7, 8, 9)

Table 3 Linguistic terms for tourist satisfaction

Linguistic term	Crisp number	Fuzzy rating
Absolutely dissatisfied (VD)	1	(0, 1, 3)
Dissatisfied (D)	3	(1, 3, 5)
Neither satisfied nor dissatisfied (NSND)	5	(3, 5, 7)
Satisfied (S)	7	(5, 7, 9)
Very much satisfied (VS)	9	(7, 9, 9)
Intermittent values	2, 4, 6, 8	(1, 2, 3), (3, 4, 5), (5, 6, 7), (7, 8, 9)

Table 4 Linguistic terms for perceived quality

Linguistic term	Crisp number	Fuzzy rating
Very low (VL)	1	(0, 1, 3)
Low quality (LQ)	3	(1, 3, 5)
Neither high nor low (NHNL)	5	(3, 5, 7)
High quality (HQ)	7	(5, 7, 9)
Very high quality (VHQ)	9	(7, 9, 9)
Intermittent values	2, 4, 6, 8	(1, 2, 3), (3, 4, 5), (5, 6, 7), (7, 8, 9)

Table 5 Linguistic terms for brand attitude

Linguistic term	Crisp number	Fuzzy rating
Worse than expectations (WE)	1	(0, 1, 3)
Below expectation (BeE)	3	(1, 3, 5)
Moderately expected (ME)	5	(3, 5, 7)
As expected (AE)	7	(5, 7, 9)
Beyond expectation (BE)	9	(7, 9, 9)
Intermittent values	2, 4, 6, 8	(1, 2, 3), (3, 4, 5), (5, 6, 7), (7, 8, 9)

Table 6 Linguistic terms for tourist loyalty

Linguistic term	Crisp number	Fuzzy rating
Definitely not (DN)	1	(0, 1, 3)
Probably not (PN)	3	(1, 3, 5)
Probably (P)	5	(3, 5, 7)
Definitely (D)	7	(5, 7, 9)
Very definitely (VD)	9	(7, 9, 9)
Intermittent values	2, 4, 6, 8	(1, 2, 3), (3, 4, 5), (5, 6, 7), (7, 8, 9)

Table 7 Linguistic terms for spirituality

Linguistic term	Crisp number	Fuzzy rating
Not spiritual (NSp)	1	(0, 1, 3)
Somewhat spiritual (SSp)	3	(1, 3, 5)
Enough spiritual (ESp)	5	(3, 5, 7)
Very spiritual (VSp)	7	(5, 7, 9)
Very spiritual (VVSp)	9	(7, 9, 9)
Intermittent values	2, 4, 6, 8	(1, 2, 3), (3, 4, 5), (5, 6, 7), (7, 8, 9)

Table 8 Linguistic terms for romantic

Linguistic term	Crisp number	Fuzzy rating
Not romantic (NRm)	1	(0, 1, 3)
Somewhat romantic (SRm)	3	(1, 3, 5)
Enough romantic (ERm)	5	(3, 5, 7)
Very romantic (VRm)	7	(5, 7, 9)
Very very romantic (VVRm)	9	(7, 9, 9)
Intermittent values	2, 4, 6, 8	(1, 2, 3), (3, 4, 5), (5, 6, 7), (7, 8, 9)

Table 9 Linguistic terms for adventure

Linguistic term	Crisp number	Fuzzy rating
Not adventurous (NAd)	1	(0, 1, 3)
Somewhat adventurous (SAd)	3	(1, 3, 5)
Enough adventurous (EAd)	5	(3, 5, 7)
Very adventurous (VAd)	7	(5, 7, 9)
Very very adventurous e (VVAd)	9	(7, 9, 9)
Intermittent values	2, 4, 6, 8	(1, 2, 3), (3, 4, 5), (5, 6, 7), (7, 8, 9)

Table 10 Linguistic ratings for weights

Criteria	Decision-makers			
	D_1	D_2	D_3	D_4
Brand awareness (C_1)	I	E	E	E
Tourist satisfaction (C_2)	VLI	VI	I	Somewhere b/w (E and VI)
Perceived quality (C_3)	VI	VI	VI	VI
Brand attitude (C_4)	E	E	VI	E
Tourist loyalty (C_5)	LI	VI	I	VI
Spirituality (C_6)	VI	E	I	LI
Romantic (C_7)	VLI	VLI	LI	VI
Adventure (C_8)	Somewhere b/w (LI & VLI)	VLI	LI	I

Table 11 Linguistic ratings for alternatives corresponding to each criterion

Criteria	Alternative	Decision-makers ratings			
		D_1	D_2	D_3	D_4
C_1	A_1	SA	EA	VA	EA
	A_2	VA	VA	VA	EA
	A_3	VA	EA	SA	VA
	A_4	VA	EA	VA	Somewhere b/w (VA&SA)
	A_5	NA	EA	NSA	SA
	A_6	VMA	EA	SA	VMA
	A_7	NSMA	SA	NA	SA
C_2	A_1	S	Somewhere b/w (VS, S)	NSND	S
	A_2	Dis	S	NSND	S
	A_3	V.Dis	NSND	NSND	S
	A_4	Dis	Somewhere b/w (VS, S)	S	Somewhere b/w (NSND&S)
	A_5	VS	S	NSND	Somewhere b/w (NSND&S)
	A_6	S	NSND	S	S
	A_7	S	VMS	S	S
C_3	A_1	VH	Somewhere b/w (H&NHNL)	NHNL	VH
	A_2	VL	Somewhere b/w (NHNL&L)	NHNL	HQ
	A_3	LQ	NHNL	NHNL	Somewhere b/w (VHQ&HQ)

(continued)

Table 11 (continued)

Criteria	Alternative	Decision-makers ratings			
		D_1	D_2	D_3	D_4
	A_4	VL	HQ	Somewhere b/w (NHNL&LQ)	HQ
	A_5	NHNL	HQ	Somewhere b/w (NHNL&L)	Somewhere b/w (H&NHNL)
	A_6	NHNL	Somewhere b/w (HQ, VHQ)	Somewhere b/w (NHNL, HQ)	HQ
	A_7	HQ	Somewhere b/w (HQ, VHQ)	NHNL	NHNL
C_4	A_1	BE	Somewhere b/w (AE&BE)	Somewhere b/w (ME, AE)	ME
	A_2	ME	ME	BeE	AE
	A_3	ME	BE	BeE	BeE
	A_4	BE	Somewhere b/w (ME, AE)	AE	BE
	A_5	BeE	Somewhere b/w (BeE, WE)	BeE	Somewhere b/w (ME, BeE)
	A_6	AE	Somewhere b/w (ME, AE)	AE	AE
	A_7	Somewhere b/w (ME, AE)	Somewhere b/w (ME, AE)	Somewhere b/w (ME, AE)	AE
C_5	A_1	VD	Somewhere b/w (VD & D)	D	D
	A_2	D	P	D	VD
	A_3	P	Somewhere b/w (PN, P)	P	D
	A_4	VD	D	D	Somewhere b/w (D, VD)
	A_5	Somewhere b/w (P, PN)	P	Somewhere b/w (P, D)	P
	A_6	Somewhere b/w (D, P)	D	VD	D
	A_7	D	D	D	P
C_6	A_1	VSp	Somewhere b/w (VSp&ESp)	Vsp	VVSp
	A_2	ESp	VSp	Somewhere b/w (ESp, SSp)	ESp
	A_3	SSp	ESp	Somewhere b/w (ESp, VSp)	Esp
	A_4	Somewhere b/w (NSp, SSp)	SSp	Esp	Somewhere b/w (ESp, SSp)

(continued)

Table 11 (continued)

Criteria	Alternative	Decision-makers ratings			
		D_1	D_2	D_3	D_4
	A_5	NSp	Somewhere b/w (NSp, SSp)	NSp	NSp
	A_6	ESp	Somewhere b/w (ESp, SSp)	VSp	Somewhere b/w(VSp, VVSp)
	A_7	NSp	NSp	NSp	NSp
C_7	A_1	VRm	Somewhere b/w (VRm, ERm)	SRm	NRm
	A_2	NRm	NRm	Somewhere b/w (NRm, SRm)	NRm
	A_3	NRm	NRm	NRm	NRm
	A_4	Somewhere b/w (NRm, SSRm)	SRm	NRm	Somewhere b/w (NRm, SRm)
	A_5	NRm	Somewhere b/w (ERm, SRm)	SRm	NRm
	A_6	NRm	NRm	NRm	Somewhere b/w (NRm, SRm)
	A_7	SRm	Somewhere b/w (NRm, SRm)	Somewhere b/w (ERm, SRm)	VRm
C_8	A_1	NAd	Somewhere b/w (NAd&SAd)	Sad	EAd
	A_2	Somewhere b/w (NAd&SAd)	NAd	Nad	SAd
	A_3	NAd	NAd	Somewhere b/w (NAd&SAd)	NAd
	A_4	EAd	Sad	Somewhere b/w (EAd&SAd)	EAd
	A_5	Somewhere b/w (NAd&SAd)	Somewhere b/w (SAd, EAd)	Sad	SAd
	A_6	NAd	Somewhere b/w (NAd&SAd)	Nad	Somewhere b/w (NAd&SAd)
	A_7	NAd	33NAd	Nad	NAd

$$\tilde{w}_1 = (3, 8, 9)$$

Similarly, aggregate fuzzy weights of all criteria are shown in Table 12.

The aggregate fuzzy rating $\tilde{x}_{11} = (a_{ij}, b_{ij}, c_{ij})$ for criterion C_1 (Brand awareness) is calculated with the help of the process given in step (2)

Table 12 Aggregate fuzzy weights

Criteria	Decision-makers				Aggregate fuzzy weights
	D_1	D_2	D_3	D_4	
C_1	(3, 5, 7)	(7, 9, 9)	(7, 9, 9)	(7, 9, 9)	(3, 8, 9)
C_2	(0, 1, 3)	(5, 7, 9)	(3, 5, 7)	(7, 8, 9)	(0, 5.25, 9)
C_3	(5, 7, 9)	(5, 7, 9)	(5, 7, 9)	(5, 7, 9)	(5, 7, 9)
C_4	(7, 9, 9)	(7, 9, 9)	(5, 7, 9)	(7, 9, 9)	(5, 8.5, 9)
C_5	(1, 3, 5)	(5, 7, 9)	(3, 5, 7)	(5, 7, 9)	(1, 5.5, 9)
C_6	(5, 7, 9)	(7, 9, 9)	(3, 5, 7)	(1, 3, 5)	(1, 6, 9)
C_7	(0, 1, 3)	(0, 1, 3)	(1, 3, 5)	(5, 7, 9)	(0, 3, 9)
C_8	(1, 2, 3)	(0, 1, 3)	(1, 3, 5)	(3, 5, 7)	(0, 2.75, 7)

$$a_{11} = \min_{k=4}(3, 7, 5, 7) = 3, \quad a_{12} = \frac{1}{4}\sum_{k=1}^{4}(5, 9, 7, 9) = 8,$$

$$a_{13} = \max_{k=4}(7, 9, 9, 9) = 9$$

$$\tilde{x}_{11} = (3, 7.5, 9)$$

Similarly, fuzzy aggregate ratings of all criteria are shown in Table 13.

An aggregate decision matrix is shown in Table 14.

In the next step, the evaluation of normalized fuzzy weights is performed for criterion C_1 an alternative A_1.

By using step 4,

$$c_1^+ = \max_i(9, 9, 9, 9, 9, 9, 7) = 9 \quad where\ i = 1, 2, \ldots 7$$

$$\tilde{r}_{11} = \left(\frac{a_{11}}{c_1^+}, \frac{b_{11}}{c_1^+}, \frac{c_{11}}{c_1^+}\right) = \left(\frac{3}{9}, \frac{7.5}{9}, \frac{9}{9}\right) = (0.333, 0.833, 1)$$

Similarly, other values are evaluated and shown in Table 15 and the different values of c_i^+ are given below.

c_1^+	c_2^+	c_3^+	c_4^+	c_5^+	c_6^+	c_7^+	c_8^+
9	9	9	9	9	9	9	7

By using step 5, we evaluate the weighted value for criteria C_1 and alternative A_1

$$(0.333, 0.833, 1)^*(3, 8, 9) = (1, 6.666, 9)$$

evaluated weighted fuzzy decision matrix, i.e., given below (Table 16).

Table 13 Aggregate fuzzy ratings for alternatives

Criteria	Alternatives	Decision-makers ratings				Aggregate fuzzy ratings
		D_1	D_2	D_3	D_4	
C_1	A_1	(3, 5, 7)	(7, 9, 9)	(5, 7, 9)	(7, 9, 9)	(3, 7.5, 9)
	A_2	(5, 7, 9)	(5, 7, 9)	(5, 7, 9)	(7, 9, 9)	(5, 7.5, 9)
	A_3	(5, 7, 9)	(7, 9, 9)	(3, 5, 7)	(5, 7, 9)	(3, 7, 9)
	A_4	(5, 7, 9)	(7, 9, 9)	(5, 7, 9)	(5, 6, 7)	(5, 7.25, 9)
	A_5	(0, 1, 3)	(7, 9, 9)	(1, 3, 5)	(3, 5, 7)	(0, 4.5, 9)
	A_6	(5, 7, 9)	(7, 9, 9)	(3, 5, 7)	(5, 7, 9)	(3, 7, 9)
	A_7	(1, 3, 5)	(3, 5, 7)	(0, 1, 3)	(3, 5, 7)	(0, 3.5, 7)
C_2	A_1	(1, 3, 5)	(7, 8, 9)	(5, 7, 9)	(5, 7, 9)	(1, 6.25, 9)
	A_2	(7, 9, 9)	(5, 6, 7)	(5, 7, 9)	(7, 9, 9)	(5, 7.75, 9)
	A_3	(3, 5, 7)	(3, 5, 7)	(3, 5, 7)	(5, 6, 7)	(3, 5.25, 9)
	A_4	(7, 9, 9)	(7, 8, 9)	(3, 5, 7)	(7, 8, 9)	(3, 7.5, 9)
	A_5	(1, 2, 3)	(5, 7, 9)	(1, 2, 3)	(5, 6, 7)	(1, 4.25, 9)
	A_6	(5, 7, 9)	(3, 5, 7)	(5, 7, 9)	(5, 7, 9)	(1, 4.25, 9)
	A_7	(5, 7, 9)	(7, 9, 9)	(5, 7, 9)	(5, 7, 9)	(1, 4.25, 9)
C_3	A_1	(7, 9, 9)	(5, 6, 7)	(3, 5, 7)	(7, 9, 9)	(3, 7.25, 9)
	A_2	(0, 1, 3)	(3, 4, 5)	(3, 5, 7)	(5, 7, 9)	(0, 4.25, 9)
	A_3	(1, 3, 5)	(3, 5, 7)	(3, 5, 7)	(7, 8, 9)	(1, 5.25, 9)
	A_4	(0, 1, 3)	(5, 7, 9)	(3, 4, 5)	(5, 7, 9)	(0, 4.75, 9)
	A_5	(3, 5, 7)	(5, 7, 9)	(3, 4, 5)	(5, 6, 7)	(3, 5.5, 9)
	A_6	(3, 5, 7)	(7, 8, 9)	(5, 6, 7)	(5, 7, 9)	(3, 6.5, 9)
	A_7	(5, 7, 9)	(7, 8, 9)	(5, 6, 7)	(3, 5, 7)	(3, 6.5, 9)
C_4	A_1	(5, 7, 9)	(7, 8, 9)	(3, 5, 7)	(5, 7, 9)	(3, 6.75, 9)
	A_2	(1, 3, 5)	(5, 7, 9)	(3, 5, 7)	(5, 7, 9)	(1, 5.5, 9)
	A_3	(0, 1, 3)	(3, 5, 7)	(3, 5, 7)	(5, 7, 9)	(0, 4.5, 9)
	A_4	(1, 3, 5)	(7, 8, 9)	(5, 7, 9)	(5, 6, 7)	(1, 6, 9)
	A_5	(7, 9, 9)	(5, 7, 9)	(3, 5, 7)	(5, 6, 7)	(3, 6.75, 9)
	A_6	(5, 7, 9)	(5, 6, 7)	(5, 7, 9)	(5, 7, 9)	(5, 6.75, 9)
	A_7	(5, 6, 7)	(5, 6, 7)	(5, 6, 7)	(5, 7, 9)	(5, 6.25, 9)
C_5	A_1	(0, 1, 3)	(7, 9, 9)	(5, 7, 9)	(7, 8, 9)	(0, 6.25, 9)
	A_2	(3, 5, 7)	(5, 7, 9)	(5, 7, 9)	(5, 6, 7)	(3, 6.25, 9)
	A_3	(7, 9, 9)	(5, 7, 9)	(3, 5, 7)	(7, 8, 9)	(3, 7.25, 9)
	A_4	(3, 5, 7)	(7, 8, 9)	(5, 7, 9)	(7, 9, 9)	(3. 7.25, 9)

(continued)

Table 13 (continued)

Criteria	Alternatives	Decision-makers ratings				Aggregate fuzzy ratings
		D_1	D_2	D_3	D_4	
	A_5	(1, 3, 5)	(5, 7, 9)	(3, 4, 5)	(5, 6, 7)	(1, 5, 9)
	A_6	(5, 6, 7)	(5, 7, 9)	(7, 9, 9)	(5, 7, 9)	(5, 7.25, 9)
	A_7	(5, 7, 9)	(5, 7, 9)	(5, 7, 9)	(3, 5, 7)	(3, 6.5, 9)
C_6	A_1	(0, 1, 3)	(1, 2, 3)	(1, 3, 5)	(7, 9, 9)	(5, 7.25, 9)
	A_2	(1, 2, 3)	(0, 1, 3)	(0, 1, 3)	(3, 5, 7)	(3, 5.25, 9)
	A_3	(0, 1, 3)	(0, 1, 3)	(1, 2, 3)	(3, 5, 7)	(1, 4.75, 7)
	A_4	(3, 5, 7)	(3, 5, 7)	(3, 4, 5)	(3, 4, 5)	(1, 3.5, 7)
	A_5	(1, 2, 3)	(0, 1, 3)	(1, 3, 5)	(0, 1, 3)	(0, 1.25, 3)
	A_6	(0, 1, 3)	(0, 1, 3)	(0, 1, 3)	(7, 8, 9)	(3, 6, 9)
	A_7	(0, 1, 3)	(0, 1, 3)	(0, 1, 3)	(0, 1, 3)	(0, 1, 3)
C_7	A_1	(5, 7,9)	(5, 6, 7)	(1, 3, 5)	(0, 1, 3)	(0, 4.25, 9)
	A_2	(0, 1, 3)	(0, 1, 3)	(1, 2, 3)	(0, 1, 3)	(0, 1.25, 3)
	A_3	(0, 1, 3)	(0, 1, 3)	(0, 1, 3)	(0, 1, 3)	(0, 1, 3)
	A_4	(1, 2, 3)	(1, 3, 5)	(0, 1, 3)	(1, 2, 3)	(0, 2, 5)
	A_5	(0, 1, 3)	(3, 4, 5)	(1, 3, 5)	(0, 1, 3)	(0, 2.25, 5)
	A_6	(0, 1, 3)	(0, 1, 3)	(0, 1, 3)	(1, 2, 3)	(0, 1.25, 3)
	A_7	(1, 3, 5)	(1, 2, 3)	(3, 4, 5)	(5, 7, 9)	(1, 4, 9)
C_8	A_1	(0, 1, 3)	(1, 2, 3)	(1, 3, 5)	(3, 5, 7)	(0, 2.75, 7)
	A_2	(1, 2, 3)	(0, 1, 3)	(0, 1, 3)	(1, 3, 5)	(0, 1.75, 5)
	A_3	(0, 1, 3)	(0, 1, 3)	(1, 2, 3)	(0, 1, 3)	(0, 1.25, 3)
	A_4	(3, 5, 7)	(3, 5, 7)	(3, 4, 5)	(3, 5, 7)	(3, 4.75, 7)
	A_5	(1, 2, 3)	(0, 1, 3)	(1, 3, 5)	(1, 3, 5)	(0, 2.25, 5)
	A_6	(0, 1, 3)	(0, 1, 3)	(0, 1, 3)	(1, 2, 3)	(0, 1.25, 3)
	A_7	(0, 1, 3)	(0, 1, 3)	(0, 1, 3)	(0, 1, 3)	(0, 1, 3)

Table 14 Normalized fuzzy decision matrix $\widetilde{R}$ for alternatives

	C1	C2	C3	C4	C5	C6	C7	C8
A1	(3 7.5 9)	(1 6.25 9)	(3 7.3 9)	(3 6.75 9)	(0 6.25 9)	(5 7.25 9)	(0 4.25 9)	(0 2.75 7)
A2	(5 7.5 9)	(5 7.75 9)	(0 4.3 9)	(1 5.5 9)	(3 6.25 9)	(3 5.25 9)	(0 1.25 3)	(0 1.75 5)
A3	(3 7 9)	(3 5.25 7)	(1 5.3 9)	(0 4.5 9)	(3 7.25 9)	(1 4.75 7)	(0 1 3)	(0 1.25 3)
A4	(5 7.25 9)	(3 7.5 9)	(0 4.8 9)	(1 6 9)	(3 7.25 9)	(1 3.5 7)	(0 2 5)	(3 4.75 7)
A5	(0 4.5 9)	(1 4.25 9)	(3 5.5 9)	(3 6.75 9)	(1 5 9)	(0 1.25 3)	(0 2.25 5)	(0 2.25 5)
A6	(3 7 9)	(3 6.5 9)	(3 6.5 9)	(5 6.75 9)	(5 7.25 9)	(3 6 9)	(0 1.25 3)	(0 1.25 3)
A7	(0 3.5 7)	(5 7.5 9)	(3 6.5 9)	(5 6.25 9)	(3 6.5 9)	(0 1 3)	(1 4 9)	(0 1 3)

Table 15 Normalized fuzzy decision matrix $\widetilde{R}$ for alternatives

	C1	C2	C3	C4	C5	C6	C7	C8
A1	(0.33 0.83 1)	(0.1 0.69 1)	(0.33 0.8 1)	(0.3 0.75 1)	(0 0.69 1)	(0.56 0.81 1)	(0 0.47 1)	(0 0.39 1)
A2	(0.55 0.83 1)	(0.6 0.86 1)	(0 0.5 1)	(0.1 0.61 1)	(0.33 0.69 1)	(0.33 0.58 1)	(0 0.14 0.33)	(0 0.25 0.71)
A3	(0.33 0.77 1)	(0.3 0.58 0.78)	(0.11 0.6 1)	(0 0.5 1)	(0.33 0.81 1)	(0.11 0.53 0.78)	(0 0.11 0.33)	(0 0.17 0.43)
A4	(0.55 0.80 1)	(0.3 0.83 1)	(0 0.5 1)	(0.1 0.66 1)	(0.33 0.81 1)	(0.11 0.39 0.78)	(0 0.22 0.56)	(0.43 0.67 1)
A5	(0 0.5 1)	(0.1 0.47 1)	(0.33 0.6 1)	(0.3 0.75 1)	(0.11 0.56 1)	(0 0.14 0.33)	(0 0.25 0.56)	(0 0.32 0.71)
A6	(0.33 0.77 1)	(0.3 0.72 1)	(0.33 0.7 1)	(0.6 0.75 1)	(0.55 0.81 1)	(0.33 0.67 1)	(0 0.14 0.33)	(0 0.17 0.43)
A7	(0 0.38 0.8)	(0.6 0.83 1)	(0.33 0.7 1)	(0.6 0.69 1)	(0.33 0.72 1)	(0 0.11 0.33)	(0.11 0.44 1)	(0 0.14 0.43)

Table 16 Normalized weighted fuzzy decision matrix $\widetilde{V}$ for alternatives

	C1	C2	C3	C4	C5	C6	C7	C8
A1	(1 6.66 9)	(0 3.65 9)	(1.67 5.6 9)	(1.7 6.37 9)	(0 3.82 9)	(0.56 4.83 9)	(0 1.42 9)	(0 1.08 7)
A2	(1.66 6.66 9)	(0 4.52 9)	(0 3.3 9)	(0.6 5.19 9)	(0.33 3.82 9)	(0.33 3.5 9)	(0 0.42 3)	(0 0.68 5)
A3	(1 6.22 9)	(0 3.06 7)	(0.56 4.1 9)	(0 4.25 9)	(0.33 4.43 9)	(0.11 3.17 7)	(0 0.33 3)	(0 0.49 3)
A4	(1.66 6.44 9)	(0 4.38 9)	(0 3.7 9)	(0.6 5.66 9)	(0.33 4.43 9)	(0.11 2.33 7)	(0 0.67 5)	(0 1.86 7)
A5	(0 4 9)	(0 2.48 9)	(1.67 4.3 9)	(1.7 6.37 9)	(0.11 3.06 9)	(0 0.83 3)	(0 0.75 5)	(0 0.88 5)
A6	(1 6.22 9)	(0 3.79 9)	(1.67 5.1 9)	(2.8 6.37 9)	(0.55 4.43 9)	(0.33 4 9)	(0 0.42 3)	(0 0.49 3)
A7	(0 3.11 7)	(0 4.38 9)	(1.67 5.1 9)	(2.8 5.90 9)	(0.33 3.97 9)	(0 0.67 3)	(0 1.33 9)	(0 0.39 3)

In this step, we used step 6 to obtain FPIS and FNIS for each criterion, which is shown in Table 17.

Evaluation of distance from FPIS and distance from FNIS (d_i^-) is performed with the help of step 7, i.e.,

$$\begin{aligned} d_1^+ &= d_v(\tilde{v}_{11}, \tilde{v}_1^+) + d_v(\tilde{v}_{12}, \tilde{v}_2^+) + d_v(\tilde{v}_{13}, \tilde{v}_3^+) + d_v(\tilde{v}_{14}, \tilde{v}_4^+) \\ &+ d_v(\tilde{v}_{15}, \tilde{v}_5^+) + d_v(\tilde{v}_{16}, \tilde{v}_6^+) + d_v(\tilde{v}_{17}, \tilde{v}_7^+) + d_v(\tilde{v}_{18}, \tilde{v}_8^+) \end{aligned}$$

Table 17 Fuzzy positive ideal solution and negative ideal solution

Criteria	FPIS	FNIS
C_1	(9, 9, 9)	(0, 0, 0)
C_2	(9, 9, 9)	(0, 0, 0)
C_3	(9, 9, 9)	(0, 0, 0)
C_4	(9, 9, 9)	(0, 0, 0)
C_5	(9, 9, 9)	(0, 0, 0)
C_6	(9, 9, 9)	(0, 0, 0)
C_7	(9, 9, 9)	(0, 0, 0)
C_8	(9, 9, 9)	(0, 0, 0)

$$d_v(\tilde{v}_{11}, \tilde{v}_1^+) = \sqrt{\frac{1}{3}[(1-9)^2 + (6.666-9)^2 + (9-9)^2]} = 4.81$$

$d_v(\tilde{v}_{12}, \tilde{v}_2^+) = 6.04$,	$d_v(\tilde{v}_{13}, \tilde{v}_3^+) =$ 4.657426,	$d_v(\tilde{v}_{14}, \tilde{v}_4^+) =$ 4.496977,	$d_v(\tilde{v}_{15}, \tilde{v}_5^+) =$ 5.939992

$d_v(\tilde{v}_{16}, \tilde{v}_6^+) = 5.43$, $d_v(\tilde{v}_{17}, \tilde{v}_7^+) = 6.79$, $d_v(\tilde{v}_{18}, \tilde{v}_8^+) = 4.12$.

Adding all values, we get $d_1^+ = 42.37$.

Similarly, we have evaluated all (d_i^+) and (d_i^-) which are mentioned in Table 18.

Further, the closeness coefficient CCi has been assessed by using step 8 of methodology, and according to the closeness coefficient CCi, we have ranked all the alternatives. The outcomes are shown in Table 18, i.e., Alternative Varanasi Ghats (A_1) is recommended as the first preference since it has the highest score while alternative Sarnath (A_4) is second, alternative Kashi Vishwanath Temple (A_2) is fourth, alternative Sankat Mochan Hanuman Temple (A_3) is seventh, alternative Ramnagar Fort (A_5) is the sixth preference, BHU Shiva temple (A_6) is at third, and Famous cuisine shops (A_7) have the fifth rank.

Table 18 Distances of alternatives from FPIS and FNIS

Alternatives	(d_i^+)	(d_i^-)	$(d_i^+) + (d_i^-)$	$CCi = \frac{d_i^-}{d_i^- + d_i^+}$	Ranking
A_1	42.37	45.65	88.01	0.518634	1
A_2	45.65	39.79	85.44	0.465657	4
A_3	47.12	35.94	83.07	0.432731	7
A_4	45.09	41.16	86.25	0.477198	2
A_5	47.42	36.49	83.9	0.434845	6
A_6	43.37	39.58	82.95	0.477152	3
A_7	45.99	37.11	83.1	0.446600	5

6 Conclusions

The purpose of the proposed Intelligent Information system is to provide the most likely ranking of the selected tourist destinations of Varanasi. The ranking mentioned in Table 18 regarding the tourist destinations has been designed by the system. This system will act as a referral system among tourists, tour operators, and respective tourism-related administrative bodies.

Acknowledgements We are grateful to Ms. Preeti Srivastava, Deputy Director of Tourism, Government U.P., India, for providing valuable suggestions. And we express our sincere thank to The Superintending Archaeologist, Archaeological Survey of India, Sarnath Circle, Varanasi, for allowing us to interact with tourists for their opinion regarding the current article.

Conflicts of Interest There is no conflict of interest among the authors.

References

1. Qu, H., Kim, L.H., Im, H.H.: A model of destination branding: integrating the concepts of the branding and destination image (2011)
2. Ruzzier, M.K., Gartner, W.C.: Customer-based brand equity for a destination branding: I feel Slovenia view project customer-based brand equity for destination view project. Ann. Tour. Res. (2013)
3. Cardoso, L., Dias, F., de Araújo, A.F., Andrés Marques, M.I.: A destination imagery processing model: structural differences between dream and favourite destinations. Ann. Tour. Res. **74**, 81–94 (2019)
4. Rajesh, R.: Impact of tourist perceptions, destination image and tourist satisfaction on destination loyalty: a conceptual model. **11**, 67–78 (2013)
5. Hunt, J.: Journal of travel research and undefined 1975, "Image as a factor in tourism development." journals.sagepub.com
6. Gunn, C.: Austin University of Texas and undefined 1972, "Vacationscape: designing tourist environments."
7. Mayo, E.: Association fourth annual conference proceedings and undefined 1973, "Regional images and regional travel behavior."
8. McFarland, C., Glisky, E.: Implementation intentions and imagery: individual and combined effects on prospective memory among young adults. Mem. Cogn. **40** (1), 62–69 (2012)
9. Fakeye, P.C., Crompton, J.L.: Image differences between prospective, first-time, and repeat visitors to the lower Rio Grande valley. J. Travel Res. **30**(2), 10–16 (1991)
10. Thai, N., Yuksel, U.: Annals of tourism research, and undefined 2017, "Too many destinations to visit: tourists' dilemma?" Elsevier
11. Ao, S.I.: Using fuzzy rules for prediction in tourist industry with uncertainty. In: 4th International Symposium on Uncertainty Modeling and Analysis, ISUMA 2003, pp. 213–218 (2003)
12. Göksu, A., Kaya, S.E.: Ranking of tourist destinations with multi-criteria decision making methods in Bosnia and Herzegovina (2014)
13. Chen, C.T.: Extensions of the TOPSIS for group decision-making under fuzzy environment. Fuzzy Sets Syst. **114**(1), 1–9 (2000)

Extensions of Some Known Differential and Integral Operators

Georgia Irina Oros

Classifications 30C20 · 30C45 · 30A10

1 Introduction and Preliminaries

The theory of differential subordinations was initiated by P. T. Mocanu together with Professor S. S. Miller in two articles in 1978 [6] and 1981 [7]. This theory subsequently became very popular and its development was broad and fast. Some recent results related to the subject can be seen in [2, 10–12, 15].

Studying subordination properties obtained by using differential and integral operators is a topic initiated at the beginning of this theory but still of interest at this time, many researchers still finding interesting results. Introducing new classes of analytic functions is the most common outcome of the study that involves operators as it can be seen in many recent papers. In [19], the authors have introduced a general family of bi-univalent functions using the Tremblay fractional derivative operator. In paper [20], the authors use the linear operator introduced and studied earlier by Srivastava and Attiya and investigate classes of admissible functions presenting properties of the third-order differential subordinations. In another recent paper [21], the authors study the Fekete–Szegő functional defined for a new subclass of analytic functions and the same Fekete–Szegő coefficient functional for a newly introduced class of normalized analytic functions in U was introduced in paper [22]. By using the Dziok–Srivastava convolution operator, in paper [23], the authors analyse some subclasses of univalent functions in U.

G. I. Oros (✉)
Department of Mathematics and Computer Science, University of Oradea,
1, Universit ă ţii street, 410087 Oradea, Romania

P. Srivastava et al. (eds.), *Mathematical, Computational Intelligence and Engineering Approaches for Tourism, Agriculture and Healthcare*, Lecture Notes in Networks and Systems 214, https://doi.org/10.1007/978-981-16-3807-7_2

The original results contained in this paper refer to a new differential–integral operator introduced by using the classical Ruscheweyh [18] and Alexander [1] operators and a new class of univalent functions defined by using this operator. This newly introduced operator is used for obtaining subordination results related to this class. The idea has been used recently for introducing a differential–integral operator in paper [13] where using the method of the subordination chains, some differential subordinations in their special Briot–Bouquet form were obtained regarding the differential–integral operator introduced in the paper. The best dominant of the investigated differential subordination is also given and, as a consequence, sufficient conditions for univalence are stated in two criteria. Similar work containing subordination results obtained by using an operator introduced by mixing a differential operator and an integral one has been published recently [16]. Also, following the same idea, in paper [17], the author introduces new classes obtained with a linear operator introduced with the use of Sălăgean differential operator and generalized Alexander operator.

Throughout this material, we use the well-known definitions and notations familiar to the field of complex analysis and its particular area, Geometric Function Theory. The unit disc of the complex plane is denoted by U. $\mathcal{H}(U)$ stands for the class of analytic functions in the unit disc and the classical definition for class A_n is applied, and it is known that it contains all functions from class $\mathcal{H}(U)$ which have the specific form $f(z) = z + a_{n+1}z^{n+1} + a_{n+2}z^{n+2} + \cdots$, with $z \in U$ and A_1 written simply A. All the functions from class A, univalent in U, and with the properties $f(0) = 0$, $f'(0) = 1$, form the class denoted by S. In particular, the functions in class A who have the property that $\mathrm{Re}\,\frac{zf'(z)}{f(z)} > 0$ represent the class of starlike functions denoted by S^* and let K denote the class of functions in class A with the property $\mathrm{Re}\left(\frac{zf''(z)}{zf'(z)} + 1\right) > 0$, namely, the convex functions. Let $\mathcal{H}[a, n]$, $a \in \mathbb{C}$, $n \in \mathbb{N}^*$, represent the class of functions $f \in \mathcal{H}(U)$ which have the specific form $f(z) = a + a_n z^n + a_{n+1}z^{n+1} + \cdots$, with $z \in U$.

The classical definition for subordination as it can be seen in [8] is used and denoted by $f(z) \prec F(z)$ with f and F analytic functions. The definitions of the solution of the differential subordination $\psi(p(z), zp'(z), z^2p''(z); z) \prec h(z)$ with $\psi : \mathbb{C}^3 \times U \to \mathbb{C}$ and h an univalent function in U, dominant and best dominant of it are also the regular ones introduced by Professors Miller and Mocanu and can be found in the monograph that contains the core of the theory of differential subordination published in 2000 [8] just as it can be seen in most of the papers cited at references.

For introducing the new differential–integral operator, the well-known definitions are used:

Definition 1 (*Ruscheweyh* [18]) For $f \in A$, $m \in \mathbb{N}$, the operator R^m is defined by $R^m : A \to A$,

$$R^0[f](z) = f(z) \tag{1}$$
$$R^1[f](z) = z \cdot [R^0 f(z)]' = zf'(z)$$
$$\vdots$$
$$(m+1)R^{m+1}[f](z) = z \cdot [R^m f(z)]' + m \cdot R^m f(z),\ z \in U.$$

Definition 2 ([14]) For $f \in A$, $m \in \mathbb{N} = \{0, 1, 2, \ldots\}$, the operator $I^m : A \to A$ is defined by

$$I^0[f](z) = f(z) \tag{2}$$
$$I^1[f](z) = \int_0^z \frac{I^0[f](t)}{t} dt = \int_0^z \frac{f(t)}{t} dt$$
$$\vdots$$
$$I^m[f](z) = \int_0^z \frac{I^{m-1}[f](t)}{t} dt.$$

Remark 3 For $m = 1$, $I^1[f](z) = \int_0^z \frac{f(t)}{t} dt$ which is Alexander integral operator studied in [1].

We next give some lemmas that are used in proving the original results in the Main Results section.

Lemma A. ([6, 8] Miller and Mocanu) *Let q be a convex function in U and let*

$$h(z) = q(z) + n\alpha z q'(z),\ z \in U,$$

where $\alpha > 0$ and n is a positive integer.

If $p(z) = q(0) + p_n z^n + p_{n+1} z^{n+1} + \cdots$, $z \in U$ is holomorphic in U and

$$p(z) + n\alpha z p'(z) \prec h(z),\ z \in U,$$

then $p(z) \prec q(z)$, $z \in U$ and this result is sharp.

Lemma B. ([8], Corollary 2.6g.2) *Let $f \in A$ and*

$$I[f](z) = F(z) = \frac{2}{z} \int_0^z f(t) dt,\ z \in U.$$

If $\operatorname{Re}\left[\frac{zf''(z)}{f'(z)} + 1\right] > -\frac{1}{2}$ *then $I[f](z) = F(z) \in K$.*

Lemma C. ([3, 8] Hallenbeck and Ruscheweyh) *Let h be a convex function with $h(0) = a$ and let $\gamma \in \mathbb{C}^*$ be a complex number with* $\operatorname{Re}\gamma \geq 0$. *If $p \in \mathcal{H}[a, n]$ and $p(z) + \frac{n}{\gamma} zp'(z) \prec h(z)$, $z \in U$ then*

$$p(z) \prec q(z) \prec h(z), \ z \in U,$$

where

$$q(z) = \frac{\gamma}{n \cdot z^{\frac{\gamma}{n}}} \int_0^z h(t) \cdot t^{\frac{\gamma}{n}-1} dt.$$

The function q is convex and is the best (a, n)-dominant.

2 Main Results

First the new differential–integral operator is defined, then a new class of univalent functions is introduced and then several differential subordinations obtained by using this operator and the newly introduced class are studied which leads to stating a condition for univalence of functions in class A.

Definition 4 Let $m \in \mathbb{N}$, $0 \leq \lambda \leq 1$. Denote by L^m the operator defined as $L^m : A \to A$,

$$L^m[f](z) = (1 - \lambda)R^m[f](z) + \lambda I^m[f](z), \ z \in U, \tag{3}$$

where R^m is given by (1) and I^m is given by (2).

Remark 5 (a) If $f \in A$, meaning that $f(z) = z + \sum_{k=2}^{\infty} a_k z^k$ and using (1) and (2), we have

$$L^m[f](z) = (1 - \lambda)R^m[f](z) + \lambda I^m[f](z) \tag{4}$$

$$= (1 - \lambda)\left(z + \sum_{k=2}^{\infty} \begin{bmatrix} m + k - 1 \\ m \end{bmatrix} a_k z^k\right) + \lambda\left(z + \sum_{k=2}^{\infty} \frac{a_k}{k^m} z^k\right)$$

$$= z + \sum_{k=2}^{\infty} a_k z^k \left\{(1 - \lambda)\begin{bmatrix} m + k - 1 \\ m \end{bmatrix} + \lambda \frac{1}{k^m}\right\}.$$

(b) For $\lambda = 1$, the differential–integral operator L^m coincides with Alexander integral operator.
(c) For $\lambda = 0$, the differential–integral operator L^m coincides with Ruscheweyh differential operator.

Definition 6 If $0 \le \beta < 1$, $0 \le \lambda \le 1$, $m \in \mathbb{N}$, we let $M^m(\lambda, \beta)$ denote the class of functions $f \in A$ for which the inequality

$$\text{Re}\,(L^m[f](z))' > \beta,\ z \in U, \tag{5}$$

is satisfied, where $L^m[f](z)$ is given by (3).

Remark 7 (a) For $m = 0$, $0 \le \lambda \le 1$, $\beta = 0$, the operator $L^m[f]$ becomes

$$L^0[f](z) = (1-\lambda)R^0[f](z) + \lambda I^0[f](z) = (1-\lambda)f(z) + \lambda f(z) = f(z).$$

Then $M^m(\lambda, \beta)$ becomes

$$M^0(\lambda, 0) = R = \{f \in A : \text{ Re}\, f'(z) > 0\}$$

the class of functions with bounded rotation studied by Alexander [1] proving that $R \subset S$ and for which Kurzyz [4] and Mocanu [9] have proved that $R \not\subset S^*$. A more systematic study of class R was done by MacGregor [5].

(b) For $m = 0$, $0 \le \beta < 1$, $0 \le \lambda \le 1$, we have

$$M^0(\lambda, \beta) = \{f \in A : \text{ Re}\, f'(z) > \beta\},$$

a class of univalent functions.

(c) It is easy to prove that $M^m(\lambda, \beta)$ is convex.

Theorem 8 *For* $0 \le \beta < 1$, $0 \le \lambda \le 1$ *and* $m \in \mathbb{N}$ *the following inclusion is true:*

$$M^m(\lambda, \beta) \subset M^m(\lambda, \delta),$$

where $\delta = 2\beta - 1 + 2(1-\beta)\ln 2 < 1$.

The result is sharp.

Proof For $\ln 2 \simeq 0,69$, $0 \le \beta < 1$,

$$0 \le 0,69\beta < 0,69,\ 0,31 \le 0,69\beta + 0,31 < 0,69 + 0,31,$$

$$0,31 \le 0,69\beta + 0,31 < 1,$$

and we obtain $\delta = 0,69\beta + 0,31 < 1$.

Let the convex function

$$h(z) = \frac{1 + (2\beta - 1)z}{1 + z},\ z \in U. \tag{6}$$

For all $z \in U$, $\text{Re}\, h(z) > \beta$ and $h(0) = 1$.

We let $f \in M^m(\lambda, \beta)$. From Definition 6

we have that

$$\operatorname{Re}\,(L^m[f](z))' > \beta. \tag{7}$$

We let

$$\frac{L^m[f](z)}{z} = p(z),\ z \in U. \tag{8}$$

Using property (4), we obtain

$$p(z) = \frac{L^m[f](z)}{z} = \frac{z + \sum_{k=2}^{\infty} a_k z^k \left\{(1-\lambda)\left[\frac{m+k-1}{m}\right] + \lambda \frac{1}{k^m}\right\}}{z}$$

$$= 1 + \sum_{k=2}^{\infty} a_k z^{k-1} \left\{(1-\lambda)\left[\frac{m+k-1}{m}\right] + \lambda \frac{1}{k^m}\right\} = 1 + p_1 z + p_2 z^2 + \cdots,$$

hence $p \in \mathcal{H}[1, 1]$ since $p(0) = 1$.

From (8), we deduce

$$L^m[f](z) = zp(z),\ z \in U. \tag{9}$$

By differentiation of (9), we obtain

$$(L^m[f](z))' = p(z) + zp'(z),\ z \in U. \tag{10}$$

Then (7) transforms into

$$\operatorname{Re}\,[p(z) + zp'(z)] > \beta. \tag{11}$$

Relation (11) is written in a subordination form as

$$p(z) + zp'(z) \prec h(z) = \frac{1 + (2\beta - 1)z}{1 + z},\ z \in U.$$

Taking $\gamma = 1$, $n = 1$, and by applying Lemma C, we obtain

$$p(z) \prec q(z) = \frac{1}{z} \int_0^z h(t)dt,\ z \in U,$$

i.e.

$$\frac{L^m[f](z)}{z} \prec q(z) = \frac{1}{z} \int_0^z \frac{1 + (2\beta - 1)t}{1 + t} dt.$$

Function $q(z)$ is a convex function and is the best dominant.

Since q is convex and $p(z) \prec q(z)$, $z \in U$, we have

$$\operatorname{Re} p(z) > \operatorname{Re} q(1) = 2\beta - 1 + 2(1-\beta)\ln 2 = \delta, \tag{12}$$

hence

$$\operatorname{Re} \frac{L^m[f](z)}{z} > \delta. \tag{13}$$

From Definition 4, relation (4), we can write

$$\frac{L^m[f](z)}{z} = \frac{z + \sum_{k=2}^{\infty} a_k z^k \left\{(1-\lambda)\begin{bmatrix} m+k-1 \\ m \end{bmatrix} + \lambda \frac{1}{k^m}\right\}}{z} \tag{14}$$

$$= 1 + \sum_{k=2}^{\infty} a_k z^{k-1} \left\{(1-\lambda)\begin{bmatrix} m+k-1 \\ m \end{bmatrix} + \lambda \frac{1}{k^m}\right\}$$

$$= 1 + \sum_{k=2}^{\infty} \frac{a_k}{k} z^{k-1} k \left\{(1-\lambda)\begin{bmatrix} m+k-1 \\ m \end{bmatrix} + \lambda \frac{1}{k^m}\right\}$$

$$= 1 + \sum_{k=2}^{\infty} b_k z^{k-1} k \left\{(1-\lambda)\begin{bmatrix} m+k-1 \\ m \end{bmatrix} + \lambda \frac{1}{k^m}\right\}$$

$$= \left(z + \sum_{k=2}^{\infty} b_k z^k \left\{(1-\lambda)\begin{bmatrix} m+k-1 \\ m \end{bmatrix} + \lambda \frac{1}{k^m}\right\}\right)' = (L^m[g](z))'$$

where

$$g(z) = z + \sum_{k=2}^{\infty} b_k z^k, \quad b_k = \frac{a_k}{k}.$$

By using (14) in (13), we get

$$\operatorname{Re} \frac{L^m[f](z)}{z} = \operatorname{Re} (L^m[g](z))' > \delta, \tag{15}$$

which is equivalent to $f \in M^m(\lambda, \delta)$ with δ given by (12).

Since $f \in M^m(\lambda, \beta)$ was taken arbitrary and we have proved that $f \in M^m(\lambda, \delta)$ we conclude that $M^m(\lambda, \beta) \subset M^m(\lambda, \delta)$.

The following corollary of Theorem 8 can be stated:

Corollary 9 *If $f \in M^m(\lambda, \delta)$, then*

$$\mathrm{Re}\,\frac{L^m[f](z)}{z} > 2\beta - 1 + 2(1-\beta)\ln 2 = \delta.$$

Proof Following the proof of Theorem 8, we have

$$\frac{L^m[f](z)}{z} \prec q(z) = \frac{1}{z}\int_0^z h(t)dt = \frac{1}{z}\int_0^z \frac{1+(2\beta-1)t}{1+t}dt$$
$$= 2\beta - 1 + 2(1-\beta)\frac{\ln(1+z)}{z}.$$

Since q is a convex function, we have

$$\mathrm{Re}\,\frac{L^m[f](z)}{z} > q(1) = 2\beta - 1 + 2(1-\beta)\ln 2.$$

Theorem 10 *Let $q(z) = \dfrac{1-z}{1+z}$ with $q(0) = 1$, a function that is convex and let h be given as*

$$h(z) = q(z) + zq'(z) = \frac{1-2z-z^2}{(1+z)^2},\ z \in U.$$

If $f \in A$, $0 \leq \lambda \leq 1$, $m \in \mathbb{N}$ and satisfies

$$\left(z \cdot \frac{L^m[f](z)}{L^{m-1}[f](z)}\right)' + z\left(z \cdot \frac{L^m[f](z)}{L^{m-1}[f](z)}\right)'' \prec \frac{1-2z-z^2}{(1+z)^2},\ z \in U, \tag{16}$$

then

$$\left(z \cdot \frac{L^m[f](z)}{L^{m-1}[f](z)}\right)' \prec \frac{1-z}{1+z},\ z \in U.$$

The result is sharp.

Proof Let

$$p(z) = \left(z \cdot \frac{L^m[f](z)}{L^{m-1}[f](z)}\right)',\ z \in U. \tag{17}$$

Using property (4) in (17), we have

$$p(z) = \left(z \cdot \frac{z + \sum_{k=2}^{\infty} a_k z^k \left\{ \left[\begin{matrix} m+k-1 \\ m \end{matrix} \right] (1-\lambda) + \lambda \cdot \frac{1}{k^m} \right\}}{z + \sum_{k=2}^{\infty} a_k z^k \left\{ \left[\begin{matrix} m+k-2 \\ m-1 \end{matrix} \right] (1-\lambda) + \lambda \cdot \frac{1}{k^{m-1}} \right\}} \right)' \tag{18}$$

$$= \left(z \cdot \frac{1 + \sum_{k=2}^{\infty} a_k z^{k-1} \left\{ \left[\begin{matrix} m+k-1 \\ m \end{matrix} \right] (1-\lambda) + \lambda \cdot \frac{1}{k^m} \right\}}{1 + \sum_{k=2}^{\infty} a_k z^{k-1} \left\{ \left[\begin{matrix} m+k-2 \\ m-1 \end{matrix} \right] (1-\lambda) + \lambda \cdot \frac{1}{k^{m-1}} \right\}} \right)'$$

$$= \left[z(1 + p_1 z + p_2 z^2 + \cdots) \right]' = 1 + q_1 z + q_2 z_2 + \cdots, \; z \in U.$$

From (18), we have that $p \in \mathcal{H}[1, 1]$ since $p(0) = 1$.
Differentiating (17), we obtain

$$p'(z) = \left(z \cdot \frac{L^m[f](z)}{L^{m-1}[f](z)} \right)'' \text{ and } zp'(z) = z \left(z \cdot \frac{L^m[f](z)}{L^{m-1}[f](z)} \right)''. \tag{19}$$

From (17) and (19), we have

$$p(z) + zp'(z) = \left(z \cdot \frac{L^m[f](z)}{L^{m-1}[f](z)} \right)' + z \left(z \cdot \frac{L^m[f](z)}{L^{m-1}[f](z)} \right)''. \tag{20}$$

Using (20), relation (16) becomes

$$p(z) + zp'(z) \prec q(z) + zq'(z) = \frac{1 - 2z - z^2}{(1+z)^2}, \; z \in U.$$

Taking now $n = 1$ and $\alpha = 1$ in order to apply Lemma A, using it, we obtain that $p(z) \prec q(z)$, i.e.

$$\left(z \cdot \frac{L^m[f](z)}{L^{m-1}[f](z)} \right)' \prec \frac{1-z}{1+z}$$

this being a sharp result.

Remark 11 The following condition for univalence of functions from class A is deduced using the results in Theorem 10:

Since $\text{Re}\, \frac{1-z}{1+z} > 0$, $z \in U$, from

$$\left(z \cdot \frac{L^m[f](z)}{L^{m-1}[f](z)} \right)' \prec \frac{1-z}{1+z},$$

we obtain

$$\operatorname{Re}\left(z\cdot\frac{L^m[f](z)}{L^{m-1}[f](z)}\right)' > \operatorname{Re}\frac{1-z}{1+z} > 0,\ z\in U,$$

which implies the univalence of the function

$$g(z) = z\cdot\frac{L^m[f](z)}{L^{m-1}[f](z)}.$$

Theorem 12 *Let $h\in A$ with*

$$\operatorname{Re}\left[1+\frac{zh''(z)}{h'(z)}\right]' > -\frac{1}{2},\ z\in U. \tag{21}$$

For $f\in A$, taking $0\le\lambda\le 1$ and $m\in\mathbb{N}$, if f satisfies

$$\frac{L^m[f](z)}{L^{m-1}[f](z)} + z\left(\frac{L^m[f](z)}{L^{m-1}[f](z)}\right)' \prec h(z),\ z\in U, \tag{22}$$

then

$$\frac{L^m[f](z)}{L^{m-1}[f](z)} \prec q(z),\ z\in U,$$

where q is the convex best dominant given by

$$q(z) = \frac{1}{z}\int_0^z h(t)dt,\ z\in U. \tag{23}$$

Proof We let

$$p(z) = \frac{L^m[f](z)}{L^{m-1}[f](z)},\ z\in U. \tag{24}$$

Using property (4) in relation (24), we get

$$p(z) = \frac{z+\sum_{k=2}^{\infty}\left\{\begin{bmatrix} m+k-1 \\ m\end{bmatrix}(1-\lambda)+\lambda\cdot\frac{1}{k^m}\right\}}{z+\sum_{k=2}^{\infty}a_k z^k\left\{\begin{bmatrix} m+k-2 \\ m-1\end{bmatrix}(1-\lambda)+\lambda\cdot\frac{1}{k^{m-1}}\right\}}. \tag{25}$$

From (24), we obtain

$$p'(z) = \left(\frac{L^m[f](z)}{L^{m-1}[f](z)}\right)',\ \text{and}\ zp'(z) = z\left(\frac{L^m[f](z)}{L^{m-1}[f](z)}\right)',$$

hence

$$p(z) + zp'(z) = \frac{L^m[f](z)}{L^{m-1}[f](z)} + z\left(\frac{L^m[f](z)}{L^{m-1}[f](z)}\right)', \ z \in U. \tag{26}$$

Using (26), relation (22) gives

$$p(z) + zp'(z) \prec h(z), \ z \in U.$$

Using Lemma C with $n = 1$, $\gamma = 1$, we obtain

$$p(z) \prec q(z) = \frac{1}{z}\int_0^z h(t)dt,$$

i.e.

$$\frac{L^m[f](z)}{L^{m-1}[f](z)} \prec \frac{1}{z}\int_0^z h(t)dt, \ z \in U.$$

We prove the convexity of function q given in (23) using Lemma B since function h satisfies (21).

Example 13 We consider function

$$h(z) = \frac{1}{(1+z)^2}, \ z \in U, \ \text{with } h(0) = 1, \ \text{for which } h'(z) = \frac{-2}{(1+z)^3},$$

$$h'(0) = -2 \neq 0, \ h''(z) = \frac{6}{(1+z)^4}.$$

Wc have

$$\operatorname{Re}\left[\frac{zh''(z)}{h'(z)} + 1\right] = \operatorname{Re}\left(1 - \frac{3z}{1+z}\right) = \operatorname{Re}\frac{1-2z}{1+z} > -\frac{1}{2}, \ z \in U$$

and

$$q(z) = \frac{1}{z}\int_0^z \frac{1}{(1+t)^2}dt = \frac{1}{z} - \frac{1}{z(1+z)}, \ z \in U.$$

Let $m = 2$, $\lambda = \frac{1}{4}$, $f(z) = z + \frac{1}{2}z^2$,

$$L^2[f](z) = z + \frac{37}{32}z^2,$$

$$L^1[f](z) = z + \frac{13}{16}z^2.$$

If $m = 2$, $\lambda = \frac{1}{4}$, $f(z) = z + \frac{1}{2}z^2$,

$$L^2[f](z) = z + \frac{37}{32}z^2,$$

$$L^1[f](z) = z + \frac{13}{16}z^2,$$

from Theorem 12, the differential subordination

$$\frac{1+\frac{37}{12}z}{1+\frac{13}{16}z} + z\left(\frac{1+\frac{37}{32}z}{1+\frac{13}{16}z}\right)' \prec \frac{1}{(1+z)^2},\ z \in U$$

implies

$$\frac{1+\frac{37}{12}z}{1+\frac{13}{16}z} \prec \frac{1}{z} - \frac{1}{z(1+z)},\ z \in U.$$

The function

$$q(z) = \frac{1}{z} - \frac{1}{z(1+z)}$$

is a convex function and is the best dominant.

Theorem 14 *Given q, a convex function in U satisfying $q(0) = 1$ we define*

$$h(z) = q(z) + \frac{1}{c+2}zq'(z),\ z \in U, \tag{27}$$

with $c > -2$.

If f and F are two functions from class $M^m(\lambda, \beta)$ which satisfy

$$L^m[F](z) = \frac{c+2}{z^{c+1}}\int_0^z t^c L^m[f](t)dt, \tag{28}$$

then

$$(L^m[f](z))' \prec h(z),\ z \in U \tag{29}$$

implies

$$(L^m[F](z))' \prec q(z),\ z \in U.$$

This result is sharp.

Proof From (28), we have

$$z^{c+1}L^m[F](z) = (c+2)\int_0^z t^c L^m[f](t)dt, \; z \in U, \; c > -2. \tag{30}$$

Differentiating (30) with respect to z, we obtain

$$(c+1)z^c L^m[F](z) + z^c(L^m[F](z))' = (c+2)\, z^c L^m[f](z), \; z \in U.$$

Using this and doing some calculations, we obtain

$$(c+1)L^m[F](z) + z(L^m[F](z))' = (c+2)\, L^m[f](z)), \; z \in U. \tag{31}$$

From (31), after a short calculation, we obtain

$$(L^m[F](z))' + \frac{1}{c+2} z(L^m[F](z))'' = \left(L^m[f](z)\right)', \; z \in U. \tag{32}$$

We let

$$p(z) = (L^m[F](z))', \text{ and } z \cdot p'(z) = z \cdot \left(L^m[F](z)\right)'', \; z \in U. \tag{33}$$

From (4), we deduce

$$\begin{aligned} p(z) &= \left(z + \sum_{k=2}^{\infty} a_k z^k \left\{(1-\lambda)\begin{bmatrix} m+k-1 \\ m \end{bmatrix} + \lambda \frac{1}{k^m}\right\}\right)' \\ &= 1 + \sum_{k=2}^{\infty} a_k k z^{k-1} \left\{(1-\lambda)\begin{bmatrix} m+k-1 \\ m \end{bmatrix} + \lambda \frac{1}{k^m}\right\} \\ &= 1 + p_1 z + p_2 z^2 + \cdots, \; p \in \mathcal{H}[1,1]. \end{aligned}$$

Using (33) in (32), we obtain

$$p(z) + \frac{1}{c+2} zp'(z) = \left(L^m[f](z)\right)', \; z \in U.$$

Then (29) becomes

$$p(z) + \frac{1}{c+2} zp'(z) \prec h(z) = q(z) + \frac{1}{c+2} zq'(z), \; z \in U. \tag{34}$$

By applying Lemma A with $n = 1$, $\alpha = \dfrac{1}{c+2} > 0$, we get that

$$p(z) \prec q(z),\ z \in U, \quad \text{i.e.} \quad (L^m[F](z))' \prec q(z),\ z \in U$$

with q being the best dominant.

Example 15 Let

$$q(z) = \frac{1}{1-z},\ z \in U,\ q(0) = 1,\ q'(z) = \frac{1}{(1-z)^2},$$

$$q''(z) = \frac{-2(1-z)(1-z)'}{(1-z)^4} = \frac{2}{(1-z)^3}.$$

Then

$$\frac{zq''(z)}{q'(z)} + 1 = \frac{\dfrac{2z}{(1-z)^3}}{\dfrac{1}{(1-z)^2}} + 1 = \frac{2z}{1-z} + 1 = \frac{1+z}{1-z}$$

$$\operatorname{Re}\left(\frac{zq''(z)}{q'(z)} + 1\right) = \operatorname{Re}\frac{1+z}{1-z} = \operatorname{Re}\frac{1+x+iy}{1-x-iy}$$

$$= \operatorname{Re}\frac{(1+x+iy)(1-x+iy)}{(1-x-iy)(1-x+iy)}$$

$$= \operatorname{Re}\frac{(1+x)(1-x) - y^2 + i[y(1-x) + y(1+x)]}{(1-x)^2 + y^2}$$

$$= \operatorname{Re}\frac{(1-x^2-y^2) + i \cdot 2y}{(1-x)^2 + y^2} = \frac{1-x^2-y^2}{(1-x)^2+y^2} > 0,$$

since $x^2 + y^2 > 1$ for $z \in U$. From this, we get that function q is convex and $q(0) = 1$.

Let $c = -1$. Then

$$h(z) = q(z) + zq'(z) = \frac{1}{(1-z)^2}$$

and

$$f(z) = z + \frac{2}{13}z^2,\ f \in A,\ m = 2,\ \lambda = \frac{1}{2},\ k = 2.$$

Using (4) we deduce that

$$L^2[f](z) = z + \frac{1}{4}z^2.$$

Using (28), we have

$$L^2[F](z) = \int_0^z t^{-1}\left(t + \frac{1}{4}t^2\right) dt = \int_0^z \frac{t + \frac{1}{4}t^2}{t} dt = z + \frac{1}{8}z^2.$$

If

$$f(z) = z + \frac{2}{13}z^2,\ L^2[f](z) = z + \frac{1}{4}z^2,\ L^2[F](z) = z + \frac{1}{8}z^2,$$

$$h(z) = \frac{1}{(1-z)^2},\ q(z) = \frac{1}{1-z},$$

from Theorem 14 the differential subordination

$$1 + \frac{1}{2}z \prec \frac{1}{(1-z)^2} \text{ implies } 1 + \frac{1}{4}z \prec \frac{1}{1-z}$$

where $q(z) = \dfrac{1}{1-z}$ is the best dominant.

Conclusion

A new differential–integral operator is introduced mixing Ruscheweyh differential operator and Alexander integral operator. Using it, a new class of univalent functions is defined and several differential subordinations are obtained with the differential–integral operator and this class. As a consequence of some new results proved in one of the theorems, a new criterion for univalence of functions from class A is stated which could be inspiring for future research work. Relevant examples are included after the proofs of the theorems containing the original results.

References

1. Alexander, J.W.: Functions which map the interior of the unit circle upon simple regions. Ann. Math. **17**, 12–22 (1915)
2. Bulboacă, T.: Differential subordinations and superordinations. Recent Results. Casa Cărţii de Ştiinţă, Cluj-Napoca (2005)

3. Hallenbeck, D.J., Ruscheweyh, S.: Subordination by convex functions, Proc. Am. Math. Soc. **52**, 191–195 (1975)
4. Krzyk, J.: A counterexample concerning univalent function. Folia Soc. Scient. Lubliniensis **2**, 57–58 (1962)
5. Mac Gregor, T.H., Functions whose derivative has a positive real part, Trans. Am. Math. Soc. **104**, 532–537 (1962)
6. Miller, S.S., Mocanu, P.T.: Second order differential inequalities in the complex plane. J. Math. Anal. Appl. **65**, 298–305 (1978)
7. Miller, S.S., Mocanu, P.T.: Differential subordinations and univalent functions. Michig. Math. J. **28**, 157–171 (1981)
8. Miller, S.S., Mocanu, P.T.: Differential Subordinations. Theory and Applications. Marcel Dekker Inc., New York, Basel (2000)
9. Mocanu, P.T.: Somes starlikeness conditions for analytic functions. Rev. Roum. Math. Pures Appl. **33**(1-2), 117–124 (1988)
10. Mocanu, P.T., Bulboacă, T., Sălăgean, G.: Geometric Function Theory. (Romanian), Casa Cărţii de Ştiinţă, Cluj-Napoca (1999)
11. Oros, G.I.: New differential subordinations and superordinations. Strong differential subordinations, strong differential superordinations. Lambert Academic Publishing, Saarbrucken, Germany (2011), p. 263
12. Oros, G.I.: Using differential subordinations in the study of certain classes of univalent functions. Casa Cărţii de Ştiinţă, Cluj-Napoca (2008)
13. Oros, G.I., Alb-Lupaş, A.: Sufficient conditions for univalence obtained by using Briot-Bouquet differential subordination. Math. Stat. **8**(2), 126–136 (2020)
14. Oros, G.I., Oros, Gh., Diaconu, R.: Differential subordinations obtained with some new integral operators. J. Comput. Anal. Appl. **19**(5), 904–910 (2015)
15. Oros, Gh.: Convexity and Starlikeness in Geometric Function Theory. Caretaken by the PAMM, Centre at the Bute, Budapest (2001)
16. Pall-Szabo, Á. O.: On a class of univalent functions defined by Sălăgean integro—differential operator. Miskolc Math. Notes **19**(2), 1095–1106 (2018)
17. Páll-Szabó, Á. O.: Coefficient bounds and Fekete-Szegő problem for new classes of analytic functions defined by Sălăgean integro-differential operator. Acta Univ. Apulensis, Math. Inform. **57**, 147–158 (2019)
18. Ruscheweyh, S.T.: New criteria for univalent functions. Proc. Am. Math. Soc. **49**, 109–115 (1975)
19. Srivastava, H.M., Motamednezhad, A., Adegan, E.A.: Faber polynomial coefficient estimates for bi-univalent functions defined by using differential subordination and a certain fractional derivative operator. Mathematics 8, Article ID 172, 1–12 (2020)
20. Srivastava, H.M., Prajapati, A., Gochhayat, P.: Third-order differential subordination and differential superordination results for analytic functions involving the Srivastava-Attiya operator. Appl. Math. Inform. Sci. **12**, 469–481 (2018)
21. Srivastava, H.M., Hussain, S., Raziq, A., Raza, M.: The Fekete-Szegö functional for a subclass of analytic functions associated with quasi-subordination. Carpathian J. Math. **34**, 103–113 (2018)
22. Srivastava, H.M., Răducanu, D., Zaprawa, P.: A certain subclass of analytic functions defined by means of differential subordination. Filomat **30**, 3743–3757 (2016)
23. Xu, Q.-H., Xiao, H.-G., Srivastava, H.M.: Some applications of differential subordination and the Dziok-Srivastava convolution operator. Appl. Math. Comput. **230**, 496–508 (2014)

Two-Stage Neural Network-Structural Equation Modeling Approach to Studying Drivers for Hotel Objectives: With Implications

Chai Ching Tan and M. S. Julian

1 Introduction

Efforts of organizational learning [1], performance measurement [2], and relationship networks [3] are important strategies and parts of the operating systems for organizations in dealing with rapid changes in the environments. Organizations should position themselves in a learning rate faster than the changes of the environment in order to not fall behind the competition [4]. When organizations learn, they acclimatize new behaviors [5] and provide the sources of motivation and direction for continuous improvement and performance maintenance [6]. Nevertheless, an effective organizational learning needs the precedence variables and sources of information, i.e., performance measurements [7]. References [1, 8] stress that both the disciplinary fields of organizational learning and performance measurement systems share common schools of thoughts and thus should be integrated for systemic advantages. In particular, this study deals with the leveraging role of organizational learning serving to better utilize performance measurements and external network relationships. In addition, this research also takes the opportunity to exploit the analytics simulation and prediction ability of neural network method [9], prior to subjecting the predicted preliminary theoretical structures for rigorous statistical analyses using structural equation modeling (SEM). Specifically, the purpose of this study is to, theoretically and empirically, demonstrate the role of organizational learning in leveraging the functions of both performance measurements and external network relationship, as the base, to provide the rule-system and knowledges, for the organization to improve

C. C. Tan (✉)
Rattanakosin International College of Creative Entrepreneurship (RICE), Rajamangala University of Technology Rattanakosin, Nakhon Pathom, Thailand
e-mail: chaiching.tan@rmutr.ac.th

M. S. Julian
Department of Business Administration, National Central University, Taoyuan City, Taiwan

P. Srivastava et al. (eds.), *Mathematical, Computational Intelligence and Engineering Approaches for Tourism, Agriculture and Healthcare*, Lecture Notes in Networks and Systems 214, https://doi.org/10.1007/978-981-16-3807-7_3

its performances at operations, customer perceptions, and organizational objective levels.

The remainder of this study is organized as follows. The next section is a literature review critically and systematically positioning the hypothetical roles of the governing variables, which describes the development of a conceptual framework. The research approach is detailed in the third section, with a particular emphasis narrating the sampling and the questionnaire design. Finally, we present the findings of the data analysis, in sequential two-stages, and conclude and offer the theoretical and practical implications.

2 Literature Review

Organizational learning can be defined and approached from various different angles of theoretical concepts and methodical means, and as a result, its utilities also vary accordingly. At the fundamental level, organizations activate effective learning when they can acquire, interpret, distribute, and make meaning of information [10], such as information about organizational performances [11]. Processual wise, organizations expect effective learning to generate knowledge to guide actions, and only with actions that change becomes possible to face up to the dynamicity of the environment, both internally and externally. Such an action-oriented learning is stressed in the concept of action learning, which requires learning to arise in some real, complex, and stressful problems [12]. Specifically, through a process of learning that involves a systematic programmed knowledge and questioning insight, being further stimulated and supported by the learning attitude of the learners, with some learning skills, and favorable learning environment, individuals and teams can significantly improve performance and self-esteem, recognition, quality of life, and self-confidence, and self-actualization [13]. A careful review will not be difficult to see the usefulness of programmed knowledge and questioning insight of a typical action learning strategy in another organizational learning concepts known as single-loop learning and double-loop learning. While single-loop learning aims to rectify the deficiency, whether at attitudinal, skill or other competency level, using concept of continuous improvement, double-loop learning engages in cognitive and behavioral breakthrough by undertaking a process of inquiry, which is, as explained and illustrated in [14], significantly different from the inquiry characteristic of single-loop learning, predominantly due to unfreezing of pre-existing assumptions about the governing variables. These iterative learning, at lower and higher order or cognitive level, have been shown robustly reliable in knowledge and rule-system creation [15], such as operations management.

Organizational learning is indicative of the existence of a capacity of organizations to take effective actions for success and performances [16] through making use of resources such as energy, information and matter, and exploiting the knowledge or "rule system" produced from the learning and the actions, which shares the insights advocated in complex adaptive system [17]. What, then, are the action scopes an organization should utilize its learning strategy and capacity for changes and better leveraging of organizational performances to achieve business objectives? Toward this end, [18] advocates using balanced scorecard (BSC)'s four-perspective structures as a platform for organizations to measure and review change and to manage change process, as in doing so, it not only fulfils strategy-driven commitment of the organizations, but, simultaneously, enable the continuous strategic alignment of the different perspectives. The strategic alignment of the different perspectives in the BSC platform is further developed into a concept known as strategy map, which considers a causally strategic linkage of four perspectives, namely, of learning and growth strategies and objectives, operations, customer objectives, and financial objectives [19, 20]. Basically, strategy map is "a visual representation of the cause-and-effect relationships among the components of an organization's strategy" [21], which describes how an organization creates values, by recognizing five principles, namely, (1) strategy balances contradictory forces, without sacrificing long-term investments, (2) strategy is balanced on a differentiated customer value proposition (customer outcome), (3) value is created through internal business processes (operations management), (4) strategy consists of simultaneous, complementary themes, and (5) strategic alignment determines the value of intangible assets [20]. To put the last principle into the context of this research, we argue that organizational learning is effective when the learning efforts can establish the positive associations of the different perspectives as reflected in the strategy map and balanced scorecard platform.

In view of the above background, we argue, by borrowing from a neuro scientific perspective of organizational learning [16], that a systematic usage of performance measurement (which can facilitate the provision of relevant neural patterns of information) should be mediated through organization learning for exerting more significant impact on the operations management aspect of the organization than direct means, leading to the following hypotheses, H1a and H2b:

H1a: Organizational learning can significantly mediate the relationship between performance measurements (both at employee domain and organizational level) and the perceived performance level of operations management of the organization.

H1b: Organizational learning can significantly mediate the relationship between external network relationship and the perceived performance level of operations management of the organization.

External network relationship is a key resource-improving business practices [22, 23] and has lately been treated as a source of innovation through concepts and strategies advocated in brand co-creation [24]. Relationship channels or networks are conceptualized in this research as means of transferring and exchanging knowledge between organizations [25], and when organizations make use of them, it can enhance their innovativeness and competitiveness [26]. Having possessed of a role as means of transferring and exchanging knowledge [25], network of relationships can share similar role to performance measurement [7, 8] and can thus be inferred to assist organizational learning. In short, through relationship networks associated with "sets of purposeful and connected exchange relationships" [3], it opens up dialogue and enriches the perspectives and efforts in performance measurements; thus, also contributing to organizational learning and operations managements [27].

Hypotheses H1a and H2b serve numerous important theoretical contributions. First, it is an extension to the "strategy map" concept of [19–21] by taking performance measurements from within the organization, both employees and organizational levels, and information that arises from the external networks, as the fundamental driving resources. Second, the hypotheses bring the context of organizational learning to a fundamental data- and information-level in order to facilitate the process of knowledge or rule-system generation for better actions development and improvement as advocated in complex adaptive system (CAS) concept. Third, with performance measurements as the base and organizational learning as the enabler, organizations acquire the "unfreezing" tool to move organization forward with changes as advocated in Lewin's change model [28].

To empirically support H1a and H1b, the following hypothesis H2, which establishes the direct relationship of performance measurements and operations management, and hypothesis H3, which states the positive association between performance measurements and organizational learning, are required:

H2: Collectively, performance measurements (both at employee domain and organizational level, and information that arises through external networks) and external network relationship, can significantly influence the perceived performance level of operations management.

H3: Collectively, performance measurements (both at employee domain and organizational level, and information that arises through external networks) and external network relationship, can significantly influence the level of organizational learning.

Based on the theoretical root of single- and double-loop learning of [29], performance measurement serves as the base to enable organizations to critically question the current assumptions of Model I or theory-in-use and thus enacts to equip organizational members with the mindfulness context for initiating a process of cognitive construction or re-construction of strategies to meet goals [7, 8]. By explicitly monitoring and measuring the efforts made by the organizational members, it prompts organization to close the performance gap [29] and reduces the negative impact of the unproductive group-think effect [30]. Based on H1a, H1b to H3, a revised generic strategy map structure can be shown in Fig. 1 to represent the theoretical model of this research. Figure 1 illustrates three additional hypotheses, namely:

H4: Organizational learning is positively related to the ability of the hotel to deliver its customer value proposition and yield customer satisfaction.

H5: Performance level of operations management is positively related to the ability of the hotel to deliver its customer value proposition and yield customer satisfaction.

H6: Collectively, the three generic strategy map's perspectives, namely organizational learning, operations management and customer outcomes, can significantly influence the hotel performance objective.

More specifically, customer outcome connotes the quality services delivered as expected, and thus, receiving the customer satisfaction and favorable hotel image. Hotel image is often considered as the most influential element in services [31] because of the natural uniqueness of services like perishability, inseparability, tangibility, and heterogeneity. With intangibility and other complexity of factors as the core features and components of businesses, the overall image formed becomes uniquely

Fig. 1 The conceptual model

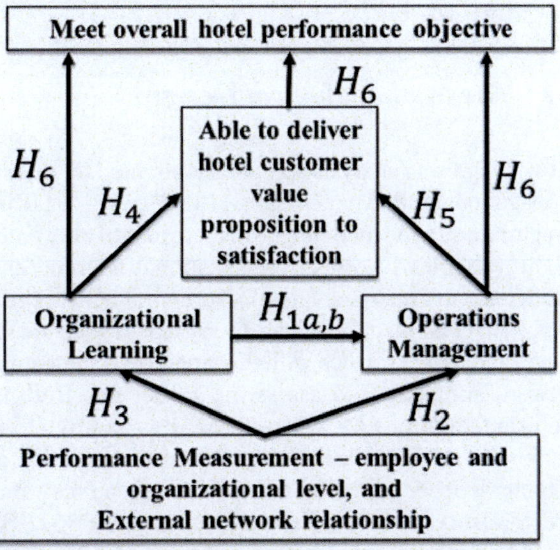

important. Image reflects the memorable experience of customers [32] and a potential value-creating force in brand-equity management [33] that characterizes a part of brand equity. In strategic and operations management context, hotels should not only pick a suitable strategic choice, but most importantly, the hotels should establish a capability to guarantee efficient service management [34], which is made possible by employees engaging in continuous learning to improve operations and service quality [7, 8].

Although Fig. 1 may seem static in nature, being constrained by the structure of the governing variables that share the fundamental shape of strategy map and balanced scorecard advocated in [20, 21], the model is actually dynamic in nature, partly driven by the double-loop and single-loop learning process that is supported by the facts tracked. In this way, the model can be continually subjected to a so-called Popperian epistemological process defined by a formula P1  TT  E  P2, where P1 is the problem in its original form, TT is a tentative solution of theory (its tentative strategy map), EE is error elimination (i.e., by means of neural network simulation, or structural equation model fit test), and P2 is the problem in a new form (i.e., in a strategic renewal state). The model shown in Fig. 1 can also connote the configuration architectural elements of generic business model, which consists of value discovery driven by performance measurements, network relationship and organizational learning [35], and value creation that is enabled by operations management and the ability to deliver hotel customer value proposition to satisfaction, and value realization being indicated by the extent of performance objectives being met. In addition, from the perspective of competence-based theory, which advocates on how competences should be developed, [36] argues that learning logic should be integrated into the operational capabilities.

3 Method

3.1 The Methodological Design

A two-stage data analysis approach is adopted for this study. First, feedforward back-propagation multilayer perceptron (MLP) is used, as a pre-run to statistical tests using structural equation modeling (SEM), to identify the preliminary relationship structure of the variables involving organizational learning and hotel objective as the two predicted outcomes (see Fig. 2). The neural network results help identify normalized important constructs, which are then used, in a second stage, in the structural equation model and in accordance with the predicted architectures, as suggested by the neural network simulation, to assess the model fits. Both incremental and absolute fits are examined in SEM, where SEM is "a family of statistical models that seek to explain the relationships among multiple variables… to estimate a series of separate, but interdependent, multiple regression equations simultaneously by specifying the structural model used by the statistical program" [37]. Both the perceptual opinions of the survey participants are used in the neural network simulation and SEM statistical tests. Although the perceptual measurements are the reflection of the consciousness, of subjective experiences and understanding of the participants (known as world 2 knowledge in Popperian epistemological terms), they are reliable mental process that intervenes in between the physical world 1 knowledge and world 3 of objective knowledge such as in the form of strategy map. These three-world epistemological insights are discussed in [38]. Popperian epistemology was also empirically shown to be a useful base for comprehending complex management issues in the discipline of strategic management in [39].

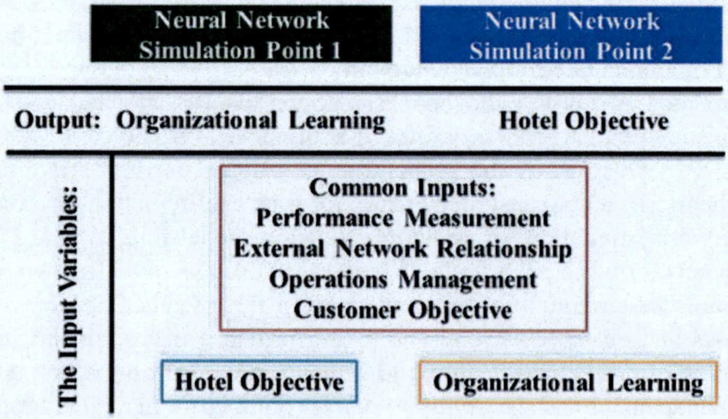

Fig. 2 The two neural network simulation points

3.2 Sampling and Data Collection

A cross-sectional survey was conducted obtaining the perceptions and attitudes of the hotel employees and management of the twenty-two hotels operated in Aceh province of Indonesia. Hotels participated include: Djawal hotel, Jeumpa hotel, Serambi Mekah, Aceh Barat, Grand Arabia, Grand Aceh Syariah, Hotel Kartika, Hotel 88, Hotel Mekkah, Hotel Kuala Radja, Hotel Lading, Sultan Hotel, Oasis Atjeh Hotel, Hotel Medan, Hotel Diana, Hotel Uin Arraniry, Meurah Mulia, Kyriad Muraya Aceh Hotel, Al-Hanafi Hotel, and The Pade and Hermes Palace Hotel. The descriptive profiles of the survey participants are given in the data analysis and discussion section. Aceh, with a population of about 4.2 million, is the westernmost province of Indonesia and has ten indigenous ethnic groups, with the largest being the Acehnese people, accounting for about 80–90% of the population. For Indonesia, in general, [40] reports that the revenue in the Indonesian hotel segments amounts to US$4133m in 2019, which ranks about top 18 in the world for the hotel industry, with 59% of the revenue through online sales in 2019, and the penetration rate is about 8%. In [41], the hotel rooms in Indonesia has been on the rise continuously, from 106,692 rooms in 2006 to 278,522 rooms in 2018. These hotel rooms were categorized as 2-star at 17.2%, 3-star at 35.9%, 4-star at 30%, and 5-star at 11.4% [41].

3.3 Measurement Instrument

The following states the definitions of the studied constructs, so as to lay the groundwork for ensuring that a measure assesses wat it purports to measure and thus conforms to the working definition of validity of construct [42], in terms of content validity, which is a formative validation that takes place prior to data collection [43], whereas the post-hoc validation of convergent validity and discriminant validity, after the research has been undertaken, is presented in the "data analysis and discussion" section. In short, measurement is designed to deliver valid measures, which depends on how the definition of the concept is designed to measure the construct [44] that also possesses reliability assurance, which describes the extent to which the measurement instrument produces similar results under constant conditions on all occasions, and thus, minimizes the errors and biases of the research study [45].

The questionnaire instrument was designed by adapting the operationalization concepts from [29] in organizational learning, as a process-oriented measure of the construct, and performance measurements, both at employee domain and organizational level [1], as the content-aspect of strategic focus, rather than as process in orientation [11]. The rationales of both organizational learning and performance measurement constructs are stated as follows:

Organizational Learning. This study operationalizes the organizational learning construct according to the single-loop and double-loop learning concept of [14], based on a definition that states the existence of organizational learning "when members of the organization act as a learning agent for the organization, responding to changes in the internal and external environment of the organization by detecting and correcting errors in organizational theory-in-use". In this definitional context, the instrumentation used are five-item statements: (1) Most of our hotel employees would raise issues and problems while at the same time suggest ways to solve them, (2) when faced with performance deficiency, i.e., lost sales, customer complaints, quality issues in food menu, our hotel will attempt to close the gaps, (3) if mistakes arise, our hotel would review the procedures and find out how we can prevent the mistakes from happening again, (4) our hotel regularly reviews and revises the policy to improve overall hotel performances, and (5) our hotel cares about employee's learning and improvement on attitude, knowledge and skills to better conduct the job role, i.e., serving customers, doing a maintenance and gardening. Item 4, in particular, involves with changes of views and assumptions, signifying some degree of double-loop learning in action.

Performance Measurement. Although there are many direct and obvious reasons for performance measurement such as to assess the performance of various processes [46], this research reinforces its stimulating effect on how organizational employees engage in learning and enhance operations improvement. Nine measurement items are employed for measuring the efforts exerted on performance measurement: (1) In our hotel, we measure employee performances in their job roles, (2) in our hotel, we measure employee performances in their attitudes to deliver the best services to their colleagues as well as the customers, (3) in our hotel, we measure employee performance in their ability to deliver quality services to our guest, (4) in our hotel, we measure customer feedback in various ways, i.e., survey forms, interviews, complaints, customer satisfaction, (5) in our hotel, we measure our ability to control costs, i.e., food and beverages, room services, (6) in our hotel, we measure and monitor the performance level of every type of operations, i.e., room service, front office, back offices and marketing and sales, (7) in our hotel, we measure continuous improvement initiatives, (8) in our hotel we monitor the perceptions of our guests in our services, and (9) in our hotel we monitor the image our guests form of our hotel. These nine items are factorized into two clusters, namely, employee-focused performance measurement (EPM) and organization-focused performance measurement (OPM).

External Network Relationship. As to the construct "external network relationship," it is operationalized to capture the relationship emphasis to benefit knowledge exchanges between organizations [25]. External network relationships with the outside stakeholders have been observed by many researchers to contribute toward information advantages [47]. A three-item measurement instrument is designed for this study, namely: (1) Our hotel maintains a network of relationship with key suppliers of raw materials, (2) our hotel maintains a network of relationship with key engineering supporting companies, and (3) our hotel maintains a network of relationship with other hotels in the same ranks. The other three constructs, namely, customer outcomes, hotel objective, and operations management, are the generic strategic focus in a company's strategy map, and thus, the concepts are adapted from [20, 21], but with application to a hotel context.

Customer Outcomes. Customer outcome is indicative of the success of the hotel operations that is able to meet the needs of the hotel customers, being represented by service quality, while fostering a positive image and customer satisfaction. Hotel image is an influential element in services and due to the uniqueness of hotel services as manifested in attributes such as perishability, inseparability, and heterogeneity [48], hotel image perception of the customers has been an important part of customer outcomes. Image reflects the memorable experience of customers [32] and is a potential value-creating force in brand-equity management [33].

Hotel Objective. Rather than focusing on only the financial perspective, this research takes a more holistic extension of the BSC concept of [19–21], which is summative or multi-dimensional in nature, which includes objectives of employee-aspect, customer-domain, financial perspective, and productivity expectation of the hotel.

Operations Management. Operations management examines the ability and efforts of the hotel in delivering according to the service standard requirement and the strategic objective of the hotel, and the questionnaire items are given as follows:

- Different departments, i.e. food, beverage, food, room service, front counters, in our hotel are working closely together.
- Our hotel uses modern communication technology and applications, i.e. LINE to communicate among each other in our team.
- Each of us can handle our job responsibilities to the standard demanded.
- Our hotel has best practices and standards to deliver quality of services in different job departments, e.g. check in.
- Our hotel continues to ensure efficiency of operations e.g. booking, room service.
- Our hotel continues to improve operational procedures to quickly respond to customer demands.
- Our hotel maintains a network of relationship with sales agents.
- Our hotel maintains a network of relationship with key suppliers of raw materials.
- Our hotel maintains a network of relationship with key engineering supporting companies from the outside.
- Our hotel maintains a network of relationship with other hotels in the same ranks.
- Our hotel maintains a network of relationship with hotel association in our region.
- We regularly develop new service ideas to increase customer expenses with us.
- Our hotel regularly participates in social responsibilities.
- Our Human Resource always focus to improve the working environment, quality of life and employee satisfaction.
- We continue to refurbish our infrastructure for the benefits of our customers and our hotel image.
- We share with other hotels on best practices such as computer programs for booking, check-in etc.

4 Data Analysis and Discussion

Data were collected from 151 employees of the hotels located in Aceh province of Indonesia. Table 1 summarizes the data profiles of the respondents, which sees majority of the hotels surveyed as 3-star or less, occupying 91.4 per cents. Majority of the hotel demographics are family-owned, at 74.2%, with the respondents holding university degree at 64.2%, and positions at supervisory level and below at 84.7%. Obviously, there is a lack of representatives at the managerial levels, but, nevertheless, in the emerging cyber-landscaping environment, where organizations continue to flatten and workforces becoming the important driving forces of the organizational success [49], the pictures revealed in the subsequent structural equation modeling (SEM) result should be representatives of the perceptions of the workforces. These

Table 1 General data profile, descriptive, and cross-comparative analyses

The Constructs										
Description		Freq	%	EPM	OPM	Network	OL	OM	HO	CO
Star	< 3 Star	78	51.7	3.81	3.80	3.76	3.76	3.74	3.70	3.76
	3 Star	60	39.7	3.88	4.01	3.93	3.94	3.85	3.88	3.96
	4 Star	13	8.6	3.77	4.06	3.86	3.92	3.67	3.48	3.69
	Sig				0.04					
Chain type	Chain	62	41.1	3.82	3.88	3.90	3.83	3.81	3.72	3.89
	Non-chain	89	58.9	3.85	3.93	3.79	3.85	3.76	3.76	3.80
Ownership	Family	112	74.2	3.87	3.90	3.83	3.86	3.79	3.78	3.85
	Public	7	4.6	3.66	3.61	3.42	3.70	3.59	3.68	3.77
	Joint V	32	21.2	3.76	4.01	3.96	3.84	3.79	3.65	3.80
Gender	Female	73	48.3	3.73	3.77	3.74	3.72	3.69	3.61	3.68
	Male	78	51.7	3.93	4.03	3.93	3.96	3.86	3.87	3.98
	Sig			0.064	0.003	0.056	0.003	0.055	0.019	0.004
Education (Degree)	Univ	97	64.2	3.95	3.98	3.88	3.92	3.79	3.76	3.85
	Other	54	36	3.63	3.77	3.75	3.71	3.76	3.72	3.81
	Sig			0.00	0.03		0.01			
Duty office	Front	77	51	3.84	3.91	3.84	3.86	3.82	3.79	3.93
	Back	74	49	3.83	3.90	3.83	3.83	3.74	3.71	3.74
	Sig									0.06
Position	<Supervisor	84	55.6	3.73	3.81	3.73	3.73	3.70	3.69	3.73
	Supervisor	44	29.1	3.93	3.99	3.93	3.93	3.88	3.81	3.97
	Assist. Mgr	13	8.6	3.82	4.06	3.87	3.95	3.68	3.75	3.83
	Manager	10	6.6	4.30	4.18	4.22	4.30	4.12	4.00	4.12
	Sig			0.054	0.073	0.067	0.002	0.063		

workforces are assumed to be result of followership and the management structures, as also highlighted in [50] that "good, skilled followers are able to nurture good leadership, by invisibly helping novice leader upright and on track". The equal distribution attributes are shown in the form of hotel operations structure, as either chain or non-chain, at respectively 41.1% and 58.9%, and also the gender, and duty distribution equally spreading between the front offices and the back offices.

Table 1 highlights the areas of significant differences, which is revealed by means of the analysis of variance (ANOVA) and *t*-tests. The areas of statistically significant differences between the different demographic groups are relating to education, duty, position, gender and star classification of the hotels. The university holders indicate a higher level of experiences with the measurement practices of the hotel operations, in areas such as using of customer feedback for service improvements. The male employees are significantly more perceptually favorable when compared to the female counterparts. Operations performance practices are more obvious for the 4-star ranked hotels than the lower tiers. The other variables, which do not show statistically significant differences, are ownership types and chain/non-chain types of the hotel operations. In note passing, the joint-venture is a value and risk co-sharing business entity that attempts to maximize the joint efforts in developing a more robust control structure for risk alleviation and performance enhancement and may thus create a performance-oriented motivation, with an emphasis on performance measurements and the external network relationships, as shown in Table 1. The different ownerships thus imply some sorts of working environment and cultural atmosphere that may influence the management efforts of the hotel operations, for instance, [51] discussed the working environment impact of joint ventures, due partly to resource and cultural compatibility of the co-investors, in organizational and operational performances.

In addition, Table 1 also shows that the front-office employees perceiving higher level on customer outcomes when compared to the back-offices. This obvious gap can be area of strains that induce the organizational members to discover and learn how to resolve the gaps or differences, in order to develop the ability, insight and understanding in regaining the trusts and loyalty of the customers. The perceptual gaps between the managers and the levels below should also be emphasized, as these can prevent misunderstanding, and upon attending to them, it could foster collaborative spirits and provide areas or opportunities for organizational learning and improvements of performances.

As indicated in Table 2, the Barlett test of sphericity is significant and that the Kaiser–Meyer–Olkin measure of sampling adequacy is sufficiently greater than 0.6. With the average variance extracted (AVEs) above the threshold of 0.50, and factor loading of the measurement items exceeding 0.70 [52], convergent validity is established. Besides, with the square-root of the AVE for each of the constructs, shown in the diagonal of the "construct" section of matrix in Table 2, as 0.82, 0.78, 0.78, 0.83, 0.81, and 0.82, in exceeding the cross-correlations below the diagonal, the discriminant validity is also maintained. Reliability of the measurement instrument is also evidenced with Cronbach alpha in exceeding the minimum threshold of 0.70

Table 2 Mean, convergent, and divergent validity, sampling adequacy for data reduction, and reliability

The constructs												
	Mean	Std. Dev	α	KMO	TVE	EPM	OPM	Net-work	OL	OM	HO	CO
EPM	3.84	0.6	0.853	0.714	0.773	0.88						
OPM	3.91	0.5	0.856	0.854	0.674	0.64	0.82					
Net-work	3.84	0.62	0.842	0.817	0.615	0.52	0.55	0.78				
OL	3.85	0.5	0.782	0.765	0.614	0.76	0.74	0.61	0.78			
OM	3.78	0.55	0.778	0.684	0.694	0.60	0.65	0.65	0.71	0.83		
HO	3.74	0.68	0.821	0.805	0.65	0.52	0.54	0.46	0.62	0.68	0.81	
CO	3.83	0.63	0.813	0.833	0.675	0.54	0.52	0.51	0.61	0.72	0.69	0.82
			Factor loadings: All > 0.70									
			Correlations Coefficients are significant at the 0.01 level (2-tailed)**									
			The diagonal = square root of TVE									

[53]. Note that ** in Table 2 indicates that the p-value of the correlations coefficients is less than the alpha threshold of 0.01 level.

Artificial neural network (ANN) is an insight-generation tool resembling the brain neuron-synapsis networks to help examine complex relationships of social and physical events or dynamics, whereby the linear models such as structural equation modeling (SEM) and multiple regressions may not be accurately feasible. The neural network configuration based on multilayer neural simulation (feedforward and backward propagation approach, with gradient descent optimization) is given in Fig. 3. The architecture illustrates a hidden layer. As a rule of thumb, the optimal number of hidden neurons is usually between the number of the input and the number of the output neurons [54].

Two staged neural networks were simulated.

First, it studies the important factors in predicting organizational learning, as shown in Table 3, which denotes of three constructs—EPM (employee performance measurement), OPM (organizational performance measurement), and external network relations (Network), as the key predictors.

Second, three key predictors considered important to explain organizational objective are OM (operations management), CO (customer outcomes), and OL (organizational learning), as shown in Table 4.

The SEM path structure is given in Fig. 4, which supports the six hypotheses stated. Based on the Maximum Livelihood (ML), it shows perfect SEM model fit, with $\chi^2/\text{df} = 0.433$, which is below the upper threshold of 5, at p not significant at 0.858. Increment model fits are also supported, with RMSEA = 0.000, Normed Fit Index (NFI) = 0.996, Tucker Lewis Index (TLI) = 1.017, Comparative Fit Index (CFI) = 1.000, and standardized root mean residual (SRMR) at 0.0097 [37]. The causal path analysis indicates that OL is an important mediator in leveraging the

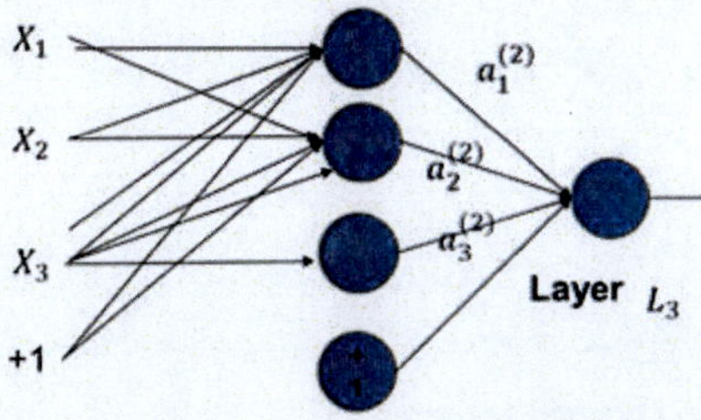

$$h_{w,b}(x) = a_1^{(3)} = f(W^T x) = f\left(\sum_{i=1}^{3} W_i a_i^{(2)} + b_1^{(2)}\right)$$

$$f: \Re \rightarrow \Re \ (activation\ function)$$

$$a_j^{(2)} = f\left(\sum_{i=1}^{3} W_{ji}^{(1)} + b_j^{(1)}\right), j = 1,2,3$$

$$J(W,b) = \frac{1}{m}\sum_{i=1}^{m}\frac{1}{2}\left\|h_{W,b}(x^{(i)}) - y^{(i)}\right\|^2 + \frac{\lambda}{2}\sum_{l=1}^{n_l-1}\sum_{i=1}^{s_l}\sum_{j=1}^{s_{l+1}}\left(W_{ji}^{(l)}\right)^2$$

$$W_{ij}^{(l)} = W_{ij}^{(l)} - \alpha\frac{\partial}{\partial W_{ij}^{(l)}}J(W,b), where\ \alpha\ is\ the\ learnign\ rate.$$

$$b_i^{(l)} = b_i^{(l)} - \alpha\frac{\partial}{\partial b_i^{(l)}}J(W,b)$$

Fig. 3 Multilayer neural network simulation structure

Table 3 Neural network simulated on organizational learning as the dependent variable

Cross-processing summary			
		N	Percent
Sample	Training	101	77.7%
	Testing	29	22.3%
Valid		130	100.0%
Excluded		21	
Total		151	
Network information			
Input layer	Factors	1	EPM
		2	OPM
		3	Network
		4	OM
		5	Overall hotel performance
		6	CO

(continued)

three predictors (EPM, OPM, Network) to influence the performances of operations management at 50 per cents of variance, which fulfils the core purpose of this study. Both OL and OM are important predictors of CO (customer outcome), in explaining 54 per cents of its variance. OL, OM, and CO, collectively, explain 55 per cents of the variance of organizational objective.

In short, as revealed in the graphical representation of the SEM paths, together with the statistics in Table 5, all the relationships hypothesized in H1a, H1b to H6, were supported.

Table 3 (continued)

Cross-processing summary

		N	Percent
	Number of units a		130
Hidden layer(s)	Number of hidden layers		1
	Number of units in hidden layer 1 a		12
	Activation function		Hyperbolic tangent
Output layer	Dependent variables	1	OL
	Number of units		1
	Rescaling method for scale dependents		Standardized
	Activation function		Identity
	Error function		Sum of squares

a Excluding the bias unit

Model summary

Training	Sum of squares error	3.854
	Relative error	0.077
	Stopping rule used	1 consecutive step with no decrease in error (a)
	Training time	0:00:00:33
Testing	Sum of squares error	6.731
	Relative error	0.507

Dependent variable: OL (Organizational learning)
(a) Error computations are based on the testing sample

Independent variable importance

	Importance	Normalized importance
EPM	0.228	100.00%
OPM	0.162	71.0%
Network	0.1830.149	80.3%
OM	0.162	65.3%
Overall hotel performance	0.115	71.1%
CO		50.5%

The present study supported the mediating role assumption of organizational learning posited in the literature review, which describes the leveraging ability of organizations to make use of the functions of performance measurements and network relationship in delivering the performances as expected at operational level, customer domain and overall company perspective. The mediating role of organizational learning is shown ranging from partial to perfect level. The study supports treating organizational learning as a means to exploit the intangible resources of the

Table 4 Neural network simulated on organizational objective as the dependent variable

Cross-processing summary			
		N	Percent
Sample	Training	101	83.5%
	Testing	20	16.5%
Valid		121	100.0%
Excluded		30	
Total		151	
Network information			
Input layer	Factors	1	EPM
		2	OPM
		3	Network
		4	OL
		5	OM
		6	CO
	Number of units a		155
Hidden layer(s)	Number of hidden layers		1
	Number of units in hidden layer 1 a		10
	Activation function		Hyperbolic tangent
Output layer	Dependent variables	1	OBJ
	Number of units		1
	Rescaling method for scale dependents		Standardized
	Activation function		Identity
	Error function		Sum of squares
a Excluding the bias unit			
Model summary			
Training	Sum of squares error		5.252
	Relative error		0.105
	Stopping rule used		1 consecutive step with no decrease in error (a)
	Training time		0:00:00:66
Testing	Sum of squares error		1.669
	Relative error		0.198
Dependent Variable: OBJ (b) Error computations are based on the testing sample			
Independent variable importance			
		Importance	Normalized importance
EPM		0.131	61.2%

(continued)

Table 4 (continued)

Cross-processing summary		
	N	Percent
OPM	0.114	53.4%
Network	0.143	66.9%
OL	0.187	87.5%
OM	0.213	100.0%
CO	0.213	99.6%

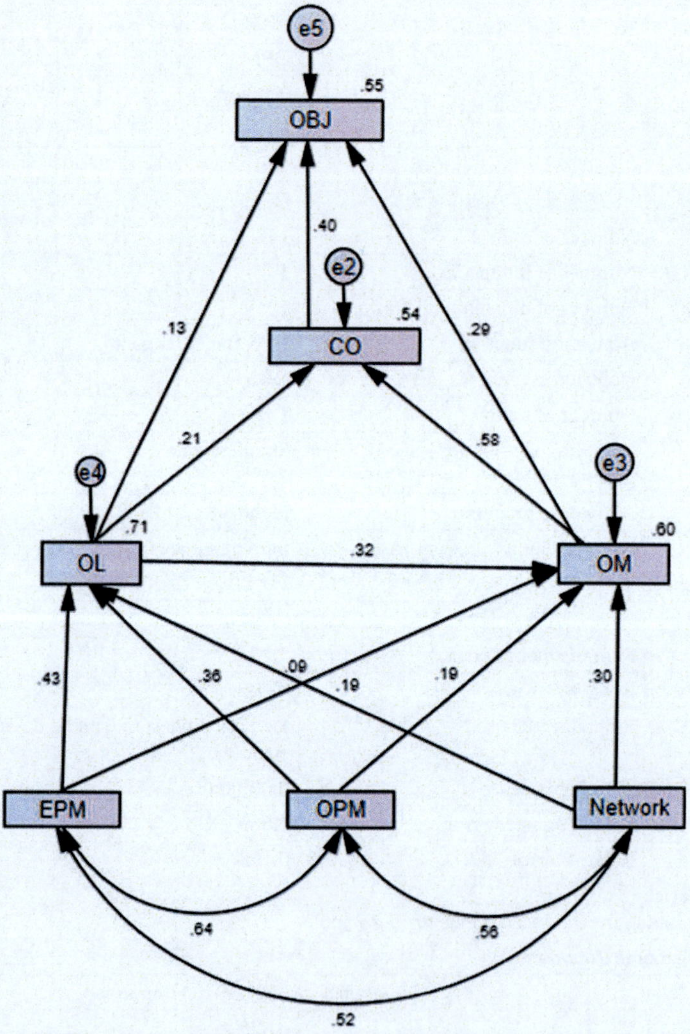

Fig. 4 Graphical representation of SEM results

Table 5 SEM results

Hypothesis	Path	Estimates	p value
H1a–H1b	OL is a perfect mediation, which has erased the significant and direct predictive power of EPM in explaining the variance of OM OL plays partial mediation role, which shows a reducing significant ability of OPM (at beta = 0.19, $p = 0.016$) and network relationship (beta = 0.30, $p = 0.0001$), as compared to the original unmediated state, at beta = 0.650	0.32	0.001
H2	EPM → OM OPM → OM Network relationship → OM	0.09 0.19 0.30	Not sig 0.0160 0.0001
H3	EPM → OL OPM → OL Network relationship → OL	0.43 0.36 0.19	0.0001 0.0001 0.0001
H4	OL → CO	0.21	0.0008
H5	OM → CO	0.58	0.0001
H6	OL → OBJ CO → OBJ OM → OBJ	0.13 0.40 0.29	0.0005 0.0001 0.0002

organizations, through performance measurements, at employee-domain and organizational level, and external network relationship. According to knowledge-based view (KBV) of firms, knowledge is often considered as the most important strategic resource of a firm in competition [55]. Fundamentally, performance measurement and external network relationship provide the data and information, and the resources, to facilitate organizational learning [56]. Considering in summative perspective, performance measurement, when aligned with other strategic and operational initiatives of organizations, such as organizational learning as empirically supported in this study, it can be reckoned to provide a cognitive and social space, sharing [57], who advocates that the appropriate use of performance measurements can create a lived space of high-energy and corporate uplifting, and organizations, as a result, can see the active and sensational interactions created among the employees and the external partners.

5 Conclusion

This research demonstrates the utility of neural network simulation to support structural equation modeling (SEM) analysis. Applying using the data from the hotel industry in Aceh, Indonesia, this research confirms the mediating role of organizational learning. The organizational learning combines single-loop and double-loop learning strategies, in targeting on continuous improvement and learning efforts to correct performance errors, by also assessing the suitability of the current mental models. Although the role of performance measurement has been widely acknowledged in terms of "what gets measured gets done", the mediating role of organizational learning cannot be ignored. Organizational learning is a principal means of achieving the strategic renewal of enterprise. This finding corresponds to a normative logic that describes the sequential flow of data (i.e., performance measurements) to information, and further, to support decision making. The model manifests as a complex adaptive system (CAS), where the organization invokes the pertaining knowledge contained in the balanced scorecard rules and exploits the stimuli the organization encounters, such as information and the external network-induced relationships, to produce effects within the system itself and externally. By embracing external network, this research makes an attempt to also extend the within-organizational effort in organizational learning, to a more open system, which contributes to a network perspective on organizational learning research.

References

1. Tan, C.C.: Stimulate changes of views, and measure to learn to transform organizational performance: an action research case. Adv. Sci. Lett. **24**(7), 5080–5085 (2018)
2. Tan, C.C., Anomasiri, S.: Attention-based view approach to the use of performance measurements to drive organizational performance. J. Accounting Profession **13**(39), 58–82 (2017)
3. Pibeam, C., Alvarez, G., Wilson, H.: The governance of supply networks: a systematic literature review. Supply Chain Manage. Int. J. **17**(4), 358–376 (2012)
4. Revans, R.W.: The origins and growth of action learning. Chartwell-Bratt, Bickley (1982)
5. Mwita, J.I.: Performance management model: a systems-based approach to public service quality. Int. J. Public Sect. Manage. **13**(1), 19–37 (2000)
6. Baker, D.F., Buckley, M.R.: A historical perspective of the impact of feedback on behavior. J. Manage. Theory **2**(4), 21–33 (1996)
7. Tan, C.C.: Double-loop learning performance measurement systems – Part I: laying the groundwork on the theories of organizational learning. IBFR-GCBF Proc. **2**(1), 91–95 (2007)
8. Tan, C.C.: Double-loop learning performance measurement systems – Part II: unifying schools of thought. IBFR-GCBF Proc. **2**(1), 96–100 (2007)
9. Delen, D., Zolbanin, H.M.: The analytics paradigm in business research. J. Bus. Res. **90**, 186–195 (2018)
10. Pokharel, M.P., Choi, S.O.: Exploring the relationships between the learning organization and organizational performance. Manage. Res. Rev. **38**(2), 126–148 (2015)
11. Henri, J.F.: Performance measurement and organizational effectiveness: bridging the gap. Manag. Finance **30**(6), 93–123 (2004)
12. Smith, P.A.C.: Q'ing action learning: more on miding our Ps and Qs. Manag. Decis. **35**(5), 365–372 (1997)
13. Koo, L.C.: Learning action learning. J. Work. Learn. **11**(3), 89–94 (1999)
14. Argyris, C., Schon, D.A.: Organizational learning: a theory of action perspective. Addison-Wesley, Harlow (1978)
15. Blackman, D., Connelly, J., Henderson, S.: Does double loop learning create reliable knowledge. Learn. Organ. **11**(1), 11–27 (2004)
16. Bennet, A., Bennet, D.: A new change model: factors for initiating and implementing personal action learning. VINE: J. Inf. Knowl. Manage. Syst. **38**(4), 378–387 (2008)
17. McElroy, M.W.: Integrating complexity theory, knowledge management, and organizational learning. J. Knowl. Manage. **4**(3), 195–203 (2000)

18. Jayashree, P., Hussain, S.J.: Aligning change deployment: a balanced scorecard approach. Meas. Bus. Excell. 63–85 (2010)
19. Kaplan, R.S., Norton, D.P.: The balanced scorecard – measures that drive performance. Harvard Bus. Rev. 71–79 (1992)
20. Kaplan, R.S., Norton, D.P.: The strategy map: guide to aligning intangible assets. Strategy Leadersh. **32**(5), 10–17 (2004)
21. Kaplan, R.S., Norton, D.P.: Strategy maps: converting intangible assets into tangible outcomes. Harvard Business School Press, Boston, Massachusetts (2004)
22. Jonwanapitakkul, P., Tan, C.C.: A resource-based view of the chili-paste business in Chiang Rai, Thailand. SELP J. Soc. Sci. **IX**(Special Issue, April), 7–12 (2018)
23. Tankam, S., Tan, C.C.: Practice-based view (PBV) of the longan industry in Chiang Rai. SELP J. Soc. Sci. **IX**(Special Issue, April), 1–6 (2017)
24. Hsieh, S.H., Chang, A.: The psychological mechanism of brand co-creation engagement. J. Interact. Mark. **33**, 13–26 (2016)
25. Beeby, M., Booth, C.: Networks and inter-organizational learning: a critical review. Learn. Organ. **7**(2), 75–88 (2000)
26. Hussain, Z., Dayan, M., Di Benedetto, C.A.: The impact of networking on competitiveness via organizational learning, employee innovativeness and innovation process: a mediation model. J. Eng. Tech. Manage. **40**, 15–28 (2016)
27. Andersen, P.H. Christensen, P.R.: Bridges over troubled water. Suppliers as connective nodes in global supply networks. J. Bus. Res. **58**(9), 1261–1273 (2005)
28. Lewin, K.: Frontiers in group dynamics: concept, method and reality in social science, social equilibria and social change. Hum. Relat. **1**(1), 5–41 (1947)
29. Argyris, C.: Double-loop learning in organizations: a theory of action perspective. In: Smith, K.G., Hitt, M.A. (eds.) Great minds in management: the process of theory development, pp. 261–279. Oxford University Press, Great Britain (2005)
30. Larsen, B.: One measurement in better than 1,000 opinions. Is it? Manag. Audit. J. **16**(2), 63–68 (2001)
31. Lahap, K., Ramli, N.S., Said, N.M., Radzi, S.M., Zain, R.A.: A study of brand image towards customer's satisfaction in the Malaysian hotel industry. Procedia Soc. Behav. Sci. **224**, 149–157 (2016)
32. Keller, K.L.: Conceptualizing, measuring and managing customer-based brand equity. J. Mark. **47**(1), 1–22 (1993)
33. Suhartanto, D., Kandampully, J.: The role of customer satisfaction and image in gaining customer loyalty in the hotel industry. J. Hosp. Leisure Mark. **10**(1/1/2), 1–24 (2003)
34. Peng, J.P., Zhao, X.Y., Mattila, A.S.: Improving service management in budget hotels. Int. J. Hosp. Manage. **49**, 139–148 (2015)
35. Tan, C.C., Sitikarn, B.: Balanced scorecard (BSC) structured business model for tribes-based community-based tourism: cases with Chiang Rai. J. MCU Peace Stud. **7**(4), 899–922 (2019)
36. Wallin, J.: Enhancing competences in business ecosystems. In: Sanchez, R., Heene, A. (eds.) A focused issue on competence perspective on new industry dynamics, pp. 83–116. Bingley, Emerald Group Publishing. https://doi.org/10.1108/S1744-2117(2012)0000006006, (2012)
37. Hair, J.F., Black, W.C., Babin, B.J., Anderson, R.E., Tatham, R.L.: Multivariate data analysis. Prentice Hall, Upper Saddle River, NJ (2006)
38. Popper, K.: All life is problem solving. Routledge, London (1999)
39. Tan, C.C.: Giving strategic management an epistemological structure: a case of community-based tourism. Natl. Acad. Manag. Staff Cult. Arts Herald **1**, 723–733 (2018)
40. Statista. https://www.statista.com/outlook/267/120/hotels/indonesia, 27 February 2020
41. CEIC, Indonesia hotel rooms. https://www.ceicdata.com/en/indonesia/number-of-hotel-rooms/number-of-hotel-rooms-total, Accessed 15 October 2019
42. Borsboom, D.: Measuring the mind: conceptual issues in contemporary psychometrics. Cambridge University Press, New York, NY (2005)
43. Brennan, L., Voros, J., Brady, E.: Paradigms at play and implications for validity in social marketing research. J. Soc. Mark. **1**(2), 100–119 (2011)

44. De Vaus, D.A.: Surveys in social research. Allen & Unwin, London (1991)
45. Amaratunga, D., Baldry, D., Sarshar, M., Newton, R.: Quantitative and qualitative research in the built environment: application of mixed research approach. Work Study **51**(1), 17–31 (2002)
46. Ahmad, S.B.S., Svalestuen, F., Andersen, B., Torp, O.: A review of performance measurement for successful concurrent construction. Procedia Soc. Behav. Sci. **226**, 447–454 (2016)
47. Galaskiewicz, J.: The new network analysis and its application to organizational theory and behavior. In: Iacobucci (ed.) Networks in marketing, pp. 19–31. Sage Publications, Thousand Oaks (1996)
48. Zeithaml, V.A., Bitner, M.J.: Services marketing. McGraw-Hill, USA (1996)
49. Brown, A.: The new followership: a challenge for leaders. Futurist **37**(2), 68 (2003)
50. Adghar, R.: Why followership is now more important than leadership. Forbes (2016)
51. Meschi, P., Roger, A.: Cultural context and social effectiveness in international joint ventures. Manag. Int. Rev. **34**(3), 197–215 (1994)
52. Fornell, C., Larcker, D.F.: Evaluating structural equation models with unobservable variables and measurement error. J. Mark. Res. **18**(1), 39–50 (1981)
53. Forza, C.: Survey research in operations management: a process-based perspective. Int. J. Oper. Prod. Manag. **22**(2), 152–194 (2002)
54. Blum, A.: Neural networks in C++. Wiley, New York, NY (1992)
55. Fang, E.A.H., Li, X.Y., Lu, J.J.: Effects of organizational learning on process technology and operations performance in mass customizers. Int. J. Prod. Econ. **174**, 68–75 (2016)
56. Srimai, S., Damsaman, N., Bangchokdee, S.: Performance measurement, organizational learning and strategic alignment: an exploratory study in Thai public sector. Meas. Bus. Excell. 57–69 (2011)
57. Tan, C.C., Arsirapongpisit, S.: Critical revisit to performance measurement from the spirituality postmodernism and quantum-relativisticism. In: The 7th Asia Pacific Decision Science Institute Proceedings, NIDA, Thailand (2002)
58. Dipboye, R.L.: Criterion development, performance appraisal, and feedback. In: Emerald Review of Industrial and Organizational Psychology, pp. 535–579. Bingley, Emerald Publishing Limited. https://doi.org/10.1108/978-1-78743-785-220181015, (2018)

Computation of Ideals in N-Topology

M. Lellis Thivagar, V. Antonysamy, and V. Ramesh

1 Introduction

The concept of ideal topological space was first studied by Kuratowski [6] and Vaidyanathaswamy [10] developed it further. The notion of semi-open sets and semi-continuity were introduced by Levine [9] in 1963. In 1990, Jankovic and Hamlett [5] investigated the topological ideals into general topology. Abd El-Monsef et al. [1] further investigated I-open sets and I-continuous functions. In 1996, Dontchev [2] extended this study to pre-I-open sets and obtained a decomposition of I-continuity. Recently, Hatir and Noiri [3, 4] have introduced the concept of semi-I-open sets and extended to the de-composition of its continuity.

Lellis Thivagar et al. [7, 8] was instrumental in extending the general topology up to N-topology. The N-topology in turn opened up the research gate and gave a lot of application to real life situations. In our study here we prove that N-topological space can be extended to ideals as well. While doing so we also discuss the possibilities of $N\tau\alpha$-local function, $N\tau$-semi and pre local functions and their continuity. As a culmination of this paper we compute an algorithm along with its programme to execute their relationship between them.

2 Preliminaries

We shall recap the necessary definitions here that will assist us to study ideal in N-topology in an elaborate manner.

Defintion 2.1 References [7, 8] let $\boldsymbol{d_1}$ and $\boldsymbol{d_2}$ be conjugate, quasi-pseudo metrics on $\boldsymbol{X}$ and define a function $\boldsymbol{d_3} : \boldsymbol{X} \times \boldsymbol{X} \to \boldsymbol{R}^+ \cup \{\boldsymbol{0}\}$ by

M. L. Thivagar · V. Antonysamy · V. Ramesh (✉)
School of Mathematics, Madurai Kamaraj University, Madurai, Tamil Nadu 625021, India

P. Srivastava et al. (eds.), *Mathematical, Computational Intelligence and Engineering Approaches for Tourism, Agriculture and Healthcare*, Lecture Notes in Networks and Systems 214, https://doi.org/10.1007/978-981-16-3807-7_4

$$d_3(x, y) = \frac{[2d_1(y, x) + d_2(y, x)]}{3} \text{ for all } x, y \in X$$

Then

(i) $d_3(x, x) = \frac{[2d_1(x,x)+d_2(x,x)]}{3}$ for all $x \in X$

(ii)
$$\begin{aligned} d_3(x, z) &= \frac{[2d_1(z, x) + d_2(z, x)]}{3} \\ &\leq \frac{[2(d_1(z, y) + d_1(y, x)) + (d_2(z, y) + d_2(y, x) + d_2(y, x))]}{3} \\ &= d_3(x, y) + d_3(y, z) \text{ for all } x, y, z \in X \end{aligned}$$

Therefore, d_3 is a quasi-pseudo metric on X and which is called a Mean Conjugate of d_1, d_2, and d_1. For each $i = 1, 2, 3$ the quasi-pseudo metric d_1 gives a topology τ_i whose base is $\{B_{d_i}(x, k_i)\}$, where $B_{d_i}(x, k_i) = \{y : d_i(x, y) < k_i\}$. Thus we define a none empty set X equipped with three arbitrary topologies τ_1, τ_2 and τ_3 is called a tri-topological space and is denoted by $(X, 3\tau)$ or $(X, \tau_1, \tau_2, \tau_3)$.

In general, let $d_1, d_2, \ldots, d_{N-1}$ be Mean Conjugate of d_1, d_2 and d_1; d_1, d_2, d_3 and d_1;...;$d_1, d_2, \ldots, d_{N-2}$ and d_1 respectively. Define a function $d_N : X \times X \to R^+ \cup \{0\}$ by.

$$d_N(x, y) = \frac{\left[d_1(y, x) + \sum_{i=1}^{N-1} d_i(y, x)\right]}{N} \text{ for all } x, y \in X$$

We can easily verify that d_N is a quasi-pseudo metric on X. We also note that for each N, $d_N(x, y) \neq d_N(y, x)$ for all $x, y \in X$ and d_N is called a Mean Conjugate of $d_1, d_2, \ldots, d_{N-1}$ and d_1. For each $i = 1, 2, \ldots, N$ the quasi-pseudo metric d_i gives a topology τ_i whose base is $\{B_{d_i}(x, k_i)\}$, where $B_{d_i}(x, k_i) = \{y : d_i(x, y) < k_i\}$. Thus, we define a non-empty set X equipped with N-arbitrary topologies $\tau_1, \tau_2, \ldots \tau_N$ is called a N-topological space and is denoted by $(X, N\tau)$ or $(X, \tau_1, \tau_2, \ldots \tau_N)$.

Definition 2.2 References [7, 8] let X be a non-empty set, τ_1 *and* τ_2 be two arbitrary topologies defined on X and the collection 2τ be defined $2\tau = \{S \subseteq X : S = (A_1 \cup A_2) \cup (B_1 \cap B_2),\ A_1, B_1 \in \tau_1$ and $A_2, B_2 \in \tau_2\}$ satisfying the following axioms:

(i) $X, \phi \in 2\tau$
(ii) $\cup_{i=1}^{\infty} S_i \in 2\tau$ for all $S_i \in 2\tau$
(iii) $\cap_{i=1}^{n} S_i \in 2\tau$ for all $S_i \in 2\tau$.

Then the pair $(X, 2\tau)$ is called a bi-topological space on X and elements of the collection 2τ are known as 2τ-open sets on X.

We can generalize the above definition as follows: Let X be a non-empty set, $\tau_1, \tau_2, \ldots \tau_N$ be N-arbitrary topologies defined on X and let the collection $N\tau$ be defined by

$N\tau = \{S \subseteq X : S = (\cup_{i-1}^{N} A_i) \cup (\cap_{i-1}^{N} B_i),\ A_i, B_i \in \tau_i\}$, satisfying the following axioms:

(i) $X, \phi \in N\tau$
(ii) $\cup_{i=1}^{\infty} S_i \in N\tau$ for all $S_i \in N\tau$
(iii) $\cap_{i=1}^{n} S_i \in N\tau$ for all $S_i \in N\tau$.

Then the pair $(X, N\tau)$ is called an N-topological space on X and the elements of the collection $N\tau$ are known as $N\tau$-open sets in X. A subset A of X is said to be $N\tau$-closed in X if the complement of A is $N\tau$-open in X. The set of all $N\tau$-open sets on X and the set of all $N\tau$-closed sets on X are respectively denoted by $N\tau O(X)$ and $N\tau C(X)$.

Definition 2.3 Vaidyanathaswamy [10] an ideal I on a topological space (X, τ) is a nonempty collection of subsets of X which satisfies the following conditions:

(i) $A \in I$ and $B \subset A \Rightarrow B \in I$;
(ii) $A \in I$ and $B \in I \Rightarrow A \cup B \in I$.

An ideal topological space is a topological space (X, τ) with an ideal I on X and if $P(X)$ is the set of all subsets of X, a set operator $(.)^* : P(X) \to P(X)$, called a local function of A with respect to τ and I is defined as follows: $A \subseteq X$, $A^*(I, \tau) = \{x \in X : U \cap A \notin I$, for every $U \in \tau(x)\}$, where $\tau(x) = \{U \in \tau : x \in U\}$. A Kuratowski [6] closure operator $cl^*(x) = A \cup A^*(I, \tau)$. When there is no chance for confusion, we will simply write A^* for $A^*(I, \tau)$ and τ^* for $\tau^*(I, \tau)$. X^* is often a proper subset of X. The hypothesis $X = X^*$ is equivalent to hypothesis $\tau \cap I = \phi$. For every ideal topological space there exists a topology $\tau^*(I)$ finer than τ generated by $\beta(I, \tau) = \{U - A : U \in \tau$ and $A \in I\}$, but in general $\beta(I, \tau)$ is not always topology.

Definition 2.4 Hatir and Noiri [4] let (X, τ, I) be an ideal topological space and A be a subset of X. Then $A_*(I, \tau) = \{x \in X : U \cap A \notin I$ for every $U \in SO((X, x)\}$ is called semi local function of A with respect to I and τ.

Definition 2.5 Levin [9] let (X, τ) be a topological space. A subset A of X is said to be semi-open if there exists an open set U in X such that $U \subseteq A \subseteq cl(U)$. The complement of a semi-open set is said to be semi-closed. The collection of all semi-open (resp. semi-closed) sets in X is denoted by $SO(X)$ (resp. $SC(X)$). The semi-closure of A in (X, τ) is denoted by the intersection of all semi-closed sets containing A and is denoted by $Scl(A)$.

Definition 2.6 A subset A of an ideal topological space (X, τ, I) is said to be

(i) α-I-open [4] if $A \subseteq \text{int}(cl^*(\text{int}(A)))$.
(ii) semi-I-open [4] if $A \subseteq cl^*(\text{int}(A)))$.
(iii) pre-I-open [2] if $A \subseteq \text{int}(cl^*(A))$.
(iv) I-open [1] if $A \subseteq \text{int}(A^*)$.
(v) β-I-open [3] if $A \subseteq cl(\text{int}(cl^*(A)))$

3 α-Local Functions in ideal N-topology

In this part of the article we try to develop the concept of $N\tau\alpha$-local functions in ideal N-topology with suitable example. Further we do discuss various topological properties that are applicable to the topic.

Definition 3.1 Let $(X, N\tau, I)$ be an ideal in N-topological space and $A \subseteq X$. Then $A^{N\alpha *}(I, N\tau) = \{x \in X : U \cap A \notin I$, for every $U \in N\tau^{\alpha}(x)\}$ is called $N\tau\alpha$–local function of A with respect to I and $N\tau$. We denote simply $A^{N\alpha *}$ for $A^{N\alpha *}(I, N\tau)$.

Example 3.2 Let $N = 4$, $X = \{a, b, c, d\}$. If $\tau_1 = \{\phi, X, \{a\}\}$, $\tau_2 = \{\phi, X, \{b\}\}$, $\tau_3 = \{\phi, X, \{a, b\}\}$ and $\tau_4 = \{\phi, X, \{c\}\}$, then $4\tau = \{\phi, X, \{a\}, \{b\}, \{c\}, \{a, b\}, \{a, c\}, \{b, c\}, \{a, b, c\}\}$ is a 4τ-topology and $4\tau\alpha O(X) = \{\phi, X, \{a\}, \{b\}, \{c\}, \{a, b\}, \{a, c\}, \{b, c\}, \{a, b, c\}\}$. Let $I = \{\phi, \{c\}, \{d\}, \{c, d\}\}$ be an ideal on X, then $(X, 4\tau, I)$ is an ideal 4-topological space on X. For $A = \{a, b\} \subset X$, we have $A^* = \{a, b, d\}$.

Definition 3.3 A subset A of an ideal N-topological space $(X, N\tau, I)$ is said to be

(i) $N\tau\alpha$-I-open if $A \subseteq N\tau - \text{int}(N\tau - cl^*(N\tau - \text{int}(A)))$.
(ii) $N\tau$-semi-I-open if $A \subseteq N\tau - cl^*(N\tau - \text{int}(A)))$.
(iii) $N\tau$-pre-I-open if $A \subseteq N\tau - \text{int}(N\tau - cl^*(A))$.
(iv) $N\tau$-I-open if $A \subseteq N\tau - \text{Int}(A^*)$.

Theorem 3.4 *Let $(X, N\tau, I)$ be an ideal N-topological space and A, B subsets of X. Then for an $N\tau\alpha$-local functions the following properties hold:*

(i) $(\phi)^{N\alpha *} = \phi$
(ii) *If $A \subset B$, then $A^{N\alpha *} \subset B^{N\alpha *}$*
(iii) *For another ideal $J \supset I$ on X, $A^{N\alpha *}(J) \subset B^{N\alpha *}(I)$.*
(iv) $A^{N\alpha *} \subset N\alpha\, cl(A)$
(v) $A^{N\alpha *}(I) = N\tau\alpha\, cl(A^{\alpha *}) \subset N\tau\alpha\, cl(A)$
(vi) $(A^{N\alpha *})^{N\alpha *} \subset A^{N\alpha *}$
(vii) $(A \cup B)^{N\alpha *} = A^{N\alpha *} \cup B^{N\alpha *}$
(viii) $A^{N\alpha *} - B^{N\alpha *} = (A - B)^{N\alpha *} - B^{N\alpha *} \subset (A - B)^{N\alpha *}$
(ix) *If $U \in N\tau^{\alpha}$, then $U \cap A^{N\alpha *} = U \cap U \cap A)^{N\alpha *} \subset (U \cap A)^{N\alpha *}$*
(x) *If $U \in I$, then $(A - U)^{N\alpha *} \subset A^{N\alpha *} = (A \cup U)^{N\alpha *}$*

Proof

(i) The proof is obvious.
(ii) Let $x \in A^{N\alpha *}$, then $U \cap A \notin I$ for every $U \in N\tau^{\alpha}(x)$. Thus $U \cap B \notin I$ for each $U \in N\tau^{\alpha}(x)$. Now because $A \subset B$ that means $U \cap A \subset U \cap B$. If $U \cap B \in I$ implies $U \cap A \in I$. Therefore $x \in B^{N\alpha *}$ and $A^{N\alpha *} \subset B^{N\alpha *}$.
(iii) Let $x \in A^{N\alpha *}(J)$. Then for every $N\tau^{\alpha}(x)$, $U \cap A \notin J$. This implies that $U \cap A \notin I$, so $x \in A^{N\alpha *}(I)$. Thus $A^{N\alpha *}(J) \subset A^{N\alpha *}(I)$.

(iv) Let $x \in A^{N\alpha*}$. Then for every $N\alpha-$ open set containing x, $U_x \cap A \notin I$. This shows that $(U_x \cap A) \neq \phi$. Thus $x \in N\alpha - cl(A)$.

(v) $A^{N\alpha*} \subset N\tau\alpha\, cl\big(A^{N\alpha*}\big)$ hold in general. Let $x \in N\tau\alpha\, cl\big(A^{N\alpha*}\big)$. Then $A^{N\alpha*} \cap U \neq \phi$ for every $U \in N\tau^{\alpha}(x)$. Hence there exist some $y \in A^{N\alpha*} \cap U$ and $U \in N\tau^{\alpha}(x)$ since $y \in A^{N\alpha*}$, $A \cap U \notin I$ and hence $x \in A^{N\alpha*}$. Thus $N\tau\alpha cl\big(A^{N\alpha*}\big) \subset A^{N\alpha*}$. Now, let $N\tau\alpha cl\big(A^{N\alpha*}\big) = A^{N\alpha*}$, then $A \cap U \notin I$ for every $U \in N\tau^{\alpha}(x)$. This shows that $A \cap U \neq \phi$ for all $U \in N\tau^{\alpha}(x)$ and so $x \in N\tau\alpha cl(X, x)$. Consequently, $A^{N\alpha*} = N\tau\alpha cl\big(A^{N\alpha*}\big) \subset N\tau\alpha cl(A)$ and $A^{N\alpha*}$ is a $N\tau\alpha-$ closed.

(vi) Let $x \in (A^{N\alpha*})^{N\alpha*}$. Then, for all $U \in N\tau^{\alpha}(x)$, $A^{N\alpha*} \cap U \notin I$ and hence $A^{N\alpha*} \cap U \neq \phi$ for every $U \in N\tau^{\alpha}(x)$. Thus we have $A \cap U \notin I$ and $x \in A^{N\alpha*}$.

(vii) $A \subset A \cup B$, and $B \subset A \cup B$ and $A^{N\alpha*} \cup B^{N\alpha*} \subset (A \cup B)^{N\alpha*}$ by (i). Conversely, let $x \in (A \cup B)^{N\alpha*}$. Then for every $U \cap (A \cup B) \notin I = (U \cap A) \cup (U \cap B) \notin I$. Hence $(U \cap A) \notin I$ or $(U \cap B) \notin I$. This implies that $x \in A^{N\alpha*}$ *or* $x \in B^{N\alpha*}$, that is, $x \in A^{N\alpha*} \cup B^{N\alpha*}$). Thus we get the required equality.

viii) Since $A - B \subset A$, by (1), $(A - B)^{N\alpha*} \subset A^{N\alpha*}$ and hence $(A - B)^{N\alpha*} - B^{N\alpha*} \subset A^{N\alpha*} - B^{N\alpha*}$. Conversely $A \subset (A - B) \cup B$, by (7) $A^{N\alpha*} \subset (A - B)^{N\alpha*} \cup B^{N\alpha*}$ and hence $A^{N\alpha*} - B^{N\alpha*} \subset \big((A - B)^{N\alpha*} \cup B^{N\alpha*}\big) - B^{N\alpha*}$. Therefore $B^{N\alpha*} - B^{N\alpha*} \subset (A - B)^{N\alpha*} - \big(B^{N\alpha*} \cup B^{N\alpha*}\big)$ and so, $A^{N\alpha*} - B^{N\alpha*} \subset (A - B)^{N\alpha*} - B^{N\alpha*}$.

(ix) Consider $U \in N\tau^{\alpha}(x)$ and $x \in U \cap A^{N\alpha*}$. Then $x \in U$ and $x \in A^{N\alpha*}$. For $V \in N\tau^{\alpha}(x)$, $U \cap V \in N\tau^{\alpha}(x)$. Thus $V \cap (U \cap A) = (U \cap V) \cap A \notin I$. So $x \in (U \cap A)^{N\alpha*}$. Therefore $U \cap A^{N\alpha*} \subset (U \cap A)^{N\alpha*}$. Also $U \cap A^{N\alpha*} \subset U \cap (U \cap A)^{N\alpha*}$, since $A \cap U \subset A$. Then by (1) $(A \cap U)^{N\alpha*} \subset A^{N\alpha*}$ and $U \cap (A \cap U)^{N\alpha*} \subset U \cap A^{N\alpha*}$. Hence we get the needed result.

(x) By (vii) we know that $(A \cup U)^{N\alpha*} = A^{N\alpha*} \cup U^{N\alpha*} = A^{N\alpha*} \cup \phi = A^{N\alpha*}$, since $A - U \subset A$ by (1), $(A - U)^{N\alpha*} \subset A^{N\alpha*}$. Hence the required result.

Theorem 3.5 *Let $(X, N\tau)$ be an N-topological space, I_1 and I_2 be ideals on X and let A be a subset of X. Then the following properties hold*

(i) *If $I_1 \subset I_2$ then $A^{N\alpha*}(I_2) \subset A^{N\alpha*}(I_1)$;*
(ii) $A^{N\alpha*}(I_1 \cap I_2) = A^{N\alpha*}(I_1) \cup A^{N\alpha*}(I_2)$.

Proof

(i) Let $I_1 \subset I_2$ and $x \in A^{N\alpha*}(I_2)$. Then $A \cap U \notin (I_2)$ for all $U \in N\tau^{\alpha*}(x)$ and hence $A \cap U \notin I_1$, Thus $x \in A^{N\alpha*}(I_1)$. This implies that $A^{N\alpha*}(I_2) \subset A^{N\alpha*}(I_1)$.

(ii) Since $I_1 \cap I_2 \subset I_1$ and $I_1 \cap I_2 \subset I_2$ by Theorem 3.4 we have $A^{N\alpha*}(I_1) \subset A^{N\alpha*}(I_1 \cap I_2)$ and $A^{N\alpha*}(I_2) \subset A^{N\alpha*}(I_1 \cap I_2)$. Hence we have $A^{N\alpha*}(I_1) \cup A^{N\alpha*} I_2 \subset A^{N\alpha*}(I_1 \cap I_2)$. Conversely let $x \in A^{N\alpha*}(I_1 \cap I_2)$. Then for every $U \in N\tau^{\alpha}(x)$, $U \cap A \notin I_1 \cap I_2$, hence $U \cap A \notin I_1$ and $U \cap A \notin I_2$. This shows that $x \in A^{N\alpha*}(I_1)$ *or* $x \in A^{N\alpha*}(I_2)$ therefore $\in A^{N\alpha*}(I_1) \cup A^{N\alpha*}(I_2)$. Hence the desired result.

Lemma 3.6 *Let $(X, N\tau, I)$ be an ideal N-topological space. If $U \in N\tau^{\alpha}(x)$, then $U \cap A^{N\alpha*} = U \cap (U \cap A)^{N\alpha*}$ for any subset A of X.*

Proof Assume that $U \in N\tau^{\alpha}(x)$ and $x \in U \cap A^{N\alpha*}$. Then $x \in U$ and $x \in A^{N\alpha*}$. Let V be any $N\alpha-$ open set containing x. Then $V \cap U \in N\tau^{\alpha}(x)$ and $V \cap (U \cap A) = (V \cap U) \cap A \notin I$. This shows that $x \in (U \cap A)^{N\alpha*}$, thus we get $U \cap A^{N\alpha*} \subseteq (U \cap A)^{N\alpha*}$. Also $U \cap A^{N\alpha*} \subseteq U \cap (U \cap A)^{N\alpha*}$ and by Theorem 3.4 $(U \cap A)^{N\alpha*} \subseteq A^{N\alpha*}$ and $U \cap (U \cap A)^{N\alpha*} \subseteq U \cap A^{N\alpha*}$. Therefore $U \cap A^{N\alpha*} = U \cap (U \cap A)^{N\alpha*}$.

Let us now show that $N\tau^{\alpha*}$ is finer than $N\tau^{\alpha}$ in terms of the closure operator $N\alpha\, cl^*(A) = A \cup A^{N\alpha*}$.

Theorem 3.7 *Let $(X, N\tau, I)$ be an ideal N-topological space, $N\alpha cl^*(A) = A^{N\alpha*} \cup A$ and A, B be subsets of X. Then*

(i) $N\tau\alpha\, cl^*(\phi) = \phi$.
(ii) $A \subseteq N\tau\alpha\, cl^*(A)$.
(iii) $N\tau\alpha cl^*(A \cup B) = N\tau\alpha\, cl^*(A) \cup N\tau\alpha\, cl^*(B)$.
(iv) $N\tau\alpha\, cl^*(A) = N\tau\alpha\, cl^*(N\tau\alpha\, cl^*(A))$.

Proof By Theorem 3.4, we get

(i) $N\tau\alpha\, cl^*(\phi) = (\phi)^{N\alpha*} = \phi$.
(ii) $A \subseteq A \cup A^{N\alpha*} = N\tau\alpha\, cl^*(A)$.
(iii) $N\tau\alpha cl^*(A \cup B) = (A \cup B)^{N\alpha*} = N\tau\alpha cl^*(A) \cup N\tau\alpha cl^*(B)$.
(iv) $N\tau\alpha\, cl^*(N\tau\alpha\, cl^*(A)) = N\tau\alpha\, cl^*\big(A^{N\alpha*} \cup A\big) = (A^{N\alpha*} \cup A)^{N\alpha*} \cup (A^{N\alpha*} \cup A) = ((A^{N\alpha*})^{N\alpha*} \cup A^{N\alpha*}) \cup \big(A^{N\alpha*} \cup A\big) = A^{N\alpha*} \cup A = N\tau\alpha\, cl^*(A)$.

Lemma 3.8 *Let $(X, N\tau, I)$ be an ideal N-topological space and A, B subsets of X. Then $A^{N\alpha*} - B^{N\alpha*} = (A - B)^{N\alpha*} - B^{N\alpha*}$.*

Proof From Theorem 3.4, we have $A^{N\alpha*} = [(A-B) \cup (A \cap B)]^{N\alpha*} = (A-B)^{N\alpha*} \cup (A \cap B)^{N\alpha*} \subseteq (A - B)^{N\alpha*} \cup B^{N\alpha*}$. Thus, $A^{N\alpha*} - B^{N\alpha*} \subseteq (A - B)^{N\alpha*} - B^{N\alpha*}$. By Theorem 3.4 $(A - B)^{N\alpha*} \subseteq A^{N\alpha*}$ and hence $(A - B)^{N\alpha*} - B^{N\alpha*} \subseteq A^{N\alpha*} - B^{N\alpha*}$. Hence $A^{N\alpha*} - B^{N\alpha*} = (A - B)^{N\alpha*} - B^{N\alpha*}$.

Lemma 3.9 *Let $(X, N\tau, I)$ be an ideal N-topological space and A, B be subsets of X then*

(i) *If $A \subseteq B$ then $N\tau\alpha\, cl^*(A) \subseteq N\tau\alpha\, cl^*(B)$.*
(ii) *$N\tau\alpha\, cl^*(A \cap B) \subseteq N\tau\alpha\, cl^*(A) \cap N\tau\alpha\, cl^*(B)$.*
(iii) *If U is $N\tau\alpha$-open, then $U \cap N\tau\alpha\, cl^*(A) \subseteq N\tau cl^*(U \cap A)$.*

Proof

(i) Since $A \subseteq B$, by Theorem 3.4 we have $N\tau\alpha cl^*(A) = A \cup A^{N\alpha*} \subseteq B \cup B^{N\alpha*} = N\tau\alpha\, cl^*(B)$.
(ii) From (i) it is obvious.

(iii) Since U is $N\tau\alpha$-open by Lemma 3.6 we have $U \cap N\tau\alpha\, cl^*(A) = U \cap (A \cup A^{N\alpha*}) = (U \cap A) \cup \big(U \cap A^{N\alpha*}\big) \subseteq (U \cap A) \cup (U \cap A)^{N\alpha*} = N\tau\alpha\, cl^*(U \cap A)$.

Example 3.10 For N $=$ 3, Let $X = \{a, b, c, d\}$, $\tau_1 = \{\phi, X, \{a, b\}\}$, $\tau_2 = \{\phi, X, \{a\}, \{a, b\}\}$ and $\tau_3 = \{\phi, X, \{c\}, \{a, c\}\}$. Then $3\tau = \{\phi, X, \{a\}, \{c\}, \{a, b\}, \{a, c\}, \{a, c, d\}, \{a, b, c\}\}$. Let $= \{\phi, \{b\}\}$. Set $A = \{a, d\}$. Then $A \in 3\tau^{\alpha*}$, but A is not $3\tau\alpha-$ open. So $A \notin 3\tau^{\alpha}(x)$.

Lemma 3.11 *Let* $(X, N\tau, I)$ *be an ideal N-topological space. A subset A of X is* $N\tau\alpha$*-I-open if and only if it is* $N\tau$*-semi-I-open and* $N\tau$*-pre-I-open.*

Proof *Necessity*: From the definition of $N\tau\alpha$-I-open and $N\tau$-semi-I-open it is clear that $N\tau\alpha$-I-open implies $N\tau$-semi-I-open.

Sufficient: Let A be $N\tau$-semi-I-open. Then, we have $A \subset N\tau - \text{int}(N\tau - cl^*(A)) \subset N\tau - \text{int}(N\tau - cl^*(N\tau - \text{int}(A)))) = N\tau - \text{int}(N\tau - cl^*(N\tau - \text{int}(A)))$. This shows that A is $N\tau\alpha$-I-open.

Proposition 3.12 *Let* $(X, N\tau, I)$ *be an ideal N-topological space.*

(1) *if* $V \in N\tau - SIO(X)$ *and* $A \in N\tau\alpha\, IO(X)$, *then* $V \cap A \in N\tau - SIO(X)$.
(2) *If* $V \in N\tau - PIO(X)$ *and* $A \in N\tau\alpha\, IO(X)$, *then* $V \cap A \in N\tau - PIO(X)$.

Theorem 3.13 *A function* $f : (X, N\tau) \to (Y, N\sigma, I)$ is $N\tau\alpha$*-I-open if and only if for each subset* $W \subset Y$ *and each closed set F of X containing* $f^{-1}(W)$, *there exists a* $N\tau\alpha$*-I-closed set* $H \subset Y$ *containing W such that* $f^{-1}(H) \subset F$.

Proof *Necessary condition*: Let $H = Y - f(X - F)$. Since $f^{-1}(W) \subset Y$, we have $f(X - F) \subset Y - W$. Since f is $N\tau\alpha$-I-open, then H is $N\tau\alpha$-I-closed and $f^{-1}(H) = X - f^{-1}(f(X - F)) \subset X - (X - F) = F$.

Sufficient condition: Let U be any $N\tau\alpha$-open set of X and $W = Y - f(U)$. Then $f^{-1}(W) = X - f^{-1}(f(U))\backslash subset\ X - U$ and $X - U$ is $N\tau\alpha-$ closed. By the hypothesis, there exists a $N\tau\alpha - I-$ closed set H of Y containing W such that $f^{-1}(H) \subset X - U$. Then we have $f^{-1}(H) \cap U = \phi$. Hence, we obtain $Y - f(U) \supset H \supset W = Y - f(U)$ and $f(U)$ is $N\tau\alpha - I-$ open *in* Y. This shows that f is $N\tau\alpha - I-$ open.

4 Semi and Pre Open in Ideal *N*-Topology

We shall define $N\tau$-semi-local function in ideal of N-topological space and develop its related properties with explicit examples. At the same time, we also introduce $N\tau$-pre-open and $N\tau$-β functions.

Definition 4.1 Let $(X, N\tau, I)$ be an ideal N-topological space and a subset of X. Then $A_*(I, N\tau) = \{x \in X : A \cap U \notin I \text{ for all } U \in N\tau - SO(X, x)\}$ is called the $N\tau$-semi-local function of A with respect to I and $N\tau$, where $N\tau - SO(X, x) = \{U \in N\tau - SO(X) : x \in U\}$.

Defintion 4.2 A subset of an ideal N-topological space $(X, N\tau, I)$ is called $N\tau$-Pre-I-open if $A \subseteq N\tau - \text{int}(N\tau - cl^*(A))$.

Theorem 4.3 *A subset A of a space* $(X, N\tau, I)$ *is* $N\tau$-semi-I-open *if and only if* $N\tau - cl^*(A) = N\tau - cl^*(N\tau - \text{int}(A))$.

Proof Let A be $N\tau-$ semi $-I-$ open, then $A \subset N\tau - cl^*(N\tau - \text{int}(A))$. This implies, $N\tau - cl^*(A) \subset N\tau - cl^*(N\tau - \text{int}(A))$ and thus $N\tau - cl^*(A) = N\tau - cl^*(N\tau - int(A))$. The converse is vivid.

Theorem 4.4 *A subset A of a space* $(X, N\tau, I)$ *is* $N\tau-$ *semi* $-I-$ open *if and only if there exists* $U \in N\tau$ *such that* $U \subset A \subset N\tau - cl^*(U)$.

Proof Let A be $N\tau$-semi-I-open then $A \subset N\tau - cl^*(N\tau - \text{int}(A))$. Considering $N\tau - \text{int}(A) = U$ we have $U \subset A \subset N\tau - cl^*(U)$. Conversely let $U \subset A \subset N\tau - cl^*(U)$ for some $U \in N\tau$. Since $U \subset A$, we know that $U \subset N\tau - \text{int}(A)$ and thus $N\tau - cl^*(U) \subset N\tau - cl^*(N\tau - \text{int}(A))$. Thus, we get $A \subset N\tau - cl^*(N\tau - \text{int}(A))$.

Theorem 4.5 *If A is a* $N\tau$*-semi-I-open set in a space* $(X, N\tau, I)$ *and* $A \subset B \subset N\tau - cl^*(A)$, *then B* is $N\tau-$ *semi* $-I-$ *open in* $(X, N\tau, I)$.

Proof Since A is $N\tau$-semi-I-open, a $N\tau\alpha$-open set U such that $U \subset A \subset N\tau - cl^*(U)$. Then we have $U \subset A \subset B \subset N\tau - cl^*(A) \subset N\tau - cl^*(N\tau - cl^*(U)) = N\tau - cl^*(U)$ and hence $U \subset B \subset N\tau - cl^*(U)$. By Theorem 4.4 we obtain $B \in SIO(X, N\tau)$.

Example 4.6 For $N = 2$, Let $X = \{a, b, c\}$, $\tau_1 = \{\phi, X, \{a\}\}$, $\tau_2 = \{\phi, X, \{c\}, \{a, c\}\}$ then $2\tau = \{\phi, X, \{a\}, \{c\}, \{a, c\}\}$. Let $I = \{\phi, \{a\}, \{b\}, \{a, b\}\}$. Then if we put $A = \{b\}$, we have $N\tau - Cl^*(N\tau - \text{int}(X - A)) = \{a\}$ and $X - N\tau - \text{int}(N\tau - cl^*(A)) = \{a, c\}$.

Theorem 4.7 *If a subset A of a space* $(X, N\tau, I)$ *is* $N\tau$*-semi-I-closed, then* $N\tau - \text{int}(N\tau - cl^*(A)) \subset A$.

Proof Since A is $N\tau$-semi-I-closed, $X - A \in N\tau - SIO(X, N\tau)$. Since $N\tau^*(I)$ is finer than $N\tau$, we have $X - A \subset N\tau - cl^*(N\tau - \text{int}(X - A)) \subset N\tau - cl(N\tau - \text{int}(X - A)) = X - (N\tau - \text{int}(N\tau - cl(A)) \subset X - N\tau - \text{int}(N\tau - cl^*(A))$.

Therefore we get $N\tau - \text{int}(N\tau - cl^*(A)) \subset A$.

Theorem 4.8 *A subset A of a space* $(X, N\tau, I)$ *is weakly* $N\tau$*-semi-I-closed if and only if* $N\tau - \text{int}^*(N\tau - cl(N\tau - \text{int}(A))) \subset A$.

Proof Let A be a weakly $N\tau$-semi-I-closed set of $(X, N\tau, I)$. Then $X - A$ is weakly $N\tau$-semi-I-open and hence $X - A \subset N\tau - cl^*(N\tau - \text{int}(N\tau - cl(X - A))) = X - N\tau - \text{int}^*(N\tau - cl(N\tau - \text{int}(A)))$. Therefore we have $N\tau - \text{int}^*(N\tau - cl(N\tau - \text{int}(A))) \subset A$.

Conversely, let $\tau - \text{int}^*(N\tau - cl(N\tau - \text{int}(A))) \subset A$. Then $X - A \subset N\tau - cl^*(N\tau - \text{int}(N\tau - cl(X - A)))$ and hence $X - A$ is weakly $N\tau$-semi-I-open. Therefore A is weakly $N\tau$-semi-I-closed.

Proposition 4.9 *Every $N\tau$-I-open set is $N\tau$-pre-I-open.*

Proof Let $(X, N\tau, I)$ be an ideal N-topological space and let $A \subseteq X$ be $N\tau - I-$ open. Then $A \subseteq N\tau - \text{int}(A^*) \subseteq (A^* \cup A) = N\tau - \text{int}(N\tau - cl^*(A))$.

Proposition 4.10 *Every $N\tau$-open set is $N\tau$-pre-I-open.*

Proof Let $A \subseteq (X, N\tau, I)$ be $N\tau$-open. Then $A \subseteq N\tau - \text{int}(A^* \cup A) = N\tau - \text{int}(N\tau - cl^*(A))$.

The following example shows that the converse of the above proposition need not be true.

Example 4.11 A $N\tau$-pre-I-open set need not be $N\tau$-I-open. For $N = 4$, let $X = \{a, b, c, d\}$, $\tau_1 = \{\phi, X,\}$, $\tau_2 = \{\phi, X, \{a, c\}\}$, $\tau_3 = \{\phi, X, \{d\}\}$ and $\tau_4 = \{\phi, X, \{a, c, d\}\}$. Then $4\tau = \{\phi, X, \{d\}, \{a, c\}, \{a, c, d\}\}$ and $I = \{\phi, \{c\}, \{d\}, \{c, d\}\}$. If the set $A = \{a, c, d\}$, then $A \in 4\tau$ and hence $A \in 4\tau - PIO(X)$ but $A \notin 4\tau - IO(X)$.

Definition 4.12 Let $(X, N\tau)$ be an N-topological space. A subset A of X is said to be $N\tau$-semi-open if there exists a $N\tau\alpha$ o-pen set U in X such that $U \subseteq A \subseteq N\tau - cl(U)$. The complement of a $N\tau$-semi-open set is said to be $N\tau$-semi-closed. The collection of all $N\tau$-semi-open (resp. $N\tau$-semi-closed) sets in X is denoted by $N\tau$-$SO(X)$ (resp. $N\tau$-$SC(X)$). The $N\tau$-semi-closure of A in (X, τ) is denoted by the intersection of all $N\tau$-semi-closed sets containing A and is denoted by $N\tau$-$Scl(A)$.

Proposition 4.13 *Let $(X, N\tau, I)$ be an ideal $N\tau-$topological space with Δ an arbitrary index set. Then*

(i) *If $\{A_\alpha : \alpha \in \Delta\} \subseteq N\tau - PIO(X)$, then $\cup\{A_\alpha : \alpha \in \Delta\} \in N\tau - PIO(X)$.*
(ii) *If $A \in N\tau - PIO(X)$ and $U \in N\tau$, then $A \cap U \in N\tau - PIO(X)$.*
(iii) *If $A \in N\tau - PIO(X)$ and $B \in N\tau^\alpha$, then $A \cap B \in N\tau - PIO(X)$.*
(iv) *If $A \in N\tau - PIO(X)$ and $B \in N\tau - SO(X)$, then $A \cap B \in N\tau - SO(X)$.*
(v) *If $A \in N\tau - PIO(X)$ and $B \in N\tau - SO(X)$, then $A \cap B \in N\tau - PIO(X)$.*

Proof

(i) Since $\{A_\alpha : \alpha \in \Delta\} \subseteq N\tau - PIO(X)$, then $A_\alpha \subseteq N\tau - \text{int}(N\tau - cl^*(A_\alpha))$ for every $\alpha \in \Delta$. Thus $\cup_{\alpha\in\Delta} A_\alpha \subseteq \cup_{\alpha\in\Delta}(N\tau - \text{int}(N\tau - cl^*(A_\alpha)) \subseteq N\tau\text{int}\ (\cup_{\alpha\in\Delta}(N\tau - cl^*(A_\alpha)) = N\tau - \text{int}(\cup_{\alpha\in\Delta}(A_\alpha \cup A_\alpha)) = N\tau - \text{int}\big((\cup_{\alpha\in\Delta} A_\alpha^*) \cup (\cup_{\alpha\in\Delta} A_\alpha)\big) \subseteq N\tau - \text{int}\big((\cup_{\alpha\in\Delta} A_\alpha)^* \cup (\cup_{\alpha\in\Delta} A_\alpha)\big) = N\tau - \text{int}(N\tau - cl^*(\cup_{\alpha\in\Delta} A_\alpha))$.

(ii) By assumption $A \subset N\tau - \text{int}(N\tau - cl^{*(A)}$ and $U \subseteq N\tau - \text{int}(U)$. This implies that $A \cap U \subseteq N\tau - \text{int}(N\tau - cl^*(A)) \cap N\tau - \text{int}(U) \subset N\tau - \text{int}(N\tau - cl^*(A) \cap U) = N\tau - \text{int}((A^* \cup A) \cap U) = N\tau - \text{int}((A^* \cap U) \cup (A \cap U)) \subseteq N\tau - \text{int}\big((A \cap U)^* \cup (A \cap U)\big) = N\tau - \text{int}(N\tau - cl^*(A \cap U))$.

(iii) Since the intersection of a $N\tau$-pre-open set and $N\tau\alpha$-set is always a $N\tau$-pre-open set and every $N\tau$-pre-I-open st is $N\tau$-pre-open the claim is true.

(iv) and (v) It is proved that the intersection of a $N\tau$-pre-open and a $N\tau$-semi-open set is a $N\tau$-pre-open subset of the $N\tau$-semi-open set and a $N\tau$-semi-open subset of the $N\tau$-pre-open set. Thus the claim follows from the fact that every $N\tau$-pre-I-open st is $N\tau$-pre-open.

Theorem 4.14 *Let $(X, N\tau, I)$ be an ideal N-topological space. Then the following holds:*

(1) *If A is a $N\tau$-pre-open, then A is weakly $N\tau$-semi-I-open.*

(2) *If $N\tau\alpha$-open sets are *–closed, then every weakly $N\tau$-semi-I-open set is $N\tau$-pre-open.*

Proof

(1) If A is $N\tau$-pre-open, then $A \subset N\tau - \text{int}(N\tau - cl(A))$ and so $A \subset N\tau - cl^*(N\tau - \text{int}(N\tau - cl(A)))$ which implies that A is weakly $N\tau$-semi- I-open.

(2) If A is weakly $N\tau$-semi-I-open, then $A \subset N\tau - cl^*(N\tau{-}\text{int}(N\tau - cl(A)))$. Since $N\tau - \text{int}(N\tau - cl(A))$ is $N\tau\alpha-$ open, by hypothesis $N\tau - \text{int}(N\tau - cl(A)) = N\tau - cl^*(N\tau - \text{int}(N\tau - cl(A)))$ and so $A \subset N\tau - \text{int}(N\tau - cl(A))$ which implies that A is $N\tau-$ pre-open.

Theorem 4.15 *Let $(X, N\tau, I)$ be an ideal space. If A is $N\tau\alpha$-I-open and B is weakly $N\tau$-semi-I-open, then $A \cap B$ is weakly $N\tau$-semi-I-opens.*

Proof Since A is $N\tau\alpha$-I-open, $A \subset N\tau - \text{int}(N\tau - cl^*(N\tau{-}\text{int}(A)))$ and B is weakly $N\tau$-semi-I-open, $B \subset N\tau - cl^*(N\tau - \text{int}(N\tau - cl(B)))$. Now $A \cap B \subset N\tau - \text{int}(N\tau - cl^*(N\tau - \text{int}(A))) \cap N\tau - cl^*(N\tau - \text{int}(N\tau - cl(B))) \subset N\tau - cl^*(N\tau - \text{int}(N\tau{-}cl^*(N\tau - \text{int}(A))) \cap N\tau - \text{int}(N\tau - cl(B))) = N\tau - cl^*(N\tau - \text{int}(N\tau - cl^*(N\tau - \text{int}(A)) \cap N\tau - \text{int}(N\tau - cl(B)))) \subset N\tau - cl^*(N\tau - \text{int}(N\tau - cl^*(N\tau - \text{int}(A) \cap N\tau - \text{int}(N\tau - cl(B))))) = N\tau - cl^*(N\tau - \text{int}(N\tau - cl^*(N\tau - \text{int}(N\tau - \text{int}(A) \cap N\tau - cl(B))))) \subset N\tau - cl^*(N\tau - \text{int}(N\tau - cl^*(N\tau - \text{int}(N\tau - cl(N\tau - \text{int}(A) \cap B))))) \subset N\tau - cl^*(N\tau - \text{int}(N\tau - cl^*(N\tau - \text{int}(N\tau - cl(A \cap B))))) = N\tau - cl^*(N\tau - \text{int}(N\tau - cl(A \cap B)))$, then $N\tau - cl^* (N\tau - \text{int}(N\tau - cl^*(N\tau - \text{int}(A)))) = N\tau - cl^*(N\tau - \text{int}(A))$ which implies that $A \cap B$ is weakly $N\tau$-semi-I-open.

Theorem 4.16 *Let $(X, N\tau, I)$ be an ideal space and $A \subset X$ be a weakly $N\tau$-semi-I-open. If A is either $N\tau$-semi-closed or $N\tau$-I-locally closed, then A is $N\tau$-semi-I-open.*

Proof Suppose A is $N\tau - I-$ locally closed implies that $A = U \cap A^*$ for some $N\tau\alpha$-open set U. A is weakly $N\tau-$semi$-I-$open implies that $A \subset N\tau - cl^*(N\tau - \text{int}(N\tau - cl(A)))$. Now $A = U \cap A^* \subset U\cap(N\tau - cl^*(N\tau - \text{int}(N\tau - cl(A))))^* \subset U \cap N\tau - cl^*(N\tau - cl^*(N\tau - \text{int}(N\tau - cl(U \cap A^*)))) = U \cap N\tau - cl^*(N\tau - \text{int}(N\tau - cl(U \cap A))) \subset N\tau - cl^*(U \cap N\tau - \text{int}(N\tau - cl(U \cap A^*))) = N\tau - cl^*(N\tau - \text{int}(U \cap (N\tau - cl(U \cap A^*))) \subset N\tau - cl^*(N\tau - \text{int}(U \cap N\tau - cl(U) \cap N\tau - cl(A)^*))) = N\tau - cl^*(N\tau - \text{int}(U \cap A^*) = N\tau - cl^*(N\tau - \text{int}(A))$. Hence A is $N\tau$-semi-I-open. Suppose A is $N\tau$-semi-closed. Then $(N\tau - \text{int}(N\tau - cl(A)) = N\tau - \text{int}(A)$. Since A is weakly $N\tau$-semi-I-open, $A \subset N\tau - cl^*(N\tau - \text{int}(N\tau - cl(A))) = N\tau - cl^*(N\tau - \text{int}(A))$. Hence A is $N\tau$-semi-I-open.

Lemma 4.17 *Let $(X, N\tau, I)$ be an ideal N-topological space. If A is $N\tau$-semi-I-open in $(X, N\tau, I)$, then it is $N\tau\beta$-I-open.*

Proof Let A be $N\tau$-semi-I-open, then $A \subset N\tau - cl^*(N\tau - \text{int}(A)) \subset N\tau - cl(N\tau - \text{int}(A)) \subset N\tau - cl(N\tau - int(N\tau - cl^*(A)))$. This shows that A is $N\tau\beta$-I-open.

Theorem 4.18 *The union of two $N\tau\beta$-I-open sets of $(X, N\tau)$ is also an $N\tau\beta$-I-open set.*

Proof Let A and B be $N\tau\beta$-I-open sets, then there exists $U, V \in N\tau$ such that $A - N\tau - \text{int}(N\tau - cl(U)) \in I$, $(U - A) \in I$, $(B - N\tau - \text{int}(N\tau - cl(V)) \in I$ and $(V - B) \in I$. Choose $W - U \cup V$, hence $(U - (A \cup B)) \subseteq (U - A) \in I, (V - (A \cup B)) \in (V - A) \in I$ and so, $(U - (A \cup B)) \cup (V - (A \cup B)) \in I$. Consequently, $W - (A \cup B)) \in I$. Also $(A - N\tau - \text{int}(N\tau - cl(W))) \subseteq (A - N\tau - \text{int}(N\tau - cl(U))) \in I$ and $(B - N\tau - \text{int}(N\tau - cl(W)) \subseteq (B - N\tau - \text{int}(N\tau - cl(V))) \in I$. Then, $(A - N\tau - \text{int}(N\tau - cl(W)) \cup (B - N\tau - \text{int}(N\tau - cl(W))) \subset I$ and so $(A \cup B) - N\tau - \text{int}(N\tau - cl(W)) \subseteq (A - N\tau - \text{int}(N\tau - cl(U)) \cup (B - N\tau - \text{int}(N\tau - cl(V)) \in I$. Thus $(A \cup B)$ is $N\tau\beta$-I-open set.

Theorem 4.19 *Let I be an ideal on $N\tau\alpha$-local indiscrete N–topological space $(X, N\tau)$ then the following statements hold:*

(1) *Every $N\tau$-semi-I-open is $N\tau$-pre-I-open.*
(2) *Every $N\tau$-semi-I-open is $N\tau\alpha$-I-open.*
(3) *Every $N\tau\beta$-I-open is $N\tau$-pre-I-open.*
(4) *Every $N\tau\alpha$-I-open* is *$N\tau$-pre-I-open.*
(5) *If $A \subseteq X$, then A is $N\tau$-pre-I-closed.*

Proof We shall prove here (1) and (5) only and the rest of the proof is obvious.

(1) Let A be a $N\tau$-semi-I-open, then there exists a $N\tau\alpha$-open set U such that $(A - (N\tau - cl(U))) \in I$ and $(U - A) \in I$. Since X is $N\tau\alpha-$locally indiscrete, then A is a $N\tau$-pre-I-open set.

(5) Let $A \subseteq X$. Choose $F = N\tau - \text{int}(A)$. Since X is $N\tau\alpha-$ locally indiscrete, then F is closed, $(N\tau - \text{int}(A) - F) = \phi \in I$ and $(F - A) = \phi \in I$. Consequently A is $N\tau$-pre-I-closed.

Example 4.20 For $N = 3$, let $X = \{a, b, c, d\}$, $\tau_1 = \{\phi, X, \{a, c\}\}$, $\tau_2 = \{\phi, X, \{a, c, d\}\}$ and $\tau_3 = \{\phi, X, \{a, c\}, \{a, b, c\}\}$, then $3\tau = \{\phi, X, \{a, c\}, \{a, c, d\}, \{a, b, c\}\}$ and $I = \{\phi, \{d\}\}$. Then

(i) $\{a\}$ is $N\tau$-pre-I-open; however, it is not $N\tau\alpha$-I-open set
(ii) $\{a\}$ is $N\tau\beta$-I-open; however, it is not $N\tau$-semi-I-open set.

Example 4.21 For $N = 3$, let $X = \{a, b, c\}$, $\tau_1 = \{\phi, X, \{b\}\}$, $\tau_2 = \{\phi, X, \{c\}\}$ and $\tau_3 = \{\phi, X, \{b\}, \{b, c\}\}$ implies $3\tau = \{\phi, X, \{b\}, \{c\}, \{b, c\}\}$ and $I = \{\phi, \{c\}\}$. Then:

(i) $\{a\}$ is $N\tau$-semi-I-open; however, it is not $N\tau\alpha$-I-open set.
(ii) $\{a, c\}$ is $N\tau\beta$-I-open; however, it is not $N\tau$-pre-I-open set.

Remark 4.22 From the above theorems, lemmas, propositions, remarks and examples, we can form the following diagram:

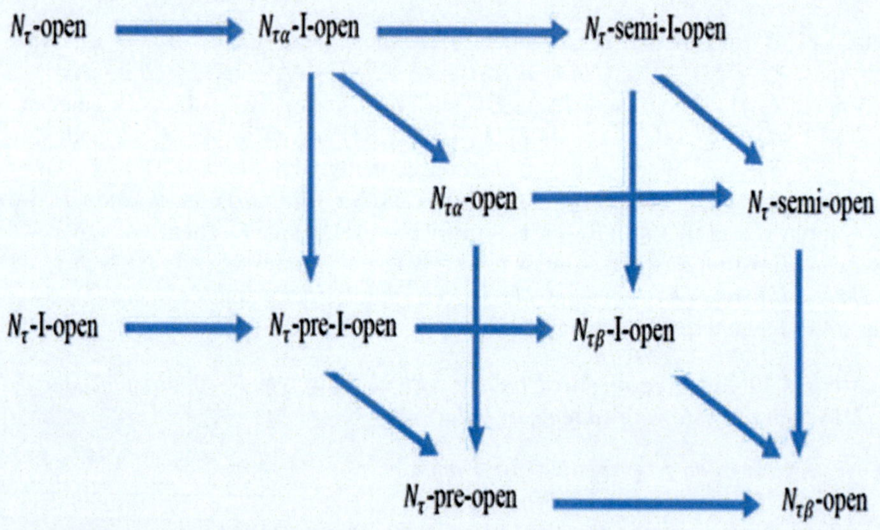

Computation From definition 3.3, theorem 4.19 and remark 4.22 we define the following algorithm.

Algorithm:

Step 1: Let A be an open set of the given N-topology.
Step 2: $N\tau$-open $\subseteq N\tau\alpha$-ideal-open $\subseteq N\tau$-semi-ideal-open $\subseteq N\tau$-semi-open $\subseteq N\tau\beta$-open.

Step 3: Also we note that $N\tau\alpha$-ideal-open $\subseteq$ $N\tau\alpha$-open $\subseteq$ $N\tau$ semi-open $\subseteq$ $N\tau\beta$-open.
Step 4: Further we find that $N\tau\alpha$-ideal-open $\subseteq$ $N\tau$ pre-ideal-open $\subseteq$ $N\tau pre$-open $\subseteq$ $N\tau\beta$-open.
Step 5: Similarly we can prove that $N\tau$-ideal-open $\subseteq$ $N\tau$-pre-ideal-open $\subseteq$ $N\tau\beta$-ideal-open $\subseteq$ $N\tau\beta$-open.
Step 6: Finally it is shown that $N\tau$-pre-ideal-open $\subseteq$ $N\tau$ pre-open $\subseteq$ $N\tau\beta$-open.

Notations and Conditions:

$N\tau\, int(A)$: Largest N-open set contained in A.
$N\tau\, cl(A)$: Smallest N-closed set containing A.
$N\tau$-open if $A = N\tau - \text{int}(N\tau - cl(A))$
$N\tau\alpha$-I-open if $A \subseteq N\tau - \text{int}(N\tau - cl^*(N\tau - \text{int}(A)))$
$N\tau$-semi-I-open if $A \subseteq N\tau - cl^*(N\tau - \text{int}(N\tau - (A)))$
$N\tau\alpha$-open if $A \subseteq N\tau - \text{int}(N\tau - cl(N\tau - \text{int}(A)))$
$N\tau$-semi-open if $A \subseteq N\tau - cl(N\tau - \text{int}(N\tau - cl(A)))$
$N\tau$-I-open if $A = N\tau - \text{int}(N\tau - cl(A^*))$
$N\tau$-pre-I-open if $A \subseteq N\tau - \text{int}(N\tau - cl^*(A))$
$N\tau\beta$-I-open if $A \subseteq N\tau - cl(N\tau - \text{int}(N\tau - cl^*(A)))$
$N\tau$-pre-open if $A \subseteq N\tau - \text{int}(N\tau - cl(A))$
$N\tau\beta$-open if $A \subseteq N\tau - cl(N\tau - \text{int}(N\tau - cl(A)))$

Conclusion. It is noteworthy to study ideals and extend it to N-topology as illustrated above. It is interesting to find that many of N-topological concepts hold good for the ideal N-topology too. Further this concept can be applied to pure and applied mathematical fields. Also this paper gives a scope for topological investigations by means of computer encoding of 3D images and digitization in computer using topological space, etc.

References

1. Abd El-monsef, M.E., Lashien, E.F., Nasef, A.A.: On I-open sets and I-continuous functions. Kyungpook Math. **32**, 21–30 (1992)
2. Dontchev, J.: On pre-I-open sets and a decomposition of I-continuity. Banyan Math. J. **2** (1996)
3. Hatir, E., Noiri, T.: on decompositions of continuity via idealization. Acta Math. Hungar. **96**(4), 341–349 (2002)
4. Hatir, E., Noiri, T.: On semi-I-open sets and semi-I-continuous functions. Acta Math. Hungar. **107**(4), 345–353 (2005)
5. Jankovic, D., Hamlett, T.R.: New topologies from old via ideals. Am. Math. Mon. **97**(4), 295–310 (1990)
6. Kuratowski, K.: Topology, vol. 1. Academic Press, New York (1966)
7. Lellis Thivagar, M., Reilly, I.L., Arockia Dasan, M., Ramesh, V.: Generalized open sets in grill N–topology. Appl. Gen. Topol. (Spain) **18**(2), 289–299 (2017)
8. Lellis Thivagar, M., Ramesh, V., Arockia Dasan, M.: On new structure of N–Topology. Cogent Math. **3**, 1–10 (2016)

9. Levin, N.: Semi-open sets and semi-continuity in topological spaces. Am. Math. Mon. **70**, 36–41 (1963)
10. Vaidyanathaswamy, R.: The localization theory in set-topology. Proc. Indian Acad. Sci. **20**, 51–61 (1945)

Arithmetic Functions, Convolution and Finite Fourier Series

Vichian Laohakosol

1 Introduction

Denote by $A = \{f : \mathbb{N} \to \mathbb{C}\}$ the ring of arithmetic functions equipped with addition, multiplication and (Dirichlet) convolution defined, for $m \in \mathbb{N}$, by

$$\begin{aligned}(f + g)(m) &= f(m) + g(m),\\ (f \cdot g)(m) &= f(m)g(m),\\ (f * g)(m) &= \sum_{d|m} f(d)g(m/d).\end{aligned}$$

The following arithmetic functions will be frequently referred to in this article.

- The *convolution identity* $I(n) = \begin{cases} 1 \ if \ n = 1 \\ 0 \ if \ n > 1. \end{cases}$
- The *Souriau–Hsu–Möbius function* [4, 13] is defined, for $\alpha \in \mathbb{C}$, by

$$\mu_\alpha(n) = \prod_{p|n} \binom{\alpha}{\nu_p(n)} (-1)^{\nu_p(n)}, \quad \mu_0(n) = I(n),$$

where $n = \prod p^{\nu_p(n)}$ denotes the unique prime factorization of n, with the convention that $\nu_p(1) := 0$. It is easily checked that

$$\mu_{\alpha+\beta} = \mu_\alpha * \mu_\beta \quad (\alpha, \beta \in \mathbb{C}).$$

V. Laohakosol (✉)
Department of Mathematics, Faculty of Science, Kasetsart University, Bangkok 10900, Thailand
e-mail: fscivil@ku.ac.th

P. Srivastava et al. (eds.), *Mathematical, Computational Intelligence and Engineering Approaches for Tourism, Agriculture and Healthcare*, Lecture Notes in Networks and Systems 214, https://doi.org/10.1007/978-981-16-3807-7_5

For some particular values of α, the corresponding Souriau–Hsu–Möbius functions represent certain well-known arithmetic functions, namely,

(i) when $\alpha = 0$, this corresponds to the convolution identity $\mu_0 = I$;

(ii) when $\alpha = 1$, this is the classical Möbius function $\mu_1 = \mu$;

(iii) when $\alpha = -1$, this is the inverse of the Möbius function $\mu_{-1} = \mu^{-1} =: u$, where $u(n) = 1 \quad (n \in \mathbb{N})$ is the *constant 1 function*;

(iv) when $\alpha = -2$, this is the number of divisors function, [4, p. 75],

$$\mu_{-2}(n) = \sigma_0(n) =: \sum_{d|n} 1.$$

- The kth-power function $\zeta_k(n) = n^k \quad (k \in \mathbb{Z}, n \in \mathbb{N})$.

For a fixed $k \in \mathbb{N}$, an arithmetic function f is said to be *periodic mod k* if

$$f(m + k) = f(m) \quad (m \in \mathbb{N}).$$

It is well-known [3, Theorem 8.4] that each arithmetic function f which is periodic mod k is uniquely representable by a finite Fourier series of the form

$$f(m) = \sum_{n \bmod k} a_n \exp(2\pi i mn/k) \quad (m \in \mathbb{N}).$$

The finite Fourier series is an important and useful tool in the study of arithmetic functions as witnessed in [3, Chap. 8], [10, Chap. 2], [12, Chaps. IX, XVI]. Of particular interest is the finite Fourier series of the greatest common divisor (gcd) function whose forms and applications have recently been treated in [1, 2, 7, 11, 15].

In this semi-expository work, we investigate the general definition and basic properties of convolution and (finite) Fourier series. As a prototype, the Fourier series representation of the gcd function is considered. Most existing identities as well as new ones are proved. Several applications related to Fourier series are subsequently obtained.

2 Generalized Convolution

For $f, g \in \mathrm{A}$, $a \in \mathbb{Z}$, and $k \in \mathbb{N}$, the *generalized convolution* of f, g is defined as the sum

$$\sum_{d|\gcd(a,k)} f(d)g(k/d).$$

We use two notational symbols for this sum. The first, taken from [2],

$$S_{f,g}(a,k) = \sum_{d|\gcd(a,k)} f(d)g(k/d),$$

is used when both a and k are considered, unless otherwise stated, as integer variables. The second, taken from [15] and referred to as *a-convolution*,

$$(f *_a g)(k) = \sum_{d|\gcd(a,k)} f(d)g(k/d),$$

is used when a is kept fixed, while k varies. The a-convolution generalizes the Dirichlet convolution because $(f *_a g)(k) = (f * g)(k)$ when $k|a$. In addition, when $\gcd(a,k) = 1$, we easily see that

$$S_{f,g}(a,k) = (f *_a g)(k) = f(1)g(k). \tag{1}$$

Some basic properties of the generalized convolution are gathered in the next theorem [2], [15, Sect. 3].

Theorem 2.1 *For* $f, g, h \in \mathrm{A},\ a \in \mathbb{Z}$, *we have*

(i) [15, Sect. 3] $(f *_a I)(k) = \begin{cases} f(k) & \text{if } k|a \\ 0 & \text{otherwise;} \end{cases}$

in particular [1] $\rho_a(k) := (\zeta_1 *_a I)(k) = \begin{cases} k & \text{if } k|a \\ 0 & \text{otherwise;} \end{cases}$

(ii) ii. (1) [2] $S_{f,g}(a,k) = \sum_{d|a} \alpha_{k,d} f(d)g(k/d) \quad (a, k \subset \mathbb{N})$

ii. (2) [2], [15, Sect. 3]

$$S_{f,g}(a,k) = \sum_{d|k} \alpha_{a,d} f(d)g(k/d) = ((f *_a I) * g)(k),$$

where $\alpha_{n,d} = \begin{cases} 1 & \text{if } d|n \\ 0 & \text{otherwise;} \end{cases}$

(iii) [15, Sect. 3] $(f *_a g) * h = f *_a (g * h)$;

(iv) [15, Sect. 3] $f *_a g = (f * g)_a = f * g_a$, *where* $f_a = u *_a f$;

(v) [15, Sect. 3] if f *and* g *are multiplicative, then so is* $f *_a g$, *i.e., for fixed* $a \in \mathbb{Z}$, *if* f, g *are multiplicative, then* $S_{f,g}(a.k)$ *is multiplicative in* k;

(vi) [2]$S_{f,g}(a,k) = (H_k * u)(a) \quad (a, k \in \mathbb{N})$, *where* $H_k(d) = \alpha_{k,d} f(d)g(k/d)$.

(vii) [2] $\sum_{d|n} S_{f,g}(d,k) = \sum_{d|\gcd(n,k)} f(d)g(k/d)\sigma_0(n/d) \quad (n, k \in \mathbb{N})$, *where* $\sigma_0(n)$ *denotes the number of the divisors of* n;

(viii) [2] $\sum_{d|n} S_{f,g}(d,k)h(n/d) = \sum_{d|\gcd(n,k)} f(d)g(k/d)(h*u)(n/d) \quad (n,k \in \mathbb{N})$;

(ix) [2] $\sum_{d|n} S_{f,g}(a,d)h(n/d) = \sum_{d|\gcd(a,n)} f(d)(g*h)(n/d) \quad (a,n \in \mathbb{N})$.

Proof Assertion (i) follows from $(f *_a I)(k) = \sum_{d|\gcd(a,k)} f(d)I(k/d)$ and the definition of the convolution identity.

Assertion ii. (1) and the first equality in ii. (2) are immediate, while the remaining relation follows, using (i), from

$$(f *_a g)(k) = \sum_{d|\gcd(a,k)} f(d)g(k/d) = \sum_{d|k} (f *_a I)(d)g(k/d).$$

To prove (iii), using ii. (2), we get $(f *_a g) * h = (f *_a I) * g * h = f *_a (g * h)$.
To prove (iv), using (iii), we have

$$f_a * g = (u *_a f) * g = u *_a (f * g) = (f * g)_a = (g * f)_a = g_a * f.$$

If f is multiplicative, then (i) shows that so is $f *_a I$. Using this together with ii. (2), we have (v). Note that (v) can also be proved directly as follows: for $k_1, k_2 \in \mathbb{N}$ with $\gcd(k_1, k_2) = 1$, using $\gcd(a, k_1k_2) = \gcd(a, k_1)\cdot\gcd(a, k_2)$ and the multiplicativity of f and g, we have

$$\begin{aligned}(f *_a g)(k_1k_2) &= \sum_{d|\gcd(a,k_1k_2)} f(d)g(k_1k_2/d)\\ &= \sum_{d_1|\gcd(a,k_1)} \sum_{d_2|\gcd(a,k_2)} f(d_1d_2)g(k_1k_2/d_1d_2)\\ &= \sum_{d_1|\gcd(a,k_1)} f(d_1)g(k_1/d_1) \sum_{d_2|\gcd(a,k_2)} f(d_2)g(k_2/d_2)\\ &= (f *_a g)(k_1)(f *_a g)(k_2).\end{aligned}$$

Assertion (vi) is immediate from the definition.
To prove (vii), we use (vi) (noting that $u = \mu^{-1}$ and $\sigma_0 = \mu^{-2}$) to get

$$\sum_{d|n} S_{f,g}(d,k) = \left(S_{f,g}(.,k) * \mu^{-1}\right)(n) = (H_k * \mu^{-1} * \mu^{-1})(n) = (H_k * \sigma_0(n)$$

$$= \sum_{d|n} \alpha_{k,d} f(d)g(k/d)\sigma(n/d).$$

To prove (viii), using (vi) we get

$$\sum_{d|n} S_{f,g}(d,k)h(n/d) = (S_{f,g}(.,k) * h)(n) = (H_k * \mu^{-1} * h)(n)$$
$$= \sum_{d|n} H_k(d)(h * u)(n/d) = \sum_{d|n} \alpha_{k,d} f(d) g(k/d)(h * u)(n/d).$$

To prove (ix), using ii. (2) and (i) we get,

$$\sum_{d|n} S_{f,g}(a,d)h(n/d) = (S_{f,g}(a,.) * h)(n) = (((f *_a I) * g) * h)(n)$$
$$= \sum_{d|n} (f *_a I)(d)(g * h)(n/d) = \sum_{d|n, d|a} f(d)(g * h)(n/d).$$

□

Remark Regarding the multiplicativity of $S_{f,g}(a,k)$, Theorem 2.1 (v) informs us that for multiplicative f, g, it is multiplicative in k; however, it is not multiplicative in a considered as element in $\mathbb{N}$, as seen from the example. Taking $f = g = \zeta_1, k = 10$, we have $S_{\zeta_1,\zeta_1}(6,10) \neq S_{\zeta_1,\zeta_1}(2,10) \times S_{\zeta_1,\zeta_1}(3,10)$ because

$$S_{\zeta_1,\zeta_1}(6,10) = \sum_{d|\gcd(6,10)=2} \zeta_1(d)\zeta_1(10/d) = \zeta_1(1)\zeta_1(10) + \zeta_1(2)\zeta_1(5) = 10 + 10 = 20$$
$$S_{\zeta_1,\zeta_1}(2,10) = \sum_{d|\gcd(2,10)=2} \zeta_1(d)\zeta_1(10/d) = \zeta_1(1)\zeta_1(10) + \zeta_1(2)\zeta_1(5) = 10 + 10 = 20$$
$$S_{\zeta_1,\zeta_1}(3,10) = \sum_{d|\gcd(3,10)=1} \zeta_1(d)\zeta_1(10/d) = \zeta_1(1)\zeta_1(10) = 10.$$

2.1 *Finite Fourier Series*

We begin by recalling a result, whose proof uses an interpolation argument, about finite Fourier series of a periodic arithmetic function and its inverse [3, Theorem 8.4].

Theorem 2.2 ([3, Theorem 8.4]) *Let $f \in \mathrm{A}$ be periodic mod k. Then there is a uniquely determined $g \in \mathrm{A}$, also periodic mod k, such that the following (finite) Fourier series holds*

$$f(m) = \sum_{n=0}^{k-1} g(n) e^{-2\pi i mn/k}.$$

The Fourier coefficient (function) g is given by the formula

$$g(n) = \tfrac{1}{k} \sum_{m=0}^{k-1} f(m) e^{2\pi i m n / k}.$$

Since both f and g are periodic mod k, we can write Theorem 2.2 as

$$f(m) = \sum_{n \bmod k} g(n) e^{-2\pi i m n / k} \quad \Leftrightarrow \quad g(n) = \frac{1}{k} \sum_{m \bmod k} f(m) e^{2\pi i m n / k}.$$

The finite Fourier series is related to generalized convolution through the following theorem.

Theorem 2.3 ([3, Theorem 8.5]) *For* $a \in \mathbb{Z}$ *and* $f, g \in \mathrm{A}$, *the function*

$$S_{f,g}(a, k) = \sum_{d \mid \gcd(a,k)} f(d) g(k/d) \quad (k \in \mathbb{N}),$$

has a finite Fourier expansion

$$S_{f,g}(a, k) = \sum_{m \bmod k} b_k(m) e^{-2\pi i m a / k}, \tag{2}$$

where

$$b_k(m) = \sum_{d \mid \gcd(m,k)} g(d) f(k/d) \frac{d}{k} \quad \left(= g *_m (f/\zeta_1)(k) = S_{g, f/\zeta_1}(m, k)\right). \tag{3}$$

Proof We adopt the proof given in [3, Theorem 8.5]. The function $S_{f,g}(\bullet, k)$ is periodic mod k, with respect to $\bullet$, because

$$S_{f,g}(a+k, k) = \sum_{d \mid \gcd(a+k,k)} f(d) g(k/d) = \sum_{d \mid \gcd(a,k)} f(d) g(k/d) = S_{f,g}(a, k).$$

By Theorem 2.2, $S_{f,g}(a, k)$ has a finite Fourier expansion of the form (2), with the coefficients $b_k(m)$ given by

$$b_k(m) = \frac{1}{k} \sum_{n \bmod k} S_{f,g}(n, k) e^{2\pi i m n / k} = \frac{1}{k} \sum_{n=1}^{k} \sum_{d \mid n, d \mid k} f(d) g(k/d) e^{2\pi i m n / k}.$$

Now write $n = cd$, and note that for each fixed d, the index c runs from 1 to k/d. Thus,

$$b_k(m) = \frac{1}{k} \sum_{d \mid k} f(d) g(k/d) \sum_{c=1}^{k/d} e^{2\pi i m c d / k}.$$

Now replace d by k/d in the sum on the right hand side to get

$$b_k(m) = \frac{1}{k}\sum_{d|k} f(k/d)g(d)\sum_{c=1}^{k/d} e^{2\pi imc/d}.$$

But the sum on c is 0 unless $d|m$, in which case the sum has the value d. Hence,

$$b_k(m) = \frac{1}{k}\sum_{d|k, d|m} f(k/d)g(d)d.$$

□

2.2 Ramanujan Sums

The classical Ramanujan sum is a Fourier series expansion defined for fixed $a \in \mathbb{Z}$ by

$$c_k(a) = \sum_{\substack{m \mod k \\ \gcd(m,k)=1}} e^{2\pi ima/k} \quad (k \in \mathbb{N}).$$

A remarkable fact connecting this series with generalized convolution is contained in the next theorem, which also shows that it takes only integral values.

Theorem 2.4 ([3, Theorem 8.6]) *For $a \in \mathbb{Z}$, we have*

$$c_k(a) = \sum_{d|\gcd(a,k)} d\mu(k/d) \quad (= S_{\zeta_1,\mu}(a,k)) \quad (k \in \mathbb{N}).$$

Proof Taking $f(k) = k$, $g(k) = \mu(k)$ in Theorem 2.3, we find

$$\sum_{d|\gcd(a,k)} d\mu(k/d) = \sum_{m \mod k} b_k(m)e^{-2\upsilon ima/k},$$

where

$$b_k(m) = \sum_{d|\gcd(m,k)} \mu(d) = I(\gcd(m,k)) = \begin{cases} 1 \; if \; \gcd(m,k) = 1 \\ 0 \; if \; \gcd(m,k) > 1. \end{cases}$$

Hence,

$$\sum_{d|\gcd(a,k)} d\mu(k/d) = \sum_{\substack{m \bmod k \\ \gcd(m,k)=1}} e^{-2\pi ima/k} = \sum_{\substack{m \bmod k \\ \gcd(m,k)=1}} e^{-2\pi im(k-a)/k}$$

$$= \sum_{\substack{m \bmod k \\ \gcd(m,k)=1}} e^{2\pi ima/k} = c_k(a).$$

□

In [8], noting the representation in Theorem 2.4, the generalized Ramanujan sum of order $\alpha \in \mathbb{C}$ is defined by

$$c_k^{(\alpha)}(a) := \sum_{d|\gcd(a,k)} d\mu_\alpha(k/d) \quad (= (\zeta_1 *_a \mu_\alpha)(k) = S_{\zeta_1,\mu_\alpha}(a,k)), \tag{4}$$

which by Theorem 2.1 (v) is a multiplicative function in k. Using Theorem 2.3 for the right-hand expression, the Fourier series representation of the generalized Ramanujan sum takes the form

$$c_k^{(\alpha)}(a) = \sum_{m \bmod k} \mu_{\alpha-1}(\gcd(m,k))e^{-2\pi ima/k}.$$

Corollary 2.5 *We have*

(i) *if* $k|a$, *then* $c_k^{(\alpha)}(a) = (\zeta_1 * \mu_\alpha)(k)$, *in particular,*

$$c_k^{(1)}(0) = c_k(0) = \varphi(k);$$

(ii) *if* $\gcd(a,k) = 1$, *then* $c_k^{(1)}(\alpha) = c_k(a) = \mu(k)$ [7, Corollary 2.5];

(iii) *for* $a, n \in \mathbb{N}$, *we have*

$$\sum_{d|n} c_k^{(\alpha)}(a) = n \prod_{p|n} \left(\sum_{t=0}^{\nu_p(n)} \binom{\alpha}{t} \left(\frac{-1}{p}\right)^t (t+1) \right);$$

(iv) *for* $a, n \in \mathbb{N}$, *we have* $\sum_{d|n} c_k^{(\alpha)}(a) = \sum_{d|\gcd(a,n)} d\mu_{\alpha-1}(n/d)$; *in particular, when* $\alpha = 1$, *we have* $\sum_{d|n} c_d(n) = \begin{cases} n & if\ n|a \\ 0 & otherwise; \end{cases}$

(v) $\sum_{d|n} c_{n/d}^{(\alpha)}(a) = \sum_{d|\gcd(a,n)} d\varphi_1^{(\alpha)}(n/d)$.

Proof Assertion (i) follows from $c_k^{(\alpha)}(a) = \sum_{d|k} d\mu_\alpha(k/d) = (\zeta_1 * \mu_\alpha)(k)$.

and $\zeta_1 * \mu = \varphi$ [3, Theorem 2.3]. Assertion (ii) is straightforward from (4).

To prove (iii), we take $k = n$, $g = \mu_\alpha$, $f = \zeta_1$ in Theorem 2.1 (vii) to get

$$\sum_{a|n} c_k^{(\alpha)}(a) = \sum_{a|n} \sum_{t|\gcd(a,k)} t\mu_\alpha(k/t) = \sum_{a|n} S_{\zeta_1,\mu_\alpha}(a,k)$$
$$= \sum_{d|n} \mu_\alpha(d)\sigma(d)n/d = n\prod_{p|n}\left(\sum_{t=0}^{\nu_p(n)} \frac{\mu_\alpha(p^t)\sigma(p^t)}{p^t}\right).$$

To prove (iv), we take $f = \zeta_1$, $g = \mu_\alpha$, $h = u(= \mu^{-1})$ in Theorem 2.1 (ix) to get

$$\sum_{d|n} c_k^{(\alpha)}(a) = \sum_{d|n} \sum_{t|\gcd(a,d)} t\mu_\alpha(d/t) = \sum_{d|n} S_{\zeta_1,\mu_\alpha}(a,d)$$
$$= \sum_{d|\gcd(a,n)} d\big(\mu_\alpha * \mu^{-1}\big)(n/d);$$

the second assertion follows from $\mu_0 = I$.

To prove (v), we take $f = h = \zeta_1$, $g = \mu_\alpha$ in Theorem 2.1 (ix) to get

$$\sum_{d|n} c_{n/d}^{(\alpha)}(a) = \sum_{d|n} \sum_{t|\gcd(a,n/d)} t\mu_\alpha((n/d)/t) = \sum_{d|n} S_{\zeta_1,\mu_\alpha}(a,n/d)$$
$$= \sum_{d|\gcd(a,n)} d(\mu_\alpha * \zeta_1)(n/d),$$

and the result follows by noting that $\varphi_1^{(\alpha)} = \mu_\alpha * \zeta_1$. □

3 Fourier Series of the gcd-Function

For $k \in \mathbb{Z}$, the gcd-function $\gcd(m,k)$, [14], considered as an arithmetic function in $m \in \mathbb{N}$ is periodic mod k, and so has a finite Fourier series representation.

Theorem 3.1 *For* $g \in \mathrm{A}$, $a \in \mathbb{Z}$, *let*

$$G_g(a;k) := \sum_{m \bmod k} (g * u)(\gcd(m,k))e^{-2\pi ma/k} \quad (k \in \mathbb{N}).$$

We have

(i) $G_g(a;k) := \sum_{d|\gcd(a,k)} dg(k/d) \quad (= S_{\zeta_1,g}(a,k))$;

in particular, if $\gcd(a,k)=1$, *then* $G_g(a;k)=g(k)$;

(ii) if g *is multiplicative, then* $G_g(a;k)$ *is a multiplicative function in* $k\in\mathbb{N}$.

Proof (i) Taking $f(k)=k$ in Theorem 2.3, we get

$$\sum_{d|\gcd(a,k)} dg(k/d) = \sum_{m \bmod k}\sum_{d|\gcd(m,k)} g(d)e^{-2\pi ima/k}$$
$$= \sum_{m \bmod k} (g*u)(\gcd(m,k))e^{-2\pi ima/k}.$$

Assertion (ii) is an immediate consequence of Theorem 2.1 (vi). □

For $f\in A$, $a\in\mathbb{Z}$, following [11], define

$$F_f(a;k) := \sum_{m \bmod k} f(\gcd(m,k))e^{-2\pi ima/k} \quad (= S_{\zeta_1, f*\mu}(a,k)) \quad (k\in\mathbb{N}),$$

and generalizing the discussion in [1], for $a\in\mathbb{N}$, define

$$J_a^f(k) := \sum_{\substack{m \bmod k\\ \gcd(m,k)|a}} f(\gcd(m,k)) \quad (k\in\mathbb{N}).$$

Theorem 3.2 *We have*

(I) I. (1) [11]

$$F_f(a:k) := \sum_{m \bmod k} f(\gcd(m,k))e^{-2\pi ima/k} = (\zeta_1 *_a (f*\mu))(k)$$
$$= (f * c_{.}(a))(k),$$

in particular, $F_{\zeta_1}(a;k) = (\zeta_1 *_a \varphi)(k) \quad (= S_{\zeta_1,\varphi}(a,k))$;

I. (2) if $\gcd(a,k)=1$, *then* $F_f(a;k)=\varphi(k)$ [7, Corollary 1];

I. (3) [7, Corollary 3] *for* $f,t\in A$ *such that* $f(n)=\sum_{d|n} t(d)=(t*u)(n)$, *if* $\gcd(a,k)=1$, *then* $F_f(a;k) := \sum_{m \bmod k} f(\gcd(m,k))e^{-2\pi ima/k} = t(k)$;

(II) [11] $f(\gcd(a,k)) = \frac{1}{k}\sum_{m \bmod k} (f*c_{\bullet}(m))(k)e^{2\pi ima/k}$;

(III) [11] If f *is multiplicative, then* $F_f(a;k)$ *is multiplicative in* k;

(IV) [11]

$$\gcd(a,k) = \sum_{m \bmod k} e^{2\pi ima/k}\left(\sum_{d|k} c_d(m)/d\right) = \frac{1}{k}\sum_{m \bmod k} e^{2\pi ima/k}(c_{\bullet} * \zeta_1)(k);$$

(V) $\sum_{m \bmod k} (f * \mu_{\alpha-1})(\gcd(m,k))e^{-2\pi i m a/k} = \big(f * c_\bullet^{(\alpha)}(a)\big)(k)$ *(This is a generalization of part (I);*

(VI) $J_a^f(k) := (f *_a \varphi)(k) = ((f *_a I) * \varphi)(k) \quad (a, k \in \mathbb{N});$

In particular ([7, p. 3]) $J_a^{\zeta_1}(k) := (\zeta_1 *_a \varphi)(k) = ((\zeta_1 *_a I) * \varphi)(k) = (\rho_a * \varphi)(k).$

Proof I. (1) Taking $f = g * u$,, i.e., $g = f * \mu$, in Theorem 3.1, we get

$$\sum_{m \bmod k} f(\gcd(m,k))e^{-2\pi i m a/k} = \sum_{d|\gcd(a,k)} d(f * \mu)(k/d) \quad (= (\zeta_1 *_a (f * \mu))(k))$$

$$= \sum_{d|a, d|k} d(f * \mu)(k/d) = \sum_{d|a,d|k} d \sum_{c|k/d} \mu(c) f\left(\frac{k/d}{c}\right) = \sum_{s|k} f(s) \sum_{d|a,d|k/s} d\mu\left(\frac{k/s}{d}\right)$$

$$= \sum_{s|k} f(s) c_{k/s}(a) = (f * c_\bullet(a))(k).$$

Using the former part and Corollary 2.5 (i), we get

$$F_{\zeta_1}(a;k) = (\zeta_1 *_a (\zeta_1 * \mu))(k) = \big(\zeta_1 *_a c_\bullet^{(1)}(0)\big)(k) = (\zeta_1 *_a \varphi)(k).$$

I. (2) follows from (1). To prove I. (3), from I. (1) with $f = t * u$ and $\gcd(a,k) = 1$, using also Corollary 2.5 (ii), we have

$$F_f(a;k) = \big(f * c_\bullet(a)\big)(k) = (t * u * c_\bullet(a))(k) = (t * u * \mu)(k) = t(k).$$

(II) follows from the formula for the inverse Fourier transform given by Theorem 2.2.

To prove (III), from Theorem 2.4 and Theorem 3.1 (ii), we see that $c_s(u)$ is a multiplicative function in s. The result is immediate from part (I).

(IV) follows from part (II) by taking $f(n) = \zeta_1(n) = n$.

To prove (V), we note from Theorem 3.1 (i) that

$$\big(f * c_\bullet^{(\alpha)}(a)\big)(k) = \sum_{s|k} f(s) c_{k/s}^{(\alpha)}(a) = \sum_{s|k} f(s) \sum_{d|\gcd(a,k/s)} d\mu_\alpha\left(\frac{k/s}{d}\right)$$

$$= \sum_{s|k} \sum_{d|a,d|k/s} f(s) d\mu_\alpha\left(\frac{k/s}{d}\right) = \sum_{d|a,d|k} d \sum_{c|k/d} \mu_\alpha(c) f\left(\frac{k/d}{c}\right)$$

$$= \sum_{d|a,d|k} d(f * \mu_\alpha)(k/d) = \sum_{d|\gcd(a,k)} d \cdot (f * \mu_\alpha)(k/d)$$

$$= \sum_{m \bmod k} (f * \mu_\alpha * u)(\gcd(m,k))e^{-2\pi i m n/k}.$$

To prove (VI), from its definition, we have

$$J_a^f(k) := \sum_{\substack{m \bmod k \\ \gcd(m,k)|a}} f(\gcd(m,k)) = \sum_{w|a, w|k} f(w)\varphi(k/w)$$

$$= \sum_{w|\gcd(a,k)} f(w)\varphi(k/w) = (f *_a \varphi)(k),$$

which is the first equality. The second equality follows from Theorem 2.1 (iii). □

Remark From part (VI) of the last theorem, we know that

$$\sum_{\substack{m \bmod k \\ \gcd(m,k)|a}} \gcd(m,k) = (\zeta_1 *_a \varphi)(k).$$

This last relation together with Theorem 3.2 (I) yields an interesting identity

$$\left(J_a^{\zeta_1}(k) =\right) \sum_{\substack{m \bmod k \\ \gcd(m,k)|a}} \gcd(m,k) = \sum_{m \bmod k} \gcd(m,k) e^{-2\pi i m a/k}.$$

4 Applications of the gcd-Function Fourier Series

In this section, we derive a number of applications from the Fourier series of the gcd-function.

4.1 *Identities*

Theorem 4.1 *We have*

(i) [11]

$$\sum_{m \bmod k} \gcd(m,k)\cos(2\pi m/k) = \varphi(k), \qquad \sum_{m \bmod k} \gcd(m,k)\sin(2\pi m/k) = 0;$$

(ii) [11], [5, Theorem 2.3] $\sum_{m \bmod k} \gcd(m,k) = (\zeta_1 * \varphi)(k)$

(This result shows that the classical gcd-sum formula is a special case of the Fourier series of the gcd-function);

(iii) [11], [6, Corollary 4.1] $\sum_{m \bmod k} f(\gcd(m,k)) = (f * \varphi)(k) \quad (f \in \mathrm{A});$

([7, Corollary 5]) *if* f *is multiplicative and* $n = p_1^{s_1} \cdots p_r^{s_r}$, *then*

$$\sum_{m \bmod k} f(\gcd(m,k)) = \prod_{i=1}^{r} \left\{ f(p_i^{s_i}) + (p_i - 1) \sum_{b=1}^{s_i} p_i^{b-1} f(p_i^{s_i - b}) \right\};$$

if f *is completely multiplicative but* $\neq \zeta_1$ *and prime factorization* $n = p_1^{s_1} \cdots p_r^{s_r}$, *then*

$$\sum_{m \bmod k} f(\gcd(m,k)) = \prod_{i=1}^{r} \left\{ f(p_i^{s_i}) + (p_i - 1) f(p_i^{s_i - 1}) \frac{f(p_i^{s_i}) - p_i^{s_i}}{f(p_i^{s_i}) - p_i f(p_i^{s_i - 1})} \right\}.$$

Proof (i) Taking $f(s) = \zeta_1(s) = s, \ a = 1$ in Theorem 3.2 (I) and using Corollary 2.5 (ii), we get

$$\sum_{m \bmod k} \gcd(m,k)\, e^{-2\pi i m/k} = (\zeta_1 * c_\bullet(1))(k) = \sum_{d|k} \frac{k}{d} \mu(d) = \varphi(k),$$

and the result follows by equating the real and imaginary parts.

(ii) follows from taking $f(s) = \zeta_1(s) = s, \ a = 0$ in Theorem 3.2 (I) and using Corollary 2.5 (i). The first part of (iii) follows from taking $a = 0$ in Theorem 3.2 (I) and using Corollary 2.5 (i), while the second and third parts are immediate from multiplicative property. □

The next application deals with the logarithmic function.

Theorem 4.2 ([11]) *Let* $h : \mathbb{N} \to \mathbb{R}_{>0}$. *We have*

$$\sum_{m \bmod k} (\log h(\gcd(m,k))) \exp(-2\pi i m a/k) = \sum_{d|k} (\log h(k/d))\, c_d(a).$$

(This generalizes Theorem 5 in [9].)

Proof Taking $f(s) = \log h(s) \ \ (s \in \mathbb{N})$ in Theorem 3.2 (I), we get

$$\log\left(\prod_{m \bmod k} h(\gcd(m,k))^{\exp(-2\pi i m a/k)} \right) = \sum_{m \bmod k} \log(h(\gcd(m,k)))\, e^{-2\pi i m a/k}$$
$$= (\log h * c_\bullet(a))(k).$$ □

The next application deals with the von Mangoldt function

$$\Lambda(k) = \begin{cases} \log p \ \ if \ k = p^m \ for \ some \ prime \ p \ and \ some \ m \in \mathbb{N} \\ 0 \qquad otherwise. \end{cases}$$

It is well-known, [3, Theorem 2.11], that $\Lambda = \log * \mu$.

Theorem 4.3 ([11]) *We have*

$$\sum_{m \bmod k} \log(\gcd(m,k)) e^{-2\pi i m/k} = \Lambda(k).$$

Proof Taking $f(s) = \log s,\ a = 1$ in Theorem 3.2 (I), we get

$$\sum_{m=1}^{k} \log(\gcd(m,k)) e^{-2\pi i m/k} = (\log * c_{\bullet}(1))(k) = (\log * \mu)(k) = \Lambda(k).$$

□

The next application involves the convolution identity.

Theorem 4.4 ([11]) *We have*

$$I(\gcd(a,k)) = \frac{1}{k} \sum_{m \bmod k} c_k(m) e^{2\pi i m a/k}.$$

Proof Letting $f(s) = I(s)$ in Theorem 3.2 (II) we get

$$I(\gcd(a,k)) = \frac{1}{k} \sum_{m \bmod k} (I * c_{\bullet}(m))(k) e^{2\pi i m a/k} = \frac{1}{k} \sum_{m \bmod k} \sum_{d|k} I(d) c_{k/d}(m)\, e^{2\pi m a/k}.$$

□

This last proposition shows that

$$\gcd(a,k) = 1 \Leftrightarrow \sum_{m \bmod k} c_k(m) e^{2\pi i m a/k} = 0$$

$\Leftrightarrow$ the m-vectors $\left(e^{2\pi i m a/k}\right)_{m \in \{1,2,\ldots,k\}}$ and the m-vectors $(c_k(m))_{m \in \{1,2,\ldots,k\}}$ are orthogonal.

The next application yields the so-called Ramanujan sum identity.

Theorem 4.5 ([11]) *Let*

$$\eta_k(a) = \sum_{m \bmod k} e^{-2\pi i m a/k}.$$

Then

$\eta_k(a) = (u * c_{\bullet}(a))(k)$ and $c_s(a) = (\mu * \eta_{\bullet}(a))(s)$.

Proof Letting $f(n) = 1 = u(n)$ in Theorem 3.2 (I), we get

$$\eta_k(a) = \sum_{m \bmod k} e^{-2\pi i m a/k} = (u * c_\bullet(a)(k),$$

and the second assertion follows from the Möbius inversion formula. □

Using this last proposition and Theorem 3.2 (IV), we get

$$\gcd(a,k) = \frac{1}{k}\sum_{m \bmod k} e^{2\pi i m a/k}(c_\bullet * \zeta_1)(k) = \frac{1}{k}\sum_{m \bmod k} e^{2\pi i m a/k}(\varphi * \eta_\bullet(m))(k),$$

where the last equality follows from the identities

$$\varphi * \eta_\bullet(m) = (\zeta_1 * \mu) * (u * c_\bullet(m)) = \zeta_1 * c_\bullet(m).$$

The next application deals with Jordan's totient defined by

$$J_z := \zeta_z * \mu \qquad (z \in \mathbb{C}).$$

Theorem 4.6 ([11]) *We have*

$$\sum_{m \bmod k} \gcd(m,k)^z e^{2\pi i m/k} = J_z(k).$$

Proof Taking $f(s) = \zeta_z(s) = s^z$ in Theorem 3.2 (I), we get

$$\sum_{m \bmod k} \gcd(m,k)^z e^{2\pi i m/k} = (\zeta_z * c_\bullet(1))(k) = (\zeta_z * \mu)(k).$$

□

4.2 Combinatorial Interpretation

The first application in this part deals with the number of solutions of a diophantine equation.

Theorem 4.7 ([11]) *Let*

$$r(n) = \sum_{n_1^2+n_2^2=n} 1$$

denote the number of ways that n can be expressed as the sum of two squares and let

$$\chi_{(1,0,-1,0)}(n) := \frac{i^{n-1} + (-i)^{n-1}}{2} = \begin{cases} 1 & \text{for } n \equiv 1 \mod 4 \\ -1 & \text{for } n \equiv 3 \mod 4 \\ 0 & \text{for } n \equiv 2 \mod 4 \end{cases}$$

be a Dirichlet character. Then

$$\sum_{m=1}^{k} r(\gcd(m,k))e^{-2\pi i m/k} = (r * \mu)(k) = 4\chi_{(1,0,-1,0)}(k).$$

Proof Taking $f(s) = r(s)$, $a = 1$ in Theorem 3.2 (I), we get

$$\sum_{m \bmod k} r(\gcd(m,k))e^{-2\pi i m/k} = (r * c_{\bullet}(1))(k) = (r * \mu)(k) = 4\chi_{(1,0,-1,0)}(k),$$

where the last equality is found in [12, Theorem 103] or [10, p. 27]. □

An interesting application to congruences, referred to as the number of factorizations of a mod k, is the next result due to van der Kamp.

Theorem 4.8 ([15, Sect. 2]) *For* $a \in \mathbb{Z}$, $k \in \mathbb{N}$, $\mathbb{Z}_k := \{0, 1, \ldots, k-1\}$, *we have*

$$F_{\zeta_1}(a;k) := \sum_{m \bmod k} \gcd(m,k)e^{-2\pi i m a/k} = \#\{(i,j) \in \mathbb{Z}_k \times \mathbb{Z}_k\,;\ i.j \equiv a \bmod k\}.$$

Proof For each fixed $i \in \mathbb{Z}_k$, let $d = \gcd(i,k)$, and note that the congruence $i.j \equiv a \mod k$ is solvable for $j \in \mathbb{Z}_k$ if and only if $d|a$ and $j \equiv \frac{a}{d}\left(\frac{i}{d}\right)^{-1} \bmod \frac{k}{d}$, where the inverse is considered mod k/d. Clearly, there are d solutions for such j mod k. Thus,

$$\#\{(i,j) \in \mathbb{Z}_k \times \mathbb{Z}_k\,;\ i \cdot j \equiv a \bmod k\} = \sum_{d|a} \sum_{\substack{i \bmod k \\ \gcd(i,k)=d}} d.$$

For $d \nmid k$ the inner sum is empty and is interpreted as 0. For $d|k$, the integers i mod k taken in the inner sum are of the form sd with $\gcd(s, k/d) = 1$, and there are exactly $\varphi(k/d)$ such i. Thus,

$$\#\{(i,j) \in \mathbb{Z}_k \times \mathbb{Z}_k \mid i \cdot j \equiv a \mod k\} = \sum_{\substack{d|a \\ d|k}} d\varphi(k/d) = (\zeta_1 *_a \varphi)(k) = F_{\zeta_1}(a;k),$$

by Theorem 3.2 (I). □

Theorem 4.8 is a special case of [10, Theorem 3.8] that we now recall.

Theorem 4.9 *Let $a, k, s \in \mathbb{N}$ and let $S(a, k, s)$ denote the number of solutions $(x_1, \ldots, x_s, y_1, \ldots, y_s) \in \mathbb{Z}_k^{2s}$ of the congruence*

$$a \equiv a_1x_1y_1 + \cdots + a_sx_sy_s \pmod{k}, \tag{5}$$

where $\gcd(a_i, k) = 1 \ \ (i = 1, \ldots, s)$. *Then*

$$S(a, k, s) = k^{2s-1} \sum_{d|k} \frac{c_d^{(1)}(a)}{d^s}. \tag{6}$$

Remarks Using (4), Theorem 2.1, (iii) and Theorem 3.2 (I), the right-hand expression of (6) is

$$k^{s-1} \sum_{d|k} \left(\frac{k}{d}\right)^s (\zeta_1 *_a \mu)(d) = k^{s-1}(\zeta_1 * (\zeta_1 *_a \mu))(k) = k^{s-1}(\zeta_1 *_a (\zeta_1 * \mu))(k)$$
$$= k^{s-1}(\zeta_1 *_a \varphi)(k) = k^{s-1} F_{\zeta_1}(a, k). \tag{7}$$

Theorem 4.8 is obtained by simply taking $s = 1, a_1 = 1$ in (7).

Closely related to Theorem 4.9 are the formulae for the number of solutions of linear congruences, [10, Theorems 3.3, 3.5], which are given, in terms of Ramanujan sums and generalized convolution, in the next theorem.

Theorem 4.10 (A) *Let $a, k, s \in \mathbb{N}$ and let $N(a, k, s)$ denote the number of solutions $(x_1, \ldots, x_s) \in \mathbb{Z}_k^s$ of the congruence*

$$a \equiv x_1 + \cdots + x_s \pmod{k}, \tag{8}$$

where $\gcd(x_i, k) = 1 \ \ (i = 1, \ldots, s)$. *Then*

$$N(a, k, s) = \frac{1}{k} \sum_{d|k} c_k(k/d)^s \ c_d(a) = \frac{1}{k} \big(c_k(\bullet)^s * c_\bullet(a)\big)(k). \tag{9}$$

(B) *Let $a, k, s \in \mathbb{N}$ and let $a_1, \ldots, a_s \in \mathbb{Z}$ be such that* $\gcd(a_1, \ldots, a_s, k) = 1$. *If $N'(a, k, s)$ is the number of solutions $(x_1, \ldots, x_s) \in \mathbb{Z}_k^s$ of the congruence*

$$a \equiv a_1x_1 + \cdots + a_sx_s \pmod{k}, \tag{10}$$

such that $\gcd(x_1, \ldots, x_s, k) = 1$, *then*

$$N'(a,k,s) = \sum_{d \mid \gcd(a,k)} \left(\frac{k}{d}\right)^{s-1} \mu(d) = (\mu *_a \zeta_{s-1})(k). \tag{11}$$

Remark Using Theorem 2.3, the formula (11) can be put into a Fourier series form as

$$N'(a,k,s) = \sum_{m \bmod k} e^{-2\pi ima/k} \sum_{d \mid \gcd(m,k)} d^{s-1}\mu\left(\frac{k}{d}\right)\frac{d}{k} = \frac{1}{k} \sum_{m \bmod k} e^{-2\pi ima/k}(\zeta_s *_m \mu)(k).$$

In [7], an application to the prime number decomposition through the use of the finite Fourier series of the gcd function is given as seen in the next theorem.

Theorem 4.11 ([7, Lemma 1]) *Let* $k = p_1^{s_1} \cdots p_r^{s_r}$, $s_i \in \mathbb{N}$ $(i = 1, \ldots, r)$ *be the prime factorization of* $k \in \mathbb{N}$, *let the prime factorization of* $a \in \mathbb{Z}$ *be* $a = up_1^{t_1} \cdots p_r^{t_r}$, $t_i \in \mathbb{N}_0 := \mathbb{N} \cup \{0\}$ *with* $\gcd(u, p_i) = 1$ $(1 \le i \le r)$, *and let* $f \in \mathrm{A}$ *be multiplicative. Then*

$$\sum_{m \bmod k} f(\gcd(m,k))e^{-2\pi ima/k} = \prod_{i=1}^{r} \left\{ f(p^{s_i}) + (p_i - 1) \sum_{d=1}^{\min(t_i,s_i)} p_1^{d-1} f(p_i^{s_i-d}) \right.$$
$$\left. - (1 - \theta_{t_i,s_i}) p_i^{t_i} f(p_i^{s_i-t_i-1}) \right\},$$

where $\theta_{t,s} = 0$ *if* $t < s$ *and* $= 1$ *if* $t \ge s$.

Proof From Theorem 3.2 (I), we know that

$$\sum_{m \bmod k} f(\gcd(m,k))e^{-2\pi ima/k} = (f * c_\bullet(a))(k).$$

Since f and $c_\bullet(a)$ are multiplicative, it suffices to evaluate the function at prime power factors p^s of k. Using [3, Example p. 164], we get

$$(f * c_\bullet(a))(p^s) = \sum_{0 \le d \le s} c_{p^d}(a) f(p^{s-d}) = \sum_{0 \le d \le s} \varphi(p^d) \frac{\mu(p^d/\gcd(p^t,p^d))}{\varphi(p^d/\gcd(p^t,p^d))} f(p^{s-d})$$

$$= f(p^s) + \sum_{d=1}^{s} \varphi(p^d) \frac{\mu(p^{d-\min(t,d)})}{\varphi(p^{d-\min(t,d)})} f(p^{s-d})$$

$$= f(p^s) + \begin{cases} (p-1) \sum\limits_{d=1}^{s} p^{d-1} f(p^{s-d}) & \text{if } s \le t \\ (p-1) \sum\limits_{d=1}^{t} p^{d-1} f(p^{s-d}) + \sum\limits_{d=t+1}^{s} \varphi(p^d) \frac{\mu(p^{d-t})}{\varphi(p^{d-t})} f(p^{s-d}) & \text{if } s > t \end{cases}$$

$$= f(p^s) + \begin{cases} (p-1)\sum_{d=1}^{s} p^{d-1} f(p^{s-d}) & \text{if } s \le t \\ (p-1)\sum_{d=1}^{t} p^{d-1} f(p^{s-d}) + \varphi(p^{t+1})\frac{\mu(p)}{\varphi(p)} f(p^{s-t-1}) & \text{if } s > t. \end{cases}$$

□

By the same proof as in the last theorem, we also have the next two theorems.

Theorem 4.12 ([7, Theorem 1], [1, Theorem 7])*Let* $k = p_1^{s_1} \cdots p_r^{s_r},\ s_i \in \mathbb{N}$ $(i = 1, \ldots, r)$ *be the prime factorization of* $k \in \mathbb{N}$, *and let the prime factorization of* $a \in \mathbb{Z}$ *be* $a = up_1^{t_1} \cdots p_r^{t_r},\ t_i \in \mathbb{N}_0 := \mathbb{N} \cup \{0\}$ *with* $\gcd(u, p_i) = 1$ $(1 \le i \le r)$. *Then*

$$\sum_{m \bmod k} \gcd(m,k) e^{-2\pi i m a/k} = \prod_{i=1}^{r} \left\{ (\min(t_i, s_i) + 1)\varphi(p_i^{s_i}) + \theta_{t_i,s_i} p_i^{s_i - 1} \right\},$$

where $\theta_{t,s} = 0$ *if* $t < s$ *and* $= 1$ *if* $t \ge s$.

Theorem 4.13 ([7, Theorem 2]) *Under the same hypotheses of Theorem 4.11 and assume that* f *is completely multiplicative. Then*

$$\sum_{m \bmod k} f(\gcd(m,k)) e^{-2\pi i m a/k}$$
$$= \prod_{i=1}^{r} \Bigg\{ f(p^{s_i}) - (1 - \theta_{t_i,s_i}) p_i^{t_i} f(p_i^{s_i - t_i - 1})$$
$$+ (p_i - 1) f(p_i^{s_i - 1}) \frac{f(p_i^{\min(t_i,s_i)}) - p_i^{\min(t_i,s_i)}}{f(p_i^{\min(t_i,s_i)}) - p_i f(p_i^{\min(t_i,s_i)-1})(1 - \delta_{0t_i})} \Bigg\}$$

where $\theta_{t,s} = 0$ *if* $t < s$ *and* $= 1$ *if* $t \ge s$ *and* δ_{0t} *is the Kronecker delta function.*

For the next application, we need some new notation. Let $\omega(n)$ be the number of distinct prime factors of $n \in \mathbb{N}$ and $\Omega(n)$ the total numbers of prime factors of n counted with multiplicity.

Theorem 4.14 ([11]) *Let* $\chi_{\wp}(n)$ *be the characteristic function of the primes. Then*

(i) $\sum_{m \bmod k} \omega(\gcd(m,k))\, e^{-2\pi i m/k} = \chi_{\wp}(k)$;
(ii) $\sum_{m \bmod k} \Omega(\gcd(m,k))\, e^{-2\pi i m/k} = \chi_{\wp^{\mathbb{N}}}(k)$.

Proof By the Möbius inversion formula, we have

$$\omega(n) = \sum_{\substack{p|n \\ p\in\wp}} 1 = \sum_{d|n} \chi_{\wp}(d) \Leftrightarrow \chi_{\wp} = \omega * \mu.$$

Taking $f(s) = \omega(s)$, $a = 1$ in Theorem 3.2 (I), we get

$$\sum_{m \bmod k} \omega(\gcd(m,k))\, e^{-2\pi i m/k} = (\omega * c_{\bullet}(1))(k) = (\omega * \mu)(k) = \chi_{\wp}(k),$$

which proves (i). The proof of (ii) is similar. □

Our next application deals with divisor functions. For $\beta \in \mathbb{R}$, define the sum of the βth powers of the divisors of n by

$$\sigma_\beta(n) = \sum_{d|n} d^\beta.$$

This function is related to the Möbius function, the constant 1-function and the power function through the identities [3, pp. 38–39]

$$\sigma_\beta(n) = \sum_{d|n} d^\beta = \zeta_\beta * u \quad \Leftrightarrow \quad \zeta_\beta = \sigma_\beta * \mu. \tag{12}$$

Theorem 4.15 ([11]) *We have*

$$\sum_{m \bmod k} \sigma_\beta(\gcd(m,n)) \cos(2\pi m/k) = k^\beta$$

$$\sum_{m \bmod k} \sigma_\beta(\gcd(m,k)) \sin(2\pi m/k) = 0.$$

Proof Taking $f(s) = \sigma_\beta(s)$, $a = 1$ in Theorem 3.2 (I) and using Theorem 2.5 (ii), we get

$$\sum_{m=1}^{k} \sigma_\beta(\gcd(m,k)) e^{-2\pi i m/k} = (\sigma_\beta * c_{\bullet}(1))(k) = \sum_{d|k} d^\beta \mu(k/d) = \zeta_\beta(k) = k^\beta.$$

□

Equating the real and imaginary parts, the two relations follow.

5 Partial Sums

The following results, taken from [2, Theorems 5–8], deal with partial sums involving the functions $S_{f,g}(a,k)$.

Theorem 5.1 *Let $f, g, h \in A$, and let $k, n \in \mathbb{N}$. Define*

$$h^*(n) = \sum_{d|n} h(d) = (h * u)(n), \quad \hat{h}(x) = \sum_{n \le x} h(n) \quad (n \in \mathbb{N}, x \in \mathbb{R}).$$

We have

I. I.1 $\sum_{m=1}^{n} S_{f,g}(m,k)\hat{h}(n/m) = \sum_{d|k, d \le n} f(d)g(k/d)\ \widehat{(h^*)}(n/m).$

I.2 $\sum_{k=1}^{n} S_{f,g}(m,k)\hat{h}(n/k) = \sum_{d|m,\ d \le n} f(d)\ \widehat{(g*h)}(n/d);$

II. if $C \in A$ *is completely multiplicative, then*

II.1 $\sum_{m=1}^{n} C(m) S_{f,g}(m,k) = \sum_{d|k,\ d\le n} C(d)\ f(d)\ g(k/d)\ \hat{C}(n/d).$

II.2 $\sum_{k=1}^{n} C(k) S_{f,g}(m,k) = \sum_{d|m,\ d \le n} C(d)\ f(d)\ \widehat{(Cg)}(n/d),$

where $\widehat{(Cg)}(r) := \sum_{d \le r} C(d) g(d).$

Examples (1) In Theorem 5.1 I.1, taking $h = I$, we get $\hat{h}(x) = 1$ for all $x \ge 1$, $h^*(n) = 1$ for all $n \in \mathbb{N}$, and so $\widehat{(h^*)}(x) = \lfloor x \rfloor$. Thus,

$$\sum_{m=1}^{n} S_{f,g}(m,k) = \sum_{d|k,\ d\le n} f(d)\ g(k/d)\ \lfloor n/d \rfloor.$$

If $k|n$, say $n = qk$, then every divisor d of k is $\le n$ and the last equation takes the form

$$\sum_{m=1}^{qk} S_{f,g}(m,k) = qk \sum_{d|k} \frac{f(d)}{d}\ g\left(\frac{k}{d}\right).$$

For the special case $f = \zeta_1$, this gives us

$$\sum_{m=1}^{qk} S_{\zeta_1,g}(m,k) = qk \sum_{d|k} g(d) = qk\ g^*(k).$$

In particular, when $g = \mu_\alpha$, we obtain the following generalized formula of Von Sterneck

$$\sum_{m=1}^{qk} c_k^{(\alpha)}(m) = \sum_{m=1}^{qk} S_{\zeta_1,\mu_\alpha}(m,k) = qk \sum_{d|k} \mu_\alpha(d) = qk\, \mu_{\alpha-1}(k).$$

(2) In Theorem 5.1 I.2, taking $h = g^{-1}$, we have $g * h = I$ and $\hat{I}(x) = 1$ for all $x \geq 1$. Thus,

$$\sum_{k=1}^{n} S_{f,g}(m,k)\widehat{(g^{-1})}(n/k) = \sum_{d|m,\ d\leq n} f(d).$$

In particular, if $g = \mu_\alpha$, then $\widehat{(g^{-1})}(x) = \sum_{n\leq x} \mu_{-\alpha}(n)$, and the last relation becomes

$$\sum_{k=1}^{n} S_{f,\mu_\alpha}(m,k)\left(\sum_{t\leq n/k} \mu_{-\alpha}(t)\right) = \sum_{d|m\ d\leq n} f(d).$$

When $f = \zeta_1$, this gives a generalization of a theorem of Nicol

$$\sum_{k=1}^{n} c_k^{(\alpha)}(m)\left(\sum_{t\leq n/k} \mu_{-\alpha}(t)\right) = \sum_{k=1}^{n} S_{\zeta_1,\mu_\alpha}(m,k)\left(\sum_{t\leq n/k} \mu_{-\alpha}(t)\right) = \sum_{d|m\ d\leq n} d = \sigma(m,n),$$

where $\sigma(m,n)$ denotes the sum of the divisors of m which are $\leq n$.

(3) In Theorem 5.1 I.2, taking $h = I$, we have

$$\sum_{k=1}^{n} S_{f,g}(m,k) = \sum_{d|m,\ d\leq n} f(d)\,\hat{g}(n/d).$$

In particular, if $f = \zeta_1,\ g = \mu_\alpha$, we obtain

$$\sum_{k=1}^{n} c_k^{(\alpha)}(m) = \sum_{k=1}^{n} S_{\zeta_1,\mu_\alpha}(m,k) = \sum_{d|m,\ d\leq n} d\,\widehat{\mu_\alpha}(n/d).$$

When $n = m$, this becomes

$$\sum_{k=1}^{m} c_k^{(\alpha)}(m) = \sum_{d|m} d\,\widehat{\mu_\alpha}(m/d) = \left(\zeta_1 * \widehat{\mu_\alpha}\right)(m).$$

Since $\zeta_1^{-1}(n) = n\mu(n)$, inversion of this last relation gives us a formula for the partial sums of the Souriau–Hsu–Möbius function

$$\sum_{n=1}^{m} \mu_\alpha(n) = \widehat{\mu_\alpha}(m) = \left(\zeta_1^{-1} * \sum_{k=1}^{\bullet} c_k^{(\alpha)}(\bullet)\right)(m) = \sum_{d|m}\left(\sum_{k=1}^{d} c_k^{(\alpha)}(d)\right)\frac{m\mu_\alpha(m/d)}{d}.$$

(4) In Theorem 5.1 II.1, taking $C = \zeta_1$ so that $\hat{C}(x) = \frac{\lfloor x\rfloor(\lfloor x\rfloor+1)}{2}$, we have

$$\sum_{m=1}^{n} mS_{f,g}(m,k) = \frac{1}{2}\sum_{d|k,\ d\le n} d\ f(d)\ g(k/d)\lfloor n/d\rfloor(\lfloor n/d\rfloor + 1).$$

If $k \mid n$, say $n = qk$, then $\lfloor n/d\rfloor = qk/d$. For $d|k$, the last relation becomes

$$\sum_{m=1}^{qk} mS_{f,g}(m,k) = \frac{1}{2}q^2k^2\sum_{d|k}\frac{f(d)}{d}\ g(k/d) + \frac{1}{2}qk\sum_{d|k} f(d)g(k/d).$$

When $g = \mu_\alpha,\ f = \zeta_1$, we obtain

$$\sum_{m=1}^{qk} mc_k^{(\alpha)}(m) = \frac{1}{2}q^2k^2\mu_{\alpha-1}(k) + \frac{1}{2}qk(\zeta_1 * \mu_\alpha)(k).$$

(5) Taking $h = \zeta_{-s}$, we have

$$\widehat{(hg)}(x) = \sum_{r\le x} g(r)r^{-s} = G(s) - \sum_{r>x} g(r)r^{-s}$$

if the series for $G(s)$ converges. Using this in Theorem 5.1 II.2, and taking $n = qm$, we find

$$\sum_{k=1}^{qm}\frac{S_{f,g}(m,k)}{k^s} = G(s)\sum_{d|m} f(d)d^{-s} - \sum_{d|m} f(d)d^{-s}\sum_{r>qm/d} g(r)r^{-s}.$$

Letting $q \to \infty$, we obtain, if the series for $G(s)$ converges,

$$\sum_{k=1}^{\infty}\frac{S_{f,g}(m,k)}{k^s} = G(s)\sum_{d|m} f(d)d^{-s}. \tag{13}$$

In the special case with $f = \zeta_{-s},\ g = \mu_\alpha$, we have [13]

$$G(s) = \sum_{r=1}^{\infty}\frac{\mu_\alpha(r)}{r^s} = \frac{1}{\zeta(s)^\alpha} \quad (\Re(s) > 1).$$

Also,

$$\sum_{d|m} \zeta_1(d)d^{-s} = \sum_{d|m} d^{1-s} = \sigma_{1-s}(m).$$

In this case (13) gives another generalization of a Ramanujan's formula

$$\sum_{k=1}^{\infty} \frac{c_k^{(\alpha)}(m)}{k^s} = \sum_{k=1}^{\infty} \frac{S_{\zeta_1,\mu_\alpha}(m,k)}{k^s} = \frac{\sigma_{1-s}(m)}{\zeta(s)^\alpha} \quad (\Re(s) > 1). \tag{14}$$

From Theorem 3.2 I. (1), we know that

$$F_f(a;k) := \sum_{m \bmod k} f(\gcd(m,k))e^{-2\pi i ma/k} = (f * c_\bullet(a))(k).$$

Taking $f = \zeta_{-s}$, for $a, k \in \mathbb{N}$, using (14), we have

$$\begin{aligned} F(s) &:= \sum_{k=1}^{\infty} \frac{F_{\zeta_1}(a;k)}{k^s} = \sum_{k=1}^{\infty} \frac{(\zeta_1 * c_\bullet(a))(k)}{k^s} = \sum_{k=1}^{\infty} \frac{\zeta_1(k)}{k^s} \cdot \sum_{k=1}^{\infty} \frac{c_k(a)}{k^s} \\ &= \sum_{k=1}^{\infty} \frac{1}{k^{s-1}} \cdot \frac{\sigma_{1-s}(a)}{\zeta(s)} = \frac{\zeta(s-1)\sigma_{1-s}(a)}{\zeta(s)}. \end{aligned}$$

References

1. Abel, U., Awan, W., Kushnirevych, V.: A generalization of the GCD-sum function. J. Integer Seq. **16** (2013). Article 13.6.7
2. Apostol, T.M.: Arithmetical properties of generalized Ramanujan sums. Pac. J. Math. **41**, 281–293 (1972)
3. Apostol, T.M.: Introduction to Analytic Number Theory. Springer, New York (1976)
4. Brown, T.C., Hsu, L.C., Wang, J., Shiue, P.J.-S.: On a certain kind of generalized number-theoretical Möbius function. Math. Sci. **25**, 72–77 (2000)
5. Broughan, K. A.: The GCD-sum function. J. Integer Seq. **4** (2001). Article 01.2.2
6. Haukkanen, P.: On a GCD-sum function. Aequ. Math. **76**, 168–178 (2008)
7. Holleboom, L. I.: Prime number decomposition of the Fourier transform of a function of the greatest common divisor. arXiv:1410.2054 [math. NT]
8. Laohakosol, V., Ruengsinsub, P., Pabhapote, N.: Ramanujan sums via generalized Möbius functions and applications. Int. J. Math. Math. Sci. **2006** (2006). Article ID 60528
9. Loveless, A. D.: The general GCD-product function. Integers **6**(A19) (2006)
10. McCarthy, P.J.: Introduction to Arithmetical Functions. Springer, New York (1986)
11. Schramm, W.: The Fourier transform of functions of the greatest common divisor. Integers **8**(A50) (2008).
12. Sivaramakrishnan, R.: Classical Theory of Arithmetic Functions. Marcel Dekker, New York (1989)

13. Souriau, J.–M.: Généralisation de certaines formules arithmétiques d' inversion. Applications. Rev. Sci. (Rev. Rose Illus.) **82**, 204–211 (1944)
14. Tóth, L.: A survey of GCD-sum functions. J. Integer Seq. **13** (2010). Article 10.8.1
15. van der Kamp, P. H.: On the Fourier transform of the greatest common divisor. Integers **13**(A24) (2013)

Problem on Coefficients of Bi-Univalent Function Class Using Chebyshev Polynomials

F. Müge Sakar and Ertuğrul Doğan

2010 AMS Classification 30C45 · 30C50

1 Introduction

Let's $\mathcal{A}$ be analytic and show Taylor expansion form

$$s(z) = z + \sum_{n=2}^{\infty} a_n z^n \tag{1}$$

in the open unit disc $\Delta = \{z : z \in \mathcal{C} \quad and \quad |z| < 1\}$. In addition, we indicate by $\mathcal{S}$ the class of all univalent functions in Δ.

For s and r, analytic functions in Δ, the function $s(z)$ is subordinate to $r(z)$ in Δ, and it can be represented by

$$s(z) \prec r(z) \qquad (z \in \Delta)$$

if there presents an analytic Schwarz function $\omega(z)$ being as

$$\omega(0) = 0 \quad and \quad |\omega(z)| < 1 \quad (z \in \Delta)$$

F. M. Sakar (✉)
Department of Business Administration, Dicle University, Diyarbakır, Turkey

E. Doğan
Institute of Science, Master's Degree in Mathematics, Batman University, Batman, Turkey

P. Srivastava et al. (eds.), *Mathematical, Computational Intelligence and Engineering Approaches for Tourism, Agriculture and Healthcare*, Lecture Notes in Networks and Systems 214, https://doi.org/10.1007/978-981-16-3807-7_6

such that

$$s(z) = r(\omega(z)) \quad (z \in \Delta).$$

Especially, if r is univalent in Δ, we can rewrite the condition of subordination as follows:

$$s(0) = r(0) \quad and \quad s(\Delta) \subset r(\Delta).$$

Every function $s \in \mathcal{S}$ is known to have an inverse s^{-1}, given by

$$s^{-1}(s(z)) = z \quad (z \in \Delta)$$

and

$$s^{-1}(s(\varpi)) = \varpi, \quad \left(|\, \varpi \,| < k_0(s); \quad k_0(s) \geq \frac{1}{4} \right),$$

where

$$s^{-1}(\varpi) = \varpi - a_2\varpi^2 + (2a_2^2 - a_3)\varpi^3 - (5a_2^3 - 5a_2a_3 + a_4)\varpi^4 + \cdots . \tag{2}$$

In order to have bi-univalent $s \in \mathcal{A}$ in Δ , both $s(z)$ and $s^{-1}(z)$ must be univalent in Δ. Let's represent all bi-univalent functions class by Σ.

Some well-known subclasses of the $\mathcal{S}$ are $\mathcal{K}(\alpha)$ and $S^*(\alpha)$. $\mathcal{K}(\alpha)$ is known as α order convex functions and $S^*(\alpha)$ is known as α order starlike functions in Δ, These subclasses are respectively shown as follows:

$$\mathcal{K}(\alpha) := \left\{ s : s \in \mathcal{A} \;\; and \;\; \mathcal{R}\left(1 + \frac{zs''(z)}{s'(z)} > \alpha \right); \quad z \in \Delta; \quad 0 \leq \alpha < 1 \right\}. \tag{3}$$

and

$$S^*(\alpha) := \left\{ s : s \in \mathcal{A} \;\; and \;\; \mathcal{R}\left(\frac{zs'(z)}{s(z)} > \alpha \right); \quad z \in \Delta; \quad 0 \leq \alpha < 1 \right\}. \tag{4}$$

For $0 \leq \alpha < 1$, if both s and s^{-1} are α order convex or α order starlike functions, respectively, $s \in \Sigma$ is in the class $\mathcal{K}_\Sigma(\alpha)$ of α order bi-convex function or $S^*_\Sigma(\alpha)$ of α order bi-starlike function.

Chebyshev polynomial is very major topic in numerical analysis in terms of both theoretic and applied approach. On the other hand, many researchers have also been dealing with orthogonal Chebyshev polynomials. One can see the details of Chebyshev polynomials which is first type $T_n(t)$, the second type $\Xi_n(t)$ and the effects of them in the referes [2, 3, 5, 6]. The well-known first and second types Chebyshev polynomials are deined as follows: $T_n(t) = \cos n\theta$ and $\Xi_n(t) = \frac{\sin(n+1)\theta}{\sin\theta}$ $(-1 \leq t < 1)$, where t is equal to $\cos\theta$ and n shows the polynomial degree.

The classical Fekete–Szegö inequality, which is studied via Loewner's chain method, contains the Taylor–Maclaurin coefficients of $s \in S$ given by Eq. (1):

$$\left|a_3 - \delta a_2^2\right| \leq 1 + 2\exp\left(\frac{-2\delta}{1-\delta}\right) \quad (0 \leq \delta < 1).$$

In its limit as $\delta \to 1^-$,we have an fundamental inequality given by

$$\left|a_3 - a_2^2\right| \leq 1.$$

In fact, for the normalized function $s \in S$, the coefficient functional $\Psi_\delta(s)$ given by the relation

$$\Psi_\delta(s) = a_3 - \delta a_2^2$$

is very important in function theory. The problem of maximizing the modulus of the functional $\Psi_\delta(s)$ is said the Fekete–Szegö problem (see [4]).

Definition 1.1 $s \in \Sigma$ given by Eq. (1) is known to be in the class $S_\Sigma(\lambda, \tau; t)$, for $0 \leq \lambda \leq 1$, $\tau \in C\backslash\{0\}$, $t \in (\frac{1}{2}, 1]$ and all $z, \varpi \in \Delta$ if the subordination conditions given below hold:

$$1 + \frac{1}{\tau}\left[\frac{zs'(z) + \lambda z^2 s''(z)}{\lambda z s'(z) + (1-\lambda)s(z)} - 1\right] \prec \Upsilon(z, t) := \frac{1}{1 - 2tz + z^2} \tag{5}$$

$$1 + \frac{1}{\tau}\left[\frac{\varpi r'(\varpi) + \lambda \varpi^2 r''(\varpi)}{\lambda \varpi r'(\varpi) + (1-\lambda)r(\varpi)} - 1\right] \prec \Upsilon(\varpi, t) := \frac{1}{1 - 2t\varpi + \varpi^2}, \tag{6}$$

where the function $r = s^{-1}$ is denoted by (2).

It is worthy that if $t = \cos\alpha$, where $\alpha \in (-\pi/3, \pi/3)$

$$\Upsilon(z, t) = \frac{1}{1 - 2\cos\alpha z + z^2} = 1 + \sum_{n=1}^{\infty} \frac{\sin(n+1)\alpha}{\sin\alpha} z^n \quad (z \in \Delta).$$

Thus

$$\Upsilon(z, t) = 1 + 2\cos\alpha z + (3\cos^2\alpha - \sin^2\alpha)z^2 + \cdots \quad (z \in \Delta).$$

From [9], $\Upsilon(z, t)$ can be written as

$$\Upsilon(z, t) = 1 + \Xi_1(t)z + \Xi_2(t)z^2 + \cdots \quad (z \in \Delta, \quad t \in (-1, 1)),$$

where

$$\Xi_{n-1} = \frac{\sin(n \arccos t)}{\sqrt{1 - t^2}} \quad (n \in N)$$

denotes Chebyshev polynomials which is second type and we get some initial forms given below:

$$\Xi_n(t) = 2t\,\Xi_{n-1}(t) - \Xi_{n-2}(t),$$

and

$$\Xi_1(t) = 2t, \quad \Xi_2(t) = 4t^2 - 1, \quad \Xi_3(t) = 8t^3 - 4t, \quad \Xi_4(t) = 16t^4 - 12t^3 + 1 \cdots . \quad (7)$$

Chebyshev polynomial which is first type $T_n(t)$, $t \in [-1, 1]$, is indicated by

$$\sum_{n=0}^{\infty} T_n(t) z^n = \frac{1 - tz}{1 - 2tz + z^2} \quad (z \in \Delta).$$

The relations between the Chebyshev polynomial which is first type $T_n(t)$ and Chebyshev polynomial which is second type $\Xi_n(t)$ are given by

$$\frac{dT_n(t)}{dt} = n\,\Xi_{n-1}(t); \quad T_n(t) = \Xi_n(t) - t\,\Xi_{n-1}(t); \quad 2T_n(t) = \Xi_n(t) - \Xi_{n-2}(t).$$

Remark 1.2 For $\tau = 1$ and $\alpha = \lambda$, we reach the class $S_\Sigma(0, 1; t) = S_\Sigma^*(t)$ occurs function $s \in \Sigma$ fulfilling the condition

$$\frac{zs'(z)}{s(z)} \prec \Upsilon(z, t) := \frac{1}{1 - 2tz + z^2}$$

and

$$\frac{\varpi r'(z)}{r(\varpi)} \prec \Upsilon(\varpi, t) := \frac{1}{1 - 2t\varpi + z^2},$$

where the function $r = s^{-1}$ is defined by (2). Bulut et al.[1]

Definition 1.3 ([7]) Let $\mathcal{P}_m(\beta)$ with the conditions of $m \geq 2$ and $0 \leq \beta < 1$ represent by analytic and univalent functions p, normalized with $p(0) = 1$ and fulfilling

$$\int_0^{2\pi} \left| \frac{\Re p(z) - \beta}{1 - \beta} \right| d\theta \leq n\pi,$$

where $z = re^{\iota\theta}$. For $\beta = 0$, we denote $\mathcal{P}_m = \mathcal{P}_m(0)$, hereby $\mathcal{P}_m$ symbolizes the class of functions $p(z)$, which are analytic and normalized with $p(0) = 1$ in Δ, also it indicated by

$$p(z) = \int_0^{2\pi} \frac{1 - ze^{\iota t}}{1 + ze^{\iota t}} d\mu(t),$$

where ζ is a function with real valued. It also satisfies the following condition of bounded variation.

$$\int_0^{2\pi} d\zeta(t) \quad and \quad \int_0^{2\pi} |d\zeta(t)| \leq m, \quad m \geq 2.$$

Note that $\mathcal{P} = \mathcal{P}_2$ is the Caratheodory function which is well-known class, i.e., the normalized functions which have the positive real part in Δ.

Definition 1.4 ([8]) For $0 \leq \lambda \leq 1, 0 \leq \beta < 1$, $s \in S_\Sigma(\lambda, \tau; \beta)$ given by (1), if the two cases given below are fulfilled.

$$1 + \frac{1}{\tau}\left[\frac{zs'(z) + \lambda z^2 s''(z)}{\lambda zs'(z) + (1-\lambda)s(z)} - 1\right] \in P_m(\beta) \quad and \tag{8}$$

$$1 + \frac{1}{\tau}\left[\frac{\varpi r'(\varpi) + \lambda \varpi^2 r''(\varpi)}{\lambda \varpi r'(\varpi) + (1-\lambda)r(\varpi)} - 1\right] \in \mathcal{P}_m(\beta), \tag{9}$$

where $\tau \in \mathcal{C} \setminus \{0\}$, the function $r = s^{-1}$ given by (2), and $z, \varpi \in \Delta$. Lemma A given below is required so as to derive our main results.

Lemma A *Let* $\Phi(z) = 1 + h_1 z + h_2 z^2 + \cdots$; $z \in \Delta$ *and* $\Phi \in \mathcal{P}_m(\beta)$. *Then*

$$|h_n| \leq m(1-\beta); \quad n \geq 1.$$

In current paper, we introduce a subclass $S_\Sigma(\lambda, \tau; t)$ and $S_\Sigma(\lambda, \tau; \beta)$ of bi-univalent-analytic functions. Also, we obtain the initial coefficient bounds and Fekete–Szegö problem by using Chebyshev polynomials expansions. In addition, we give some corollaries related to our main results.

2 Main Results

Theorem 2.1 *For* $0 \leq \lambda \leq 1$, $\tau \in \mathcal{C} \setminus \{0\}$ *and* $t \in (\frac{1}{2}, 1]$, *let's take* $s \in \Sigma$ *denoted by in the class* $S_\Sigma(\lambda, \tau; t)$. *Then*

$$|a_2| \leq \frac{2|\tau|t\sqrt{2t}}{\sqrt{|4[2\tau(1+2\lambda) - (\tau+1)(1+\lambda)^2]t^2 + (1+\lambda)^2|}}, \tag{10}$$

$$|a_3| \leq \frac{4|\tau|^2 t^2}{(1+\lambda)^2} + \frac{|\tau|}{1+2\lambda}, \tag{11}$$

and for some $\eta \in \mathcal{R}$,

$$|a_3 - \eta a_2^2| \leq \begin{cases} \frac{|\tau| t}{1+2\lambda} & , \quad |\eta - 1| \leq \frac{|4[2\tau(1+2\lambda)-(\tau+1)(1+\lambda)^2]t^2+(1+\lambda)^2|}{8|\tau|(1+2\lambda)t^2} \\ \frac{8|\tau|^2|\eta-1|t^3}{|4[2\tau(1+2\lambda)-(\tau+1)(1+\lambda)^2]t^2+(1+\lambda)^2|} & , \quad |\eta - 1| \geq \frac{|4[2\tau(1+2\lambda)-(\tau+1)(1+\lambda)^2]t^2+(1+\lambda)^2|}{8|\tau|(1+2\lambda)t^2}. \end{cases} \tag{12}$$

Proof Let $s \in S_{\Sigma}(\lambda, \tau; t)$. Then from Eqs. (5) and (6), we attain

$$1 + \frac{1}{\tau}\left[\frac{zs'(z) + \lambda z^2 s''(z)}{\lambda z s'(z) + (1-\lambda)s(z)} - 1\right] = 1 + \Xi_1(t)p(z) + \Xi_2(t)p^2(z) + \cdots \tag{13}$$

and

$$1 + \frac{1}{\tau}\left[\frac{\varpi r'(\varpi) + \lambda \varpi^2 r''(\varpi)}{\lambda \varpi r'(\varpi) + (1-\lambda)r(\varpi)} - 1\right] = 1 + \Xi_1(t)q(\varpi) + \Xi_2(t)q^2(\varpi) + \cdots \tag{14}$$

for functions which is analytic

$$p(z) = c_1 z + c_2 z^2 + c_3 z^3 + \cdots, \quad (z \in \Delta), \tag{15}$$

and

$$q(\varpi) = d_1 \varpi + d_2 \varpi^2 + d_3 \varpi^3 + \cdots, \quad (\varpi \in \Delta), \tag{16}$$

such that $p(0) = q(0) = 0$, $|p(z)| < 1$ $(z \in \Delta)$ and $|q(\varpi)| < 1$ $(\varpi \in \Delta)$. It is clear that if $|p(z)| < 1$ and $|q(\varpi)| < 1$, then

$$|c_j| \leq 1 \quad and \quad |d_j| \leq 1 \quad for \quad j \in \mathbf{N}. \tag{17}$$

From (11), (12), (13) and (14), we have

$$1 + \frac{1}{\tau}\left[\frac{zs'(z) + \lambda z^2 s''(z)}{\lambda z s'(z) + (1-\lambda)s(z)} - 1\right] = 1 + \Xi_1(t)c_1 z + [\Xi_1(t)c_2 + \Xi_2(t)c_1^2]z^2 + \cdots \tag{18}$$

and

$$1 + \frac{1}{\tau}\left[\frac{\varpi r'(\varpi) + \lambda \varpi^2 r''(\varpi)}{\lambda \varpi r'(\varpi) + (1-\lambda)r(\varpi)} - 1\right] = 1 + \Xi_1(t)d_1 \varpi + [\Xi_1(t)d_2 + \Xi_2(t)d_1^2]\varpi^2 + \cdots. \tag{19}$$

We can obtain equalities of (18), (19), (20) and (21) by equating the coefficients in (16) and (17),

$$\frac{1}{\tau}(1+\lambda)a_2 = \Xi_1(t)c_1 \tag{20}$$

$$\frac{1}{\tau}[2(1+2\lambda)a_3 - (1+\lambda)^2 a_2^2] = \Xi_1(t)c_2 + \Xi_2(t)c_1^2 \tag{21}$$

$$-\frac{1}{\tau}(1+\lambda)a_2 = \Xi_1(t)d_1 \tag{22}$$

$$\frac{1}{\tau}\left\{[4(1+2\lambda)-(1+\lambda)^2]a_2^2 - 2(1+2\lambda)a_3\right\} = \Xi_1(t)d_2 + \Xi_2(t)d_1^2. \tag{23}$$

From Eqs. (18) and (20), we have

$$c_1 = -d_1 \tag{24}$$

and

$$\frac{2}{\tau^2}(1+\lambda)^2 a_2^2 = \Xi_1^2(t)(c_1^2 + d_1^2). \tag{25}$$

Also, by using (19) and (21), we obtain

$$\frac{2}{\tau}[2(1+2\lambda)-(1+\lambda)^2]a_2^2 = \Xi_1(t)(c_2+d_2) + \Xi_2(t)(c_1^2+d_1^2). \tag{26}$$

By using (23) in (24), we get

$$\left\{\frac{2}{\tau}[2(1+2\lambda)-(1+\lambda)^2] - \frac{2\Xi_2(t)}{\tau^2\Xi_1^2(t)}(1+\lambda)^2\right\} a_2^2 = \Xi_1(t)(c_2+d_2). \tag{27}$$

From Eqs. (7), (15) and (25), we obtain the inequality (8). Afterward, we get the following equality, by subtracting (21) from (19)

$$\frac{4}{\tau}(1+2\lambda)(a_3 - a_2^2) = \Xi_1(t)(c_2-d_2) + \Xi_2(t)(c_1^2 - d_1^2). \tag{28}$$

Further, from Eq. (22), we obtain

$$a_3 = a_2^2 + \frac{\tau\,\Xi_1(t)(c_2-d_2)}{4(1+2\lambda)}. \tag{29}$$

Thus, we get desired inequality (9) by using (23) and applying (7).

Now, we get the following equalities by using (25) and (27) for some $\eta \in \mathcal{R}$

$$(a_3 - \eta a_2^2) = (1-\eta)a_2^2 + \frac{\tau\,\Xi_1(t)(c_2-d_2)}{4(1+2\lambda)}$$

and

$$(a_3 - \eta a_2^2) = \tau\,\Xi_1(t)\left[\left(h(\eta) + \frac{1}{4(1+2\lambda)}\right)c_2 + \left(h(\eta) - \frac{1}{4(1+2\lambda)}\right)d_2\right],$$

where

$$h(\eta)=\frac{\tau\Xi_1^2(t)(1-\eta)}{2\tau\Xi_1^2(t)[2(1+2\lambda)-(1+\lambda)^2]-2\Xi_2(t)(1+\lambda)^2}.$$

So, we deduce that

$$|a_3-\eta a_2^2|\le\begin{cases}\frac{|\tau|t}{1+2\lambda} & , \quad 0\le|h(\eta)|\le\frac{1}{4(1+2\lambda)}\\ 4|\tau||h(\eta)|t & , \quad |h(\eta)|\ge\frac{1}{4(1+2\lambda)}\end{cases}.$$

So, this is the completed proof of Theorem 2.1.

We can attain the following corollary by taking $\tau=1$ and $\lambda=0$ in Theorem 2.1

Corollary 2.2 *Let $s\in S_\Sigma^*$ for $t\in(\frac{1}{2},1]$. Then*

$$|a_2|\le 2t\sqrt{2t},$$

$$|a_3|\le 4t^2+t$$

and for some $\eta\in\mathcal{R}$,

$$|a_3-\eta a_2^2|\le\begin{cases}t & , \quad |\eta-1|\le\frac{1}{8t^2}\\ 8|\eta-1|t^3 & , \quad |\eta-1|\ge\frac{1}{8t^2}\end{cases}.$$

Theorem 2.3 *For $0\le\lambda\le1$ and $0\le\beta<1$, let $s\in S_\Sigma(\lambda,\tau;\beta)$ be given by (1) then one has the following*

$$|a_2|\le\begin{cases}\sqrt{\frac{|\tau|m(1-\beta)}{1+2\lambda-\lambda^2}}, & 0\le\beta<1-\frac{(1+\lambda)^2}{|\tau|m(1+2\lambda-\lambda^2)}\\ \frac{|\tau|m(1-\beta)}{1+\lambda}, & 1-\frac{(1+\lambda)^2}{|\tau|m(1+2\lambda-\lambda^2)}\le\beta<1\end{cases}, \tag{30}$$

$$|a_3|\le\begin{cases}\frac{|\tau|m(1-\beta)}{1+2\lambda-\lambda^2}, & 0\le\beta\le1-\frac{(1+\lambda)^4}{|\tau|m[2(1+2\lambda)(1+2\lambda-\lambda^2)]}\\ \frac{|\tau|m(1-\beta)}{2(1+2\lambda)}+\frac{|\tau|^2m^2(1-\beta)^2}{(1+\lambda)^2}, & 1-\frac{(1+\lambda)^4}{|\tau|m[2(1+2\lambda)(1+2\lambda-\lambda^2)]}\le\beta<1\end{cases}. \tag{31}$$

Proof Since $s\in S_\Sigma(\lambda,\tau;\beta)$ and $r=s^{-1}\in S_\Sigma(\lambda,\tau;\beta)$, by definition, there present two functions with positive real part

$$p(z)=1+\sum_{n=1}^{\infty}c_nz^n\in\mathcal{A}\quad and\quad q(\varpi)=1+\sum_{n=1}^{\infty}d_n\varpi^n\in\mathcal{A}, \tag{32}$$

where $\Re p(z))>0$ and $\Re(q(z))>0$ in Δ.

$$1+\frac{1}{\tau}\left[\frac{zs'(z)+\lambda z^2s''(z)}{\lambda zs'(z)+(1-\lambda)s(z)}-1\right]=p(z), \tag{33}$$

and

$$1+\frac{1}{\tau}\left[\frac{\varpi r'(\varpi)+\lambda\varpi^2 r''(\varpi)}{\lambda\varpi r'(\varpi)+(1-\lambda)r(\varpi)}-1\right]=q(\varpi). \tag{34}$$

So that, when we take $n = 2$ and $n = 3$ in (33) and (34), we can obtain following equalities, respectively;

$$\frac{1}{\tau}(1+\lambda)a_2 = c_1, \tag{35}$$

$$\frac{1}{\tau}[2(1+2\lambda)a_3-(1+\lambda)^2a_2^2]=c_2, \tag{36}$$

$$-\frac{1}{\tau}(1+\lambda)a_2 = d_1, \tag{37}$$

$$\frac{1}{\tau}\{[4(1+2\lambda)-(1+\lambda)^2]a_2^2-2(1+2\lambda)a_3\}=d_2. \tag{38}$$

From (35) and (37) we find (by Lemma A)

$$a_2=\frac{c_1\tau}{1+\lambda}=\frac{d_1\tau}{-(1+\lambda)},$$

by the Lemma A

$$|a_2|\leq\frac{|\tau|m(1-\beta)}{1+\lambda}. \tag{39}$$

From (36) and (38) we obtain,

$$\frac{2}{\tau}[2(1+2\lambda)-(1+\lambda)^2]a_2^2=c_2+d_2. \tag{40}$$

Using by Lemma A, we have

$$|a_2|\leq\sqrt{\frac{|\tau|m(1-\beta)}{1+2\lambda-\lambda^2}}$$

and joining this with Eq. (37), we derived the required estimate on the coefficient $|a_2|$ as given by (30). Now, we subtract (36) from (38) so as to obtain the coefficient bound on a_3. we thus get

$$\frac{1}{\tau}[4(1+2\lambda)(a_3-a_2^2)]=c_2-d_2$$

or

$$a_3=a_2^2+\frac{\tau(c_2-d_2)}{4(1+2\lambda)}, \tag{41}$$

when the value of a_2^2 from (35) takes its place into (39), the equality can be obtained as follows:

$$a_3 = \frac{\tau^2 c_1^2}{(1+\lambda)^2} + \frac{\tau(c_2 - d_2)}{4(1+2\lambda)}.$$

We thus find (by the Lemma A) that

$$|a_3| \leq \frac{|\tau|^2 m^2 (1-\beta)^2}{(1+\lambda)^2} + \frac{|\tau| m (1-\beta)}{2(1+2\lambda)}. \tag{42}$$

Moreover, when the value of a_2^2 from (38) takes its place into (39), the equality can be obtained as follows:

$$a_3 = \frac{\tau(c_2 + d_2)}{2(1+2\lambda-\lambda^2)} + \frac{\tau(c_2 - d_2)}{4(1+2\lambda)},$$

making necessary arrangements and applying Lemma A,

$$|a_3| \leq \frac{|\tau| m (1-\beta)}{1+2\lambda-\lambda^2}. \tag{43}$$

Merging Eqs. (40) and (41), we get the required estimates on the coefficient of $|a_3|$ as presented in (31). Thus, the proof of Theorem 2.3 is completed.

Note that Kant and Vays [8] derived the results in the Theorem 2.3. By setting $\tau = 1, m = 2, \lambda = 0, \beta = \alpha$ for the Theorem 2.3, we can easily obtain the consequence given below.

Corollary 2.4 *Let $s \in S_\Sigma^*(\alpha)$ $(0 \leq \alpha < 1)$ be given by (1). Then one has the following*

$$|a_2| \leq \begin{cases} \sqrt{2(1-\alpha)}, & 0 \leq \alpha < \frac{1}{2} \\ 2(1-\alpha), & \frac{1}{2} \leq \alpha < 1 \end{cases}$$

$$|a_3| \leq \begin{cases} 2(1-\alpha), & 0 \leq \alpha \leq \frac{1}{4} \\ 4(1-\alpha)^2 + (1-\alpha), & \frac{1}{4} \leq \alpha < 1 \end{cases}.$$

References

1. Bulut, S., Magesh, N., Balaji, V.K.: Initial bounds for analytic and bi-univalent functions by means of Chebyshev polynomials. J. Class. Anal. **11**(1), 83–89 (2017)
2. Doha, E.H.: The first and second kind Chebyshev coefficients of the moments of the general order derivative of an infinitely differentiable function. Int. J. Comput. Mat. **51**, 21–35 (1994)
3. Dziok, J., Raina R.K., Sokól, J.: Application of Chebyshev polynomials to classes of analytic functions. C. R. Math. Acad. Sci. Paris **353**, 433–438 (2015)
4. Fekete, M., Szegö, G.: Eine Bemerkung über ungerade schlichte funktionen. J. Lond. Math. Soc. Apr 1 **1**(2), 9–85 (1933)

5. Mason, J.C.: Chebyshev polynomial approximations for the L-membrane eigenvalue problem. SIAM J. Appl. Math. **15**, 172–186 (1967)
6. Sakar, F.M.: Estimate for initial Tschebyscheff polynomials coefficients on a certain subclass of bi-univalent functions defined by Salagean differential operator. Acta Universitatis Apulensis. **54**, 45–54 (2018)
7. Padmanabhan, K., Parvatham, R.: Properties of a class of functions with bounded boundary rotation. Ann. Polon. Math. **31**, 311–323 (1975)
8. Vays, P.P., Kant, S.: Estimates on initial coefficients of cerain subclasses of bi-univalent functions associated with the class $P_m(\beta)$. Int. J. Math. Andits Appl. **5**(1-B), 165–169 (2017)
9. Whittaker, T., Watson, G.N.: A Course of Modern Analysis. Cambridge University Press, Cambridge (1996)

Transcendental Numbers Arising from Mahler's Method

Saeree Wananiyakul, Vichian Laohakosol, and Janyarak Tongsomporn

1 Introduction

A complex number ξ is called an algebraic number if it satisfies a non-trivial polynomial equation $f(x) = 0$, where $f(x)$ is a polynomial over $\mathbb{Q}$. Any complex number that is not algebraic is said to be transcendental. It is well-known that if α is a quadratic irrationality (i.e., an algebraic number of degree 2), then there is a positive constant c such that

$$\left|\alpha - \frac{p}{q}\right| > \frac{c}{q^2},$$

for every pair of rational integers p, q with $q > 0$. Liouville [1] proved the following generalization of this result, which has since been known as Liouville's Theorem: if α is an algebraic number of degree $n \geq 2$, then there exists a positive constant c such that

$$\left|\alpha - \frac{p}{q}\right| > \frac{c}{q^n}, \tag{1}$$

for every pair of rational integers p, q with $q > 0$.

Liouville used this theorem to show the existence of transcendental numbers, known as Liouville numbers, which is made precise by the definition [2, Definition 1.1, p. 4]: let $x \in \mathbb{R}$. If for any positive real number ω there exists a rational number p/q with $p, q \in \mathbb{Z}, q > 0$, such that

S. Wananiyakul · J. Tongsomporn (✉)
School of Science, Walailak University, Nakhon Si Thammarat 80160, Thailand

V. Laohakosol
Department of Mathematics, Kasetsart University, Bangkok 10900, Thailand
e-mail: fscivil@ku.ac.th

P. Srivastava et al. (eds.), *Mathematical, Computational Intelligence and Engineering Approaches for Tourism, Agriculture and Healthcare*, Lecture Notes in Networks and Systems 214, https://doi.org/10.1007/978-981-16-3807-7_7

$$0 < \left|x - \tfrac{p}{q}\right| < \tfrac{1}{q^{\omega}},$$

then x is called a Liouville number.

In order to explain the significance of improving upon Liouville's theorem, we reproduce from [3, pp. 122–123] the following remarks: consider a hypothesis improvement of Liouville's theorem, in which the inequality (1) is replaced by

$$\left|\alpha - \tfrac{p}{q}\right| > \tfrac{c}{q^{\nu}}, \tag{2}$$

where ν is any number smaller than n. Thue [4] noticed that if such a theorem could be proved, it would have the important consequence that the Diophantine equation

$$q^n f\left(\tfrac{p}{q}\right) = a_0 p^n + a_1 p^{n-1} q + \ldots + a_n q^n = A \tag{3}$$

can have only finitely many solutions for any fixed $A \in \mathbb{Z}\backslash\{0\}$.

(To see this, let the zeros of $f(z)$ be $\alpha_1 = \alpha, \ldots, \alpha_n$, and put

$$\gamma = \min_{i \neq j}\left|\alpha_i - \alpha_j\right|.$$

Suppose that (3) has infinitely many solutions p, q, then there must be at least one α_i, which by suitable naming we can take to be α, which is a limit point of the numbers p/q, since otherwise the quantity

$$q^n f\left(\frac{p}{q}\right) = a_0 q^n \prod_{k=1}^{n}\left(\frac{p}{q} - \alpha_k\right)$$

is certainly not bounded as q increases indefinitely. There must therefore be infinitely many solutions of (3) for which $|\alpha - p/q| < \gamma/2$. But for all such solutions,

$$\left|\alpha - \tfrac{p}{q}\right| = \frac{A}{a_0 q^n \prod_{k=2}^{n}\left|\frac{p}{q} - \alpha_k\right|} \leq \frac{A}{a_0(\gamma/2)^{n-1}} \cdot \frac{1}{q^n},$$

and this is at variance with (2) if ν is a constant $< n$ and q is sufficiently large.)

Thue showed that (2) holds with

$$\nu = \tfrac{n}{2} + 1.$$

Later Siegel [5] improved Thue's result, showing that (2) holds with

$$\nu > \min_{1 \leq s \leq n-1, s \in \mathbb{Z}}\left(\tfrac{n}{s+1} + s\right),$$

and in particular with $\nu = s\sqrt{n}$. In 1947, Dyson [6] made the further improvement $\nu > \sqrt{2n}$, and finally in 1955, Roth [7] proved that (2) holds with $\nu = 2 + \varepsilon$, for each $\varepsilon > 0$, for all but a finite number of fractions p/q. This is the best theorem possible if ν is to be independent of q, since Hurwitz's theorem shows that the corresponding statement is false for every irrational algebraic number, for $\nu = 2$ and suitable c. Roth's work is similar in some respects to a simplification of Dyson's proof, published by Schneider [8].

This result has since then been known as the Thue–Seigel–Roth's theorem. This result may be stated in several equivalent forms; the form we take for our work here is: **Thue–Siegel–Roth theorem**. Let α be an algebraic number of degree $n \geq 2$. For $\rho > 2$, the inequality

$$\left|\alpha - \frac{p}{q}\right| < \frac{1}{q^\rho}$$

has at most finitely many rational solutions, and hence, there is then a constant $\gamma(\alpha, \rho)$ such that

$$\left|\alpha - \frac{p}{q}\right| \geq \frac{\gamma(\alpha, \rho)}{q^\rho}$$

for all rational numbers p/q.

Later in 1957, Ridout generalized the Thue–Siegel–Roth theorem to what is now known as **Ridout's theorem** [9]: let α be a nonzero algebraic number, let

$$P_1, P_2, \ldots, P_s, Q_1, Q_2, \ldots, Q_t$$

be distinct primes, and let μ, ν, c be real numbers satisfying

$$0 \leq \mu \leq 1, 0 \leq \nu \leq 1, c > 0.$$

Let x, y be restricted to integers of the form

$$x = x^* P_1^{\delta_1} P_2^{\delta_2} \cdots P_s^{\delta_s}, y = y^* Q_1^{\sigma_1} Q_2^{\sigma_2} \cdots Q_t^{\sigma_t},$$

where $\delta_1, \delta_2, \ldots, \delta_s, \sigma_1, \sigma_2, \ldots, \sigma_t$ are non-negative integers and x^*, y^* are integers satisfying

$$0 < |x^*| \leq c|x|^\mu, 0 < |y^*| \leq c|y|^\nu.$$

If $\kappa > \mu + \nu$, then the inequality

$$0 < \left| \alpha - \frac{x}{y} \right| < y^{-\kappa}$$

has at most a finite number of solutions in $x, y \in \mathbb{Z}$.

Mahler [10, 11] proved that the number (written in decimal representation)

$$0.123456789\ 10\ 11\ 12 \cdots$$

is a transcendental number. Later in 1976, he generalized this result in [12], using Ridout's Theorem, to the number

$$\sigma(a) = 0.d_1 d_2 d_3 \cdots ,$$

where the expansion after the point is to the base $q (\geq 2)$ of the integers 1 repeated a_1 times, 2 repeated a_2 times, 3 repeated a_3 times, etc.

Our objective here is to apply this Mahler's method to prove:

Theorem 1.1 *The decimal expansion*

$$\begin{aligned} S :=& 0.13579\ 11\ 13\ 15\ 17\ 19 \ldots \\ =& \left\{ \frac{1}{10} + \frac{3}{10^2} + \ldots + \frac{9}{10^5} \right\} \\ &+ \left\{ \left(\frac{1}{10^6} + \frac{1}{10^7} \right) + \left(\frac{1}{10^8} + \frac{3}{10^9} \right) + \ldots \right. \\ &+ \left. \left(\frac{9}{10^{92}} + \frac{7}{10^{93}} \right) + \left(\frac{9}{10^{94}} + \frac{9}{10^{95}} \right) \right\} + \left\{ \left(\frac{1}{10^{96}} + \frac{0}{10^{97}} + \frac{1}{10^{98}} \right) \right. \\ &+ \left. \left(\frac{1}{10^{99}} + \frac{0}{10^{100}} + \frac{3}{10^{101}} \right) + \ldots + \left(\frac{9}{10^{1443}} + \frac{9}{10^{1444}} + \frac{9}{10^{1445}} \right) \right\} \\ &+ \ldots \end{aligned}$$

which consists of all odd positive integers written successively after the decimal point, is a transcendental number but is not a Liouville number.

Proof of Theorem 1.1

In the expansion of S, the odd integers from 1 to 9 have exactly one digit, those from 10 to 99 have exactly two digits, those from 100 to 999 have exactly three digits, etc. Hence in the expansion of S there are after the decimal point.

- $\frac{10}{2}$ single digits which correspond to 1-digit integers;
- next, there are $2\left(\frac{10^2}{2} - \frac{10}{2}\right)$ digits in pairs which correspond to 2-digit integers;

- following this, there are $3\left(\frac{10^3}{2}-\frac{10^2}{2}\right)$ digits in sets of three which correspond to 3-digit integers, etc.

Consequently, we define A_n to be the total number of digits in S after the decimal point derived from integers which have at most n digits, so that $A_1 = 5$, $A_2 = \frac{10}{2}+2\left(\frac{10^2}{2}-\frac{10}{2}\right) = 95$, and in general, for $n \geq 2$,

$$A_n = \frac{10}{2}+2\left(\frac{10^2}{2}-\frac{10}{2}\right)+3\left(\frac{10^3}{2}-\frac{10^2}{2}\right)+\ldots+n\left(\frac{10^n}{2}-\frac{10^{n-1}}{2}\right)$$
$$= \frac{n10^n}{2}-\frac{1}{2}(10+10^2+10^3+\ldots+10^{n-1}) = \frac{n10^n}{2}-\frac{5(10^{n-1}-1)}{9}. \quad (4)$$

Now put

$$s_1 = \sum_{k=1}^{5}(2k-1)10^{-k} = \frac{1}{10}+\frac{3}{10^2}+\frac{5}{10^3}+\frac{7}{10^4}+\frac{9}{10^5}$$

$$s_2 = \sum_{k=\frac{10}{2}+1=6}^{\frac{10^2}{2}=50}(2k-1)10^{-2k} = \frac{11}{10^{12}}+\frac{13}{10^{14}}+\ldots+\frac{97}{10^{98}}+\frac{99}{10^{100}}$$

$$s_n = \sum_{k=\frac{10^{n-1}}{2}+1}^{\frac{10^n}{2}}(2k-1)10^{-nk}(n \geq 2).$$

Observe that the terms in s_2 with the lowest and highest powers of 10^{-1} are, respectively,

$$(10+1)\cdot 10^{-2\left(\frac{10}{2}+1\right)},\ \left(10^2-1\right)\cdot 10^{-2\cdot\frac{10^2}{2}}.$$

In general, for $n \geq 2$, the terms in s_n with the lowest and highest powers of 10^{-1} are, respectively,

$$\left(10^{n-1}+1\right)\cdot 10^{-n\left(\frac{10^{n-1}}{2}+1\right)},\ (10^n-1)\cdot 10^{-n\cdot\frac{10^n}{2}}.$$

Hence, the first and the last terms of the product

$$t_2{:=}10^{-\left(A_1-2\cdot\frac{10}{2}\right)}s_2$$

are, respectively,

$$(10+1)\cdot 10^{-(A_1+2)} = \frac{11}{10^7},\ \left(10^2-1\right)\cdot 10^{-A_2} = \frac{99}{10^{95}},$$

and in general, for $n \geq 2$, the first and the last terms of the product

$$t_n := 10^{-\left(A_{n-1} - n \cdot \frac{10^{n-1}}{2}\right)} s_n$$

are, respectively,

$$\left(10^{n-1} + 1\right) \cdot 10^{-(A_{n-1}+n)}, \, (10^n - 1) \cdot 10^{-A_n},$$

using also

$$A_n = A_{n-1} + \tfrac{9n}{2} \cdot 10^{n-1}. \tag{5}$$

Note that s_1 is the sum of all the contributions to S from the 1-digit integers, and for $n \geq 2$, t_n is the sum of all the contributions to S from the n-digit integers. Consequently,

$$S = s_1 + \sum_{n=2}^{\infty} t_n.$$

Here, for $n \geq 2$, by the definition of A_n, we have

$$A_{n-1} - n\left(\tfrac{10^{n-1}}{2}\right) = -\tfrac{5(10^{n-1}-1)}{9} := -e(n). \tag{6}$$

Thus,

$$S = s_1 + \sum_{n=2}^{\infty} 10^{e(n)} s_n.$$

We proceed now to simplify the sums s_n.

$$\begin{aligned} s_1 &= \sum_{k=1}^{5} (2k-1) 10^{-k} \\ &= \frac{1}{10-1}\left(1 + \frac{2}{10-1}\right) - \frac{1}{10-1}\left(\frac{20}{10-1} + (10-1)\right) 10^{-5} \end{aligned}$$

and for $n \geq 2$ with $y = 10^{n-1}/2$,

$$\begin{aligned} s_n &= \sum_{k=y+1}^{y+9y} (2k-1) 10^{-nk} \\ &= 10^{-(y+1)n} \Big\{ (2y+1) + (2y+3) 10^{-n} + (2y+5) 10^{-2n} + \ldots + (2y + (18y-1)) 10^{-(9y-1)n} \Big\} \end{aligned}$$

$$=\frac{(2y+1)+2\left(10^{-n}+\ldots+10^{-(9y-1)n}\right)-(2y+(18y-1))10^{-9yn}}{10^{(y+1)n}\left(1-10^{-n}\right)}$$
$$=\frac{1}{10^{yn}(10^n-1)}\left\{\left(10^{n-1}+1\right)+\frac{2}{10^n-1}\left(1-\frac{10^n}{10^{9yn}}\right)-\left(\frac{10^n-1}{10^{9yn}}\right)\right\}.$$

From (5) and (6), we obtain

$$A_n = A_{n-1}+\tfrac{9}{2}\left(10^{n-1}\right)n = A_{n-1}+9yn,\ A_{n-1}-yn=-e(n), \tag{7}$$

and so

$$\begin{aligned}
S &= s_1+\sum_{n=2}^{\infty}10^{e(n)}s_n \\
&= s_1+\sum_{n=2}^{\infty}\frac{1}{10^{yn-e(n)}(10^n-1)}\left\{\left(10^{n-1}+1\right)+\frac{2}{10^n-1}\left(1-\frac{10^n}{10^{9yn}}\right)-\left(\frac{10^n-1}{10^{9yn}}\right)\right\} \\
&= s_1+\sum_{n=2}^{\infty}\frac{1}{10^{A_{n-1}}(10^n-1)}\left\{\left(10^{n-1}+1\right)+\frac{2}{10^n-1}\left(1-\frac{10^n}{10^{9yn}}\right)-\left(\frac{10^n-1}{10^{9yn}}\right)\right\} \\
&= s_1+\sum_{n=2}^{\infty}\frac{1}{10^n-1}\left\{\left(10^{n-1}+1+\frac{2}{10^n-1}\right)-\left(\frac{2}{10^n-1}\cdot\frac{10^n}{10^{9yn}}+\frac{10^n-1}{10^{9yn}}\right)\right\}\frac{1}{10^{A_{n-1}}} \\
&= s_1+\sum_{n=2}^{\infty}\frac{1}{10^n-1}\left(10^{n-1}+1+\frac{2}{10^n-1}\right)\frac{1}{10^{A_{n-1}}} \\
&\quad -\sum_{n=2}^{\infty}\frac{1}{10^n-1}\left(\frac{2\cdot10^n}{10^n-1}+10^n-1\right)\frac{1}{10^{A_{n-1}+9yn}} \\
&= \frac{1}{10-1}\left(1+\frac{2}{10-1}\right)-\frac{1}{10-1}\left(\frac{20}{10-1}+(10-1)\right)10^{-5} \\
&\quad +\sum_{n=1}^{\infty}\frac{1}{10^{n+1}-1}\left(10^n+1+\frac{2}{10^{n+1}-1}\right)\frac{1}{10^{A_n}}-\sum_{n=2}^{\infty}\frac{1}{10^n-1}\left(\frac{2\cdot10^n}{10^n-1}+10^n-1\right)\frac{1}{10^{A_n}} \\
&= \frac{1}{10-1}\left(1+\frac{2}{10-1}\right)+\sum_{n=1}^{\infty}\frac{1}{10^{n+1}-1}\left(10^n+1+\frac{2}{10^{n+1}-1}\right)\frac{1}{10^{A_n}} \\
&\quad -\sum_{n=1}^{\infty}\frac{1}{10^n-1}\left(\frac{2\cdot10^n}{10^n-1}+10^n-1\right)\frac{1}{10^{A_n}} \\
&= \frac{11}{81}-\sum_{n=1}^{\infty}\left\{\left(\frac{2\cdot10^n}{(10^n-1)^2}+1\right)-\left(\frac{10^n+1}{10^{n+1}-1}+\frac{2}{(10^{n+1}-1)^2}\right)\right\}\frac{1}{10^{A_n}}.
\end{aligned}$$

From its definition, S is obviously an irrational number. We shall show that this number is transcendental. For $n\in\mathbb{N}$, $n\geq 2$, put

$$u_n=\left(\tfrac{2\cdot10^n}{(10^n-1)^2}+1\right)-\left(\tfrac{10^n+1}{10^{n+1}-1}+\tfrac{2}{(10^{n+1}-1)^2}\right)>0.$$

Hence,

$$S = \frac{11}{81} - \sum_{k=1}^{\infty} \frac{u_k}{10^{A_k}} = \frac{11}{81} - \sum_{k=1}^{n-1} \frac{u_k}{10^{A_k}} - \sum_{k=n}^{\infty} \frac{u_k}{10^{A_k}}.$$

For $n \in \mathbb{N}$, let

$$D_n = \left\{(10-1)(10^2-1)(10^3-1)\cdots(10^n-1)\right\}^2$$

and for $n \geq 2$, let

$$B_n = D_n 10^{A_{n-1}}, \; C_n = B_n\left(\tfrac{11}{81} - \sum_{k=1}^{n-1} \tfrac{u_k}{10^{A_k}}\right), \; R_n = \sum_{k=n}^{\infty} \tfrac{u_k}{10^{A_k}}. \tag{8}$$

Clearly, both B_n and C_n are positive integers for all $n \geq 2$, and

$$S = \tfrac{C_n}{B_n} - R_n. \tag{9}$$

By the definition of u_n, we see that

$$\lim_{n\to\infty} u_n = \tfrac{9}{10}, \tag{10}$$

and since A_n increases indefinitely, using (7) and (10), we have

$$\lim_{n\to\infty} \frac{R_n}{\frac{9}{10}10^{-A_n}} = \lim_{n\to\infty} \frac{1}{\frac{9}{10}10^{-A_n}}\left(\frac{u_n}{10^{A_n}} + \frac{u_{n+1}}{10^{A_{n+1}}} + \cdots\right) = 1,$$

i.e.,

$$R_n \sim \tfrac{9}{10}10^{-A_n}. \tag{11}$$

Since $0 < D_n < \left(10^{1+2+3+\cdots+n}\right)^2 = 10^{n(n+1)}$, using (4), we obtain

$$\lim_{n\to\infty} \tfrac{\log D_n}{A_n} < \lim_{n\to\infty} \frac{\log 10^{n(n+1)}}{\frac{n10^n}{2} - \frac{5\left(10^{n-1}-1\right)}{9}} = 0,$$

which implies that

$$\lim_{n\to\infty} \tfrac{\log D_n}{A_n} = 0. \tag{12}$$

Moreover, by (4), we have

$$\lim_{n\to\infty}\frac{A_n}{A_{n-1}}=\lim_{n\to\infty}\frac{\frac{n10^n}{2}-\frac{5\left(10^{n-1}-1\right)}{9}}{\frac{(n-1)10^{n-1}}{2}-\frac{5\left(10^{n-2}-1\right)}{9}}=10, \tag{13}$$

which shows that there exists an infinite strictly increasing sequence of positive integers N such that

$$A_n > 9A_{n-1} \tag{14}$$

for all $n \in N$.

Hereafter, all large integer n are taken from N.

By (9) and (11), we get

$$0<\left|S-\frac{C_n}{B_n}\right|=|R_n|<\frac{c}{10^{A_n}}<\frac{c}{10^{9A_{n-1}}}(n\in N), \tag{15}$$

where c is a positive constant independent of n. Letting

$$A_n = \alpha_n A_{n-1}, \tag{16}$$

we see that (13) implies that the positive number $\alpha_n \to 10(n \to \infty)$. Putting

$$\frac{\log D_n}{A_n} = \varepsilon_n, \tag{17}$$

the relation (12) implies that $\varepsilon_n \to 0(n \to \infty)$, and so for sufficiently large $n \in N$, we obtain $\varepsilon_n < \frac{1}{8\alpha_n}$. Consequently,

$$D_n = 10^{A_n\varepsilon_n} = 10^{\alpha_n A_{n-1}\varepsilon_n} < 10^{\frac{1}{8}A_{n-1}}. \tag{18}$$

By the definition of B_n and (18), for sufficiently large $n \in N$, we have

$$B_n = D_n 10^{A_{n-1}} < 10^{\frac{9}{8}A_{n-1}}. \tag{19}$$

Applying (19) to (15), for sufficiently large $n \in N$, we get

$$0<\left|S-\frac{C_n}{B_n}\right|<\frac{c}{10^{9A_{n-1}}}<\frac{c}{B_n^8}.$$

It follows from Ridout's Theorem with $\mu = 1$, $\nu = 1$, and $\kappa = 8 > 1+1$ that S is a transcendental number.

Finally, we verify that S is not a Liouville number. For any $B, C \in \mathbb{Z}$ with sufficiently large $B > 0$, from (11), (16), (17), and the definition of B_n and R_n in (8), it is readily shown that

$$\left|S - \frac{C}{B}\right| \geq cB^{-12},$$

where c is a positive constant independent of the sufficiently large integers n taken from N. It follows from the definition of Liouville number that S is not a Liouville number.

References

1. Liouville, J.: Sur des classes très-étendues de quantités dont la valeur n'est ni algébrique, ni même réductible à des irrationelles algébriques. J. Math. Pures Appl. **16**, 133–142 (1851)
2. Bugeau, Y.: Approximation by Algebraic Number. Cambridge Tracts in Mathematics No. 160, Cambridge University Press (2004)
3. LeVeque, W.J.: Topics in Number Theory, Vol. II. Addison-Wesley, Reading (1961)
4. Thue, A.: Über Annäherungswerte algebraischer Zahlen. J. reine angew. Math. **135**, 284–305 (1909)
5. Siegel, C.L.: Approximation algebraischer Zahlen. Math. Z. **10**, 173–213 (1921)
6. Dyson, F.J.: The approximation to algebraic number by rationals. Acta Math. **79**, 225–240 (1947)
7. Roth, K.F.: Rational approximations to algebraic numbers. Mathematika **2**, 1–20 (1955); Corrigendumm, Mathematika **2**, 168 (1955)
8. Schneider, T.: Über eine Dysonsche Verschärfung des Siegel-Thueschen Satzes. Arch. Math. **1**, 288–295 (1949)
9. Ridout, D.: Rational approximation to algebraic numbers. Mathematika **4**, 125–131 (1957)
10. Mahler, K.: Über die Dezemalbruchtwicklung gewisser Irrationalzahlem , Mathematica, 4B (Zeuthen), 15 p. (1937)
11. Mahler, K.: Arithmetische Eigenschaften einer Klasse von Dezimalbrüchen. Proc. Kon. v. Wetensch **40**, 421–428 (1937)
12. Mahler, K.: On a class of transcendental; decimal fraction, Commun. Pure Appl. Math. **XXIX**, 717–725 (1976)

Intelligent System for ECG Beat Classification

Pankaj Srivastava and Neeraja Sharma

1 Introduction

From the very beginning, the application of mathematics has been in vogue in the field of science. But its actual significance and applicability in the medical sciences have become handy in this Anthropocene epoch. No medical sciences, however, are left untouched by the influence of the concepts and theories of mathematics. Medical science is full of uncertainties and is an art of probability as well; so in the current scenario, the soft computing domain, fuzzy tools can be very fruitful and applicable tools in medical science. Fuzzy logic has been applied to many fields but it was due to Zadeh and Bellman [3] that the application of fuzzy tools in the field of medical science could become possible. Later on, Zadeh [19] proposed innumerable procedures in fuzzy mathematics including—fuzzy logic, fuzzy algorithm, fuzzy system, fuzzy semantic, etc. to design and develop the analysis of complex systems and their decision processes. In the presence of fuzzy variables, the decision-making process was introduced by Jain [5]; in the same way, Jena et al. [7] described the soft computing methods in bioinformatics. In the current hectic life, cardiovascular disease has become the leading cause of death among metropolitan and the people from hinterland areas as well in the country. As per the World Health Organization, cardiovascular diseases are the most common and non-communicable disease globally responsible for an estimated 17.8 million deaths in 2017. The ECG (electrocardiogram) is an important tool in the diagnosis of heart disease that shows the information of the heart and cardiovascular condition, which is essential to enhance the living quality and appropriate treatment of the concerned patient. Degani [4] introduced computerized electrocardiogram diagnosis on the basis of fuzzy tools.

P. Srivastava
Department of Mathematics, Motilal Nehru National Institute of Technology, Allahabad, India

N. Sharma (✉)
P.D.D.U.G.M.I.C, Kamalanagar, Allahabad, India

P. Srivastava et al. (eds.), *Mathematical, Computational Intelligence and Engineering Approaches for Tourism, Agriculture and Healthcare*, Lecture Notes in Networks and Systems 214, https://doi.org/10.1007/978-981-16-3807-7_8

Hu et al. [5] introduced the ECG beat classifier using a mixture of experts approach. Osowski and Linh [10] introduced an ECG beat recognition scheme using a fuzzy hybrid neural network. Andreao et al. [2] proposed an ECG signal analysis through hidden Markov models. Übeyli [16] introduced ECG beat classification scheme on combining recurrent neural networks with Eigenvector methods. Adeli and Neshat [1] proposed a fuzzy expert system for heart disease diagnosis. Pandey et al. [11] designed a rule base system for cardiac analysis. An approach has been made by Verma [17] to study the classification scheme of ECG beats on the basis of soft computing methods. Recently, Srivastava and Srivastava [12, 13] proposed a risk assessment scheme for cardiac analysis and hypertension. In the present paper, we have proposed a soft computing technique to design and develop a soft computing diagnostic model that classifies ECG beats and detects the various phases of cardiac normalities and abnormalities on the basis of available ECG data of the patient concerned. A number of cases have been discussed to show the effectiveness of the proposed soft computing technique.

2 Materials and Methodology

The ECG data, Einthoven triangle, hexaxial system, fuzzy tools and fuzzy expert system have been used to carry out the work (Fig. 1).

R–R Gap: The R–R gap is a part of ECG output, and further it will be used for classification on the basis of available data. It is also useful in calculating heart rate.

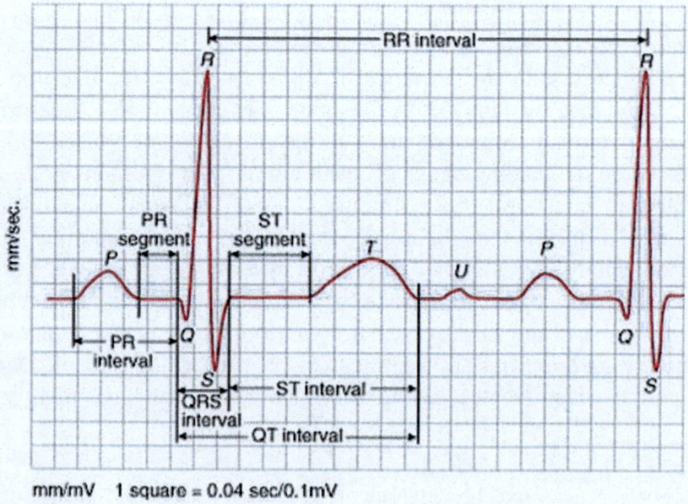

Fig. 1 ECG graph

2.1 Input Variables

1. **QRS Electrical Axis Degree**

We have classified eight regions of the hexaxial system on the basis of available ECG beats and fuzzy tools. The QRS electrical axis degree is classified in the following type of fuzzy sets:

i. Negative very large (NVL): −180° to −150°
ii. Negative large (NL): −150° to −90°
iii. Negative medium (NM): −90° to −30°
iv. Negative small (NS): −30° to 0°
v. Positive small (PS): 0° to 30°
vi. Positive medium (PM): 30° to 110°
vii. Positive large (PL): 110° to 150°
viii. Positive very large (PVL): 150° to 180°

The membership functions designed for these fuzzy sets are as follows (Fig. 2):

i. $\mu_{\mathrm{NVL}}(x) = \exp\left(\frac{-(x+180)^2}{2\times 30^2}\right)$
ii. $\mu_{\mathrm{NL}}(x) = \exp\left(\frac{-(x+150)^2}{2\times 60^2}\right)$
iii. $\mu_{\mathrm{NM}}(x) = \exp\left(\frac{-(x+90)^2}{2\times 60^2}\right)$
iv. $\mu_{\mathrm{NS}}(x) = \exp\left(\frac{-(x+30)^2}{2\times 30^2}\right)$
v. $\mu_{\mathrm{PS}}(x) = \exp\left(\frac{-(x-30)^2}{2\times 30^2}\right)$
vi. $\mu_{\mathrm{PM}}(x) = \exp\left(\frac{-(x-110)^2}{2\times 80^2}\right)$
vii. $\mu_{\mathrm{PL}}(x) = \exp\left(\frac{-(x-150)^2}{2\times 40^2}\right)$
viii. $\mu_{\mathrm{PVL}}(x) = \exp\left(\frac{-(x-180)^2}{2\times 30^2}\right)$

1. **R–R Gap**: The membership functions of R–R gap are given below:

i. $\mu_{\mathrm{Short}}(x) = \begin{cases} 1, & x \le 10 \\ \frac{(15-x)}{5}, & 10 < x < 15 \\ 0, & x \ge 15 \end{cases}$

ii. $\mu_{\mathrm{Normal}}(x) = \begin{cases} 0, & x \le 10 \text{ or } x \ge 30 \\ \frac{(x-10)}{5}, & 10 < x < 15 \\ 1, & 15 \le x \le 25 \\ \frac{(30-x)}{5}, & 25 < x < 30 \end{cases}$

iii. $\mu_{\mathrm{Large}}(x) = \begin{cases} 0, & x \le 25 \\ \frac{(x-25)}{5}, & 25 < x < 30 \\ 1, & x \ge 30 \end{cases}$

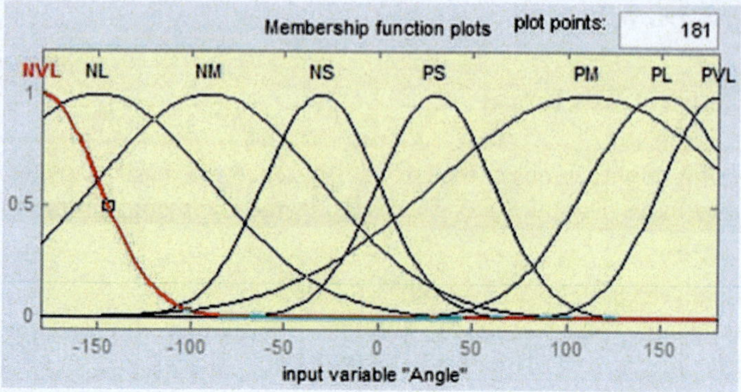

Fig. 2 Membership functions of QRS electrical axis degree

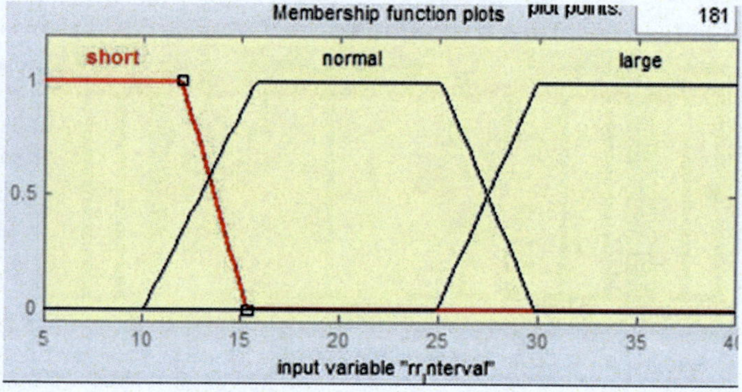

Fig. 3 Membership functions of R–R Gap

In this section, we have designed a diagnostic system, which is governed by the following fuzzy rules (Fig. 3):

S.N	QRS angle	AND/OR	R–R gap	Cardiac status
1	NVL	And	Short	Rare
2	NVL	OR	Short	Rare
3	NVL	AND	Large	Rare
4	NVL	OR	Large	Rare
5	NVL	AND	Normal	Little bit normal
6	NVL	OR	Normal	Little bit normal
7	NL	AND	Short	Rare
8	NL	OR	Short	Rare
9	NL	AND	Large	Rare
10	NL	OR	Large	Rare

(continued)

(continued)

S.N	QRS angle	AND/OR	R–R gap	Cardiac status
11	NL	AND	Normal	Little bit normal
12	NL	AND	Normal	Little bit normal
13	NM	AND	Short	Little bit normal
14	NM	OR	Short	Little bit normal
15	NM	AND	Large	Little bit normal
16	NM	OR	Large	Little bit normal
17	NM	AND	Normal	Little bit normal
18	NM	OR	Normal	Little bit normal
19	NS	AND	Short	Little bit normal
20	NS	OR	Short	Little bit normal
21	NS	AND	Large	Little bit normal
22	NS	OR	Large	Little bit normal
23	NS	AND	Normal	Normal
24	NS	OR	Normal	Normal
25	PS	AND	Normal	Normal
26	PS	OR	Normal	Normal
27	PS	AND	Large	Somewhat abnormal
28	PS	OR	Large	Somewhat abnormal
29	PS	AND	Short	Somewhat abnormal
30	PS	OR	Short	Somewhat abnormal
31	PM	AND	Short	Somewhat abnormal
32	PM	OR	Short	Somewhat abnormal
33	PM	AND	Large	Somewhat abnormal
34	PM	OR	Large	Somewhat abnormal
35	PM	AND	Normal	Normal
36	PM	OR	Normal	Normal
37	PL	AND	Short	Abnormal
38	PL	OR	Short	Abnormal
39	PL	AND	Large	Abnormal
40	PL	OR	Large	Abnormal
41	PL	AND	Normal	Somewhat abnormal
42	PL	OR	Normal	Somewhat abnormal
43	PVL	AND	Short	Abnormal
44	PVL	OR	Short	Abnormal
45	PVL	AND	Large	Abnormal
46	PVL	OR	Large	Abnormal
47	PVL	AND	Normal	Somewhat abnormal

(continued)

(continued)

S.N	QRS angle	AND/OR	R–R gap	Cardiac status
48	PVL	OR	Normal	Somewhat abnormal

In order to represent state of patient's health on the basis of QRS axis degree and R–R gap, linguistic strings are designed. As per available details, the 24 linguistic strings are framed as follows:

1. $J_1 = NVL_{QRS}Short_{RR}$
2. $J_2 = NVL_{QRS}Large_{RR}$
3. $J_3 = NVL_{QRS}Normal_{RR}$
4. $J_4 = NL_{QRS}Short_{RR}$
5. $J_5 = NL_{QRS}Large_{RR}$
6. $J_6 = NL_{QRS}Normal_{RR}$
7. $J_7 = NM_{QRS}Short_{RR}$
8. $J_8 = NM_{QRS}Large_{RR}$
9. $J_9 = NM_{QRS}Normal_{RR}$
10. $J_{10} = NS_{QRS}Short_{RR}$
11. $J_{11} = NS_{QRS}Large_{RR}$
12. $J_{12} = NS_{QRS}Normal_{RR}$
13. $J_{13} = PS_{QRS}Short_{RR}$
14. $J_{14} = PS_{QRS}Large_{RR}$
15. $J_{15} = PS_{QRS}Normal_{RR}$
16. $J_{16} = PM_{QRS}Short_{RR}$
17. $J_{17} = PM_{QRS}Large_{RR}$
18. $J_{18} = PM_{QRS}Normal_{RR}$
19. $J_{19} = PL_{QRS}Short_{RR}$
20. $J_{20} = PL_{QRS}Large_{RR}$
21. $J_{21} = PL_{QRS}Normal_{RR}$
22. $J_{22} = PVL_{QRS}Short_{RR}$
23. $J_{23} = PVL_{QRS}Large_{RR}$
24. $J_{24} = PVL_{QRS}Normal_{RR}$

Case 1

The input variables of the patient are as follows: QRS degree = 73.9088° and R–R gap = 25 mm.

1. QRS degree = {(0, NVL), (0, NL), (0.024, NM), (0.0025, NS), (0.3426, PS), (0.9032, PM), (0.1638, PL), (0.0019, PVL)}
2. R-R gap = {(0, Short), (0, Large), (1, Normal)}

This is the fuzzy set that represents the state of the concerned patient (Fig. 4):

$$X = \{(0.024, J_9), (0.0025, J_{12}), (0.3426, J_{15}), (0.9032, J_{18}), (0.1638, J_{21}), (0.0019, J_{24})\}$$

Fig. 4 ECG graph of Case 1

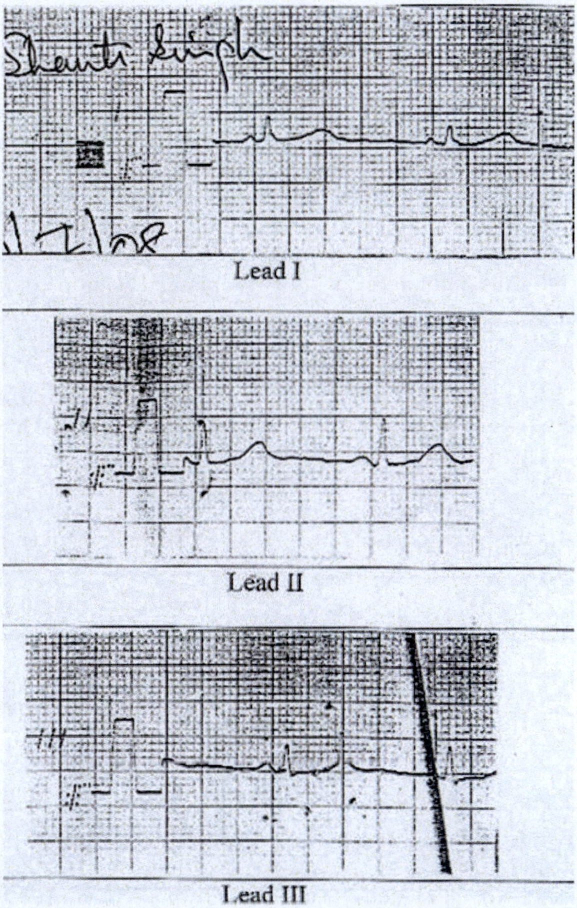

The utility matrix U of order 5 × 24 by using the fuzzy rule base designed is as follows:

$$U = \begin{pmatrix} 9&9&7&8&8&6&7&7&6&6&6&1&3&3&1&4&4&1&7&7&4&8&8&7 \\ 6&6&9&7&7&9&9&9&9&8&8&5&3&3&2&2&2&2&5&5&3&4&4&3 \\ 1&1&2&1&1&3&4&4&4&5&5&9&5&5&9&4&4&9&1&1&3&1&1&2 \\ 4&4&2&3&3&3&2&2&3&2&2&4&8&8&4&9&9&5&7&7&6&8&8&7 \\ 7&7&5&6&6&4&4&4&3&3&3&1&6&6&1&7&7&1&9&9&7&9&9&8 \end{pmatrix}$$

The five fuzzy utilities are:

1. $U_1 = \{(0.9363, 1), (0.1638, 4), (0.024, 6), (0.0019, 7)\}$
2. $U_2 = \{(0.9364, 2), (0.1654, 3), (0.0025, 5), (0.024, 9)\}$
3. $U_3 = \{(0.0019, 2), (0.1638, 3), (0.024, 4), (0.9363, 9)\}$
4. $U_4 = \{(0.024, 3), (0.3442, 4), (0.9032, 5), (0.1638, 6), (0.0019, 7)\}$

5. $U_5 = \{(0.9363, 1), (0.024, 3), (0.1638, 7), (0.0019, 8)\}$

The maximizing sets corresponding to each alternative are:

1. $U_{1M} = \{(0.1111, 1), (0.4444, 4), (0.6666, 6), (0.7777, 7)\}$
2. $U_{2M} = \{(0.0494, 2), (0.1111, 3), (0.3086, 5), (1, 9)\}$
3. $U_{3M} = \{(0.011, 2), (0.0370, 3), (0.0878, 4), (1, 9)\}$
4. $U_{4M} = \{(0.0123, 3), (0.039, 4), (0.0953, 5), (0.1975, 6), (0.3660, 7)\}$
5. $U_{5M} = \{(0, 1), (0.0041, 3), (0.2846, 7), (0.5549, 8)\}$

Now the optimal fuzzy utility is obtained by union of fuzzy utility and maximizing sets are:

1. $U_{1O} = \{(0.1111, 1), (0.1638, 4), (0.024, 6), (0.0019, 7)\}$
2. $U_{2O} = \{(0.0494, 2), (0.1111, 3), (0.0025, 5), (0.024, 9)\}$
3. $U_{3O} = \{(0.0019, 2), (0.0370, 3), (0.024, 4), (0.9363, 9)\}$
4. $U_{4O} = \{(0.0123, 3), (0.039, 4), (0.0953, 5), (0.1638, 6), (0.0019, 7)\}$
5. $U_{5O} = \{(0, 1), (0.0041, 3), (0.1638, 7), (0.0019, 8)\}$

Using these utilities, the optimal alternative is given as

$$A_O = \{(0.1638, \textbf{rare}), (0.1111, \textbf{little bit normal}), (0.9363, \textbf{normal}), (0.1638, \textbf{some} - \textbf{what abnormal}), (0.1638, \textbf{abnormal})\}$$

This optimal alternative indicates the normal health status of the patient.

Case 2

The input variables of the patient are as follows: QRS degree = −55.9540° and R–R gap = 23 mm.

1. QRS degree = {(0, NVL), (0.2928, NL), (0.8513, NM), (0.6878, NS), (0.0165, PS), (0.1163, PM)}
2. R-R gap = {(0, Short), (0, Large), (1, Normal)}

The fuzzy set of the state of the patient is represented as follows:

$$X = \{(0.2928, J_6), (0.8513, J_9), (0.6878, J_{12}), (0.0165, J_{15}), (0.1163, J_{18})\}$$

The fuzzy utilities are:

1. $U_1 = \{(0.7287, 1), (0.8948, 6)\}$
2. $U_2 = \{(0.1309, 2), (0.6878, 5), (0.8948, 9)\}$
3. $U_3 = \{(0.2928, 3), (0.8513, 4), (0.7287, 9)\}$
4. $U_4 = \{(0.8948, 3), (0.6930, 4), (0.1163, 5)\}$
5. $U_5 = \{(0.7287, 1), (0.8513, 3), (0.2928, 4)\}$

Proceed as above we obtained as below:

A_O = {(0.6666, **rare**), (0.8948, **little bit normal**), (0.7287, **normal**), (0.0953, **somewhat abnormal**), (0.0173, **abnormal**)}

The optimal alternative indicates that the health status of the patient is a little bit normal (Fig. 5).

Case 3

The input variables of the patient are as below: 72.9794° and R–R gap = 31 mm.

1. QRS degree = {(0, NVL), (0.001, NL), (0.025, NM), (0.0028, NS), (0.3584, PS), (0.8985, PM), (0.1566, PL), (0.0017, PVL)}
2. R-R gap = {(0, Short), (1, Large), (0, Normal)}

The fuzzy set of the state of the patient is represented as follows:

X = {(0.001, J_5), (0.025, J_8), (0.0028, J_{11}), (0.3584, J_{14}), (0.8985, J_{17}), (0.1566, J_{20}), (0.0017, J_{23})}

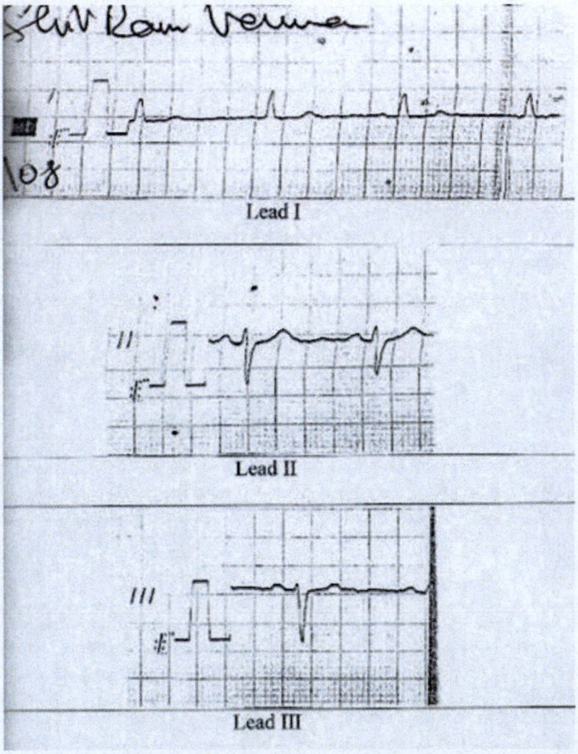

Fig. 5 ECG graph of Case 2

The fuzzy utility sets are:

1. $U_1 = \{(0.3584, 3), (0.8985, 4), (0.0028, 6), (0.1777, 7), (0.0027, 8)\}$
2. $U_2 = \{(0.8985, 2), (0.3584, 3), (0.0017, 4), (0.1566, 5), (0.001, 7), (0.0028, 8), (0.025, 9)\}$
3. $U_3 = \{(0.1589, 1), (0.9010, 4), (0.3602, 5)\}$
4. $U_4 = \{(0.0277, 2), (0.001, 3), (0.1566, 7), (0.3595, 8), (0.8985, 9)\}$
5. $U_5 = \{(0.0028, 3), (0.025, 4), (0.3590, 6), (0.8985, 7), (0.1580, 9)\}$

The optimal alternative for this case is

A_O = {(0.4444, **rare**), (0.1566, **little bit normal**), (0.1715, **normal**), (0.8985, **somewhat abnormal**), (0.2846, **abnormal**)}

The optimal alternative indicates that the health status of the patient is somewhat abnormal (Fig. 6).

Fig. 6 ECG graph of Case 3

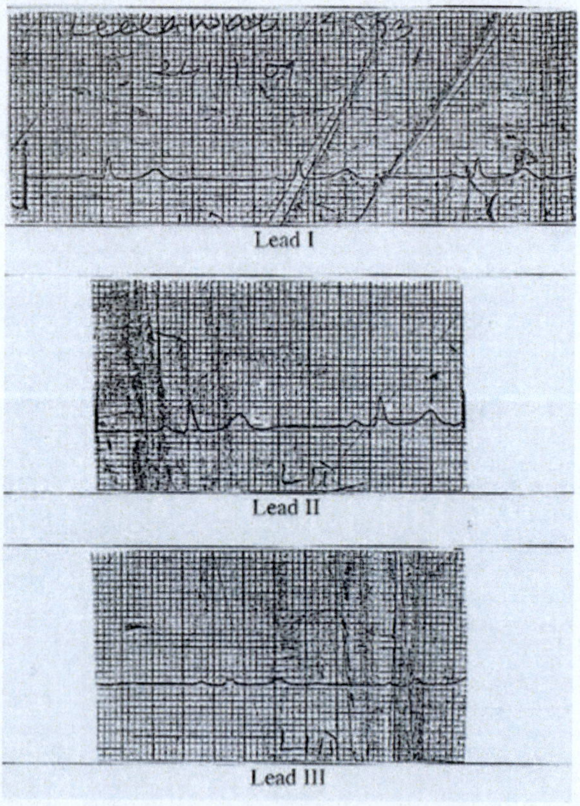

3 Conclusion

The output of the proposed system for various patients has been verified with the help of cardiac experts and gives details of different phases of cardiac abnormalities.The application of this diagnostic model will be helpful in designing and developing a machine that on scanning the ECG graph provide a spectrum of cardiac normalities and abnormalities as per the soft computing techniques. This process will sharpen the diagnostic skills of medical practitioner and will work as an effective referral system between patients and medical experts.

Acknowledgements We sincerely acknowledge Dr. Geeta Shukla, cardiologist, Nazarth Hospital, Allahabad and Dr. R.P. Singh, Prabha Medical Center, Shantipuram, Phaphamau for providing different ECGs data and regarding their useful suggestions and discussion on the validation of results in the present article.

References

1. Adeli, A., Neshat, M.: A fuzzy expert system for heart disease diagnosis. In: Proceedings of the International Multi Conference of Engineers and Computer Scientists, pp. 134–139 (2010)
2. Andreao, R.V., Dorizzi, B., Boudy, J.: ECG Signal Analysis through Hidden Markov Models. IEEE Trans. Biomed. Eng. **53**(8), 1541–1549 (2006)
3. Bellman, R.E., Zadeh, L.A.: Decision making in a fuzzy environment. Manage. Sci. **17**(4), B141–B164 (1970)
4. Degani, R.: Computerized electrocardiogram diagnosis: fuzzy approach. Methods Inf. Med. **31**, 225–233 (1992)
5. Hu, Y.H., Palreddy, S., Tompkins, W.J.: Patient adaptable ECG beat classifier using a mixture of experts approach. IEEE Trans. Biomed. Eng. **44**(9), 891–897 (1997)
6. Jain, R.: Decision making in the presence of fuzzy variables. IEEE Trans. Syst. Man Cybern. **6**(10), 698–703 (1967)
7. Jena, R.K., et al.: Soft computing methodologies in bioinformatics. Eur. J. Sci. Res. **26**, 189–203 (2004)
8. Luthra, A.: ECG Made Easy, 3rd edn, Jaypee Brothers, New Delhi(2007)
9. Klire, G.J., Yuan, B.: Fuzzy Sets and Fuzzy Logic: Theory and Applications, PHI Learning Private Limited, New Delhi (2009)
10. Osowski, S., Linh, T.H.: ECG beat recognition using fuzzy hybrid neural network. IEEE Trans. Biomed. Eng. **48**(11), 1265–1271 (2001)
11. Pandey, D., Mahajan, V., Srivastava, P.: Rule based system for cardiac analysis. Natl. Acad. Sci. Lett. **29**(7 & 8), 299–309 (2006)
12. Srivastava, P., Srivastava, A.: A note on soft computing approach for cardiac analysis. J. Basic. Appl. Sci. Res. **2**(1), 376–385 (2012)
13. Srivastava, P., Srivastava, A.: Spectrum of soft computing risk assessment scheme for hypertension. Int. J. Comput. Appl. (0975–8887) **44**(17), 23–30 (2012)
14. Srivastava, P., Sharma, N.: A spectrum of soft computing model for medical diagnosis. Appl. Math. Inf. Sci. **8**(3), 1225–1230 (2014)
15. Srivastava, P., Sharma, N., Singh, R.: Soft computing diagnostic system for diabetes. Int. J. Comput. Appl. (0975–888) **47**(18), 22–27 (2012)
16. Übeyli, E.D.: Combining recurrent neural networks with eigen vector methods for classification of ECG beats. Dig. Signal Process. **19**(2) (2009)

17. Verma, P.: A soft computing approach for classification scheme of ECG beats, M.Sc. thesis Environment Science Allahabad University (2009)
18. Zadeh, L.A.: Fuzzy sets. Inf. Control **8**, 338–353 (1965)
19. Zadeh, L. A.: On fuzzy algorithms, Memo No. ERL-M325, Electronics Research Laboratory, University of California (1972)
20. Zadeh, L.A.: Outline of a new approach to the analysis of complex systems and decision process. IEEE Trans. Syst. Man Cybern. SMC **3**(1), 28–44 (1973)

Efficient Algorithm in IoT for Drip Irrigation System

S. Raji Reddy and M. V. Ramana Murthy

1 Introduction

About IoT: The Internet of Things (IoT) is a system of interrelated computing devices, mechanical and digital machines, objects, animals or people that are provided with unique identifiers (UIDs) and the ability to transfer data over a network without requiring human-to-human or human-to-computer interaction.

It is the application of the latest ICT (information and communication technologies) into agriculture. In IoT-based smart farming, a system is built for monitoring the crop field with the help of sensors (light, humidity, temperature, soil moisture, etc.) and automating the irrigation system [1].

About drip irrigation system: It is an application concerned with farming related to different crops. This application can be used in three ways to support farmers in maintaining proper water level at fields in reducing human interference at different seasons. The application is suitable to work in the following conditions in the absence of farmer:

- Manual watering to fields and automatic water level control.
- Sensing the temperature and start/stop watering to fields.
- Automatic water level control in any situation while watering to fields.

2 Problem Analysis

Manual Process: The manual process needs human interaction. If human interaction is mission, then the entire work cannot continue. There is a disadvantage of human interaction in water pumping to crops. Most of the time the water resources are

S. R. Reddy (✉) · M. V. R. Murthy
Mahatma Gandhi Institute of Technology, Hyderabad, Telangana 500075, India

P. Srivastava et al. (eds.), *Mathematical, Computational Intelligence and Engineering Approaches for Tourism, Agriculture and Healthcare*, Lecture Notes in Networks and Systems 214, https://doi.org/10.1007/978-981-16-3807-7_9

wasted, and they will not be available when they require. Another problem with manual watering is that sometimes the water we pour may be more or less. It is not possible to provide exactly the required water. In this scenario, we can adopt an automated watering system for crops, called drip irrigation system. It will not pour water into the crop, it sprinkles the water so that the water will not be wasted and less quantity of water is provided to the field for wetness to maintain in the field.

Even in the drip irrigation system, we can minimize the human interaction with the system by introducing an IoT (Internet of Things) system in the agriculture field [2–4].

This paper provides the solution into two phases. The first phase is to identify the available water resources and the number of drip points for water sprinkling. The second phase checks the water utilized from each drip point to analyze, whether the allotted water is consumed from each point. If there are some points from which water is released in less quantity or not released, it helps the system to analyze the successful utilization of water in the field or any leakage/seepage points that exist in the drip irrigation system.

3 Solution Implementation

The system offers the required water to field i.e. more water or limited water as depicted in Fig. 1. It is the simulated environment for drip irrigation system in crop fields. It saves a lot of water from wastage and water pump energy. The devices used in designing the system are:

- Home gateway
- Smartphone

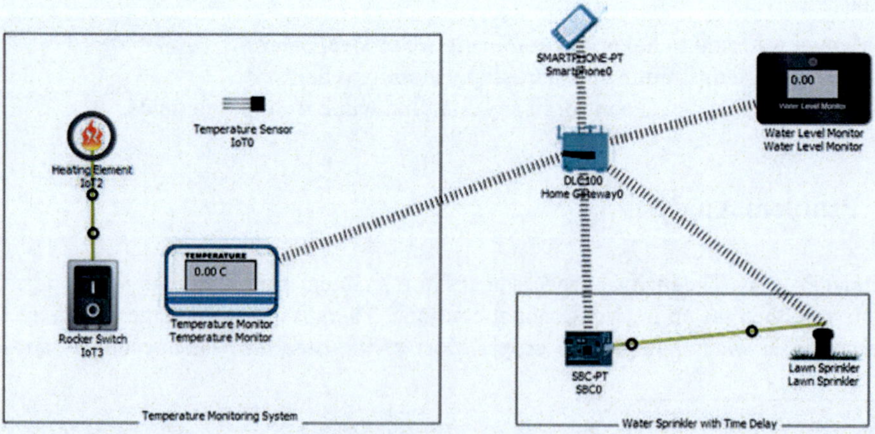

Fig. 1 Drip irrigation system with IoT control with sensors

- Water level monitor
- Temperature sensor
- Temperature monitor

At the back end, the system uses an algorithm to sense the weather condition, water resource availability, and also the water utilized at every drip point in the crop field. The system uses the following Divide and Conquer algorithm to assess the data from the IoT system and weather conditions in the field.

The algorithm Divide and Control is suitable to implement in Drip Irrigation System. It works in identifying the available water resources, the existing number of drip points and requirement of water in the field according to the sensed weather. The Divide and Conquer algorithm [5] is used to divide available water resources equally to each drip point and release water for a particular period. While providing water for the next time, the system senses the weather status and takes the decision [6, 7] that how much water is required to support the fields. Following are the steps to implement the algorithm:

Implementation of Divide and Conquer Algorithm for Drip Irrigation System

Phase I:

Step 1: Select number of drip points available for water sprinkling in the field.
Step 2: Identify the drip points require water by sensing the weather status/moisture status in the field.
Step 3: Identify the available water resource in the field to supply.
Step 4: Apply equal water quantity sprinkling from each drip point. Use divide process of algorithm to divide equally the available water resource (use only divide process of Divide and Conquer algorithm).

Phase II:

Step 1: Conquer (Implementation of Conquer process in Divide and Conquer algorithm) the water utilized from each drip point according to the allocated water from each drip point to sprinkle.
Step 2: Assess the total water consumed from all drip points and compare with the quantity of water proposed to pump to the field.
Step 3: If the sum of water pumped from all drip points is equal to the quantity pumped from the source, then the system is perfect and no correction is required.
Step 4: If there is a difference, then we can identify that there is a seepage or leakage in the pipeline. There is an indicator that shows the status of water utilization in the field.
Step 5: We can still go in deep to identify at which drip point the assigned water is not released and decide the preceding point that drip point is having a leakage/seepage.
Step 6: A decision can be taken to repair the defective system and repair it for regular functionality.

The above algorithm can be inserted into the smartphone device. The algorithm code can be written using popular platforms like Python, Java, etc.

4 Conclusion and Future Scope

In the present scenario, the optimum resource utilization and minimizing wastage are required. In this situation, the agriculturist can test the functionality of drip irrigation system and implement it in the agriculture fields with the consultation of the farmer [8]. The technical factors and maintenance are to be trained to the farmers. A security feature [1, 9] is to be inculcated in the system to monitor or give an alert to the organizer about water resource utilization or seepage. There is a problem that arises if the system is open to all users. For this purpose, it is recommended to provide access rights to certain users and implement the system.

The farmer can't totally depend on rainwater resources. It is always available. In non-available situations, the drip irrigation system can preserve rainwater and use it in rains unavailable situations. The efficiency of water resource utilization is possible if we use weather and moisture sensors used in the field. It is one-time investment and can be used in any weather situation and for any crop.

References

1. Zhou, W., Zhang, Y., Liu, P.: The effect of IoT new features on security and privacy: new threats, existing solutions, and challenges, member, IEEE
2. Suchitra, C., Vandana, C.: Internet of things and security issues. IJCSMC. **5**(1), 133–139 (2016). ISSN 2320–088X
3. Sharma, V., Tiwari, R.: A review paper on IOT & it's smart applications. IJSETR. **5**(2) (2016). ISSN 2278–7798
4. Sarika, A., Korade, A.: Review paper on internet of things (IoT) and its applications
5. Paira, S., Chandra, S., Alam, S.S., Dey, P.S.: A review report on Divide and Conquer sorting algorithm. In: IEEE CALCON2014
6. Kotak, V., Durafe, A.: Int. Res. J. Eng. Technol., e-ISSN 2395-0056, p-ISSN 2395-0072
7. Mudga, P.K.: Study of applications of divide and conquer approach in algorithm and complexity. In: 2nd International Conference on Science, Technology and Management conference, ICSTM-978-81-931039-6-8 (2015)
8. Dhanalakshmi, S., Krishna Kishore, D., Prabakaran, T.: Analysis and performance of Divide and Conquer methodology. Int. J. Adv. Res. Comp. Eng. Technol. **6**(8) (2017). ISSN 2278–1323
9. Security in the Internet of Things, 2012. In: International Conference on Computer Science and Electronics Engineering, IEEE Computer Society
10. Miraz, M.H., Ali, M., Excell, P.S., Picking, R.: Internet of nano-things, things and everything: future growth trends. Future Internet **10**, 68 (2018). https://doi.org/10.3390/fi10080068 www.mdpi.com/journal/futureinternet

Computational Techniques on Totally Continuous Functions Via Čech Rough and Nano Soft Topological Spaces

M. Lellis Thivagar, B. Tamilarasan, and G. Kabin Antony

1 Introduction

Čech closure spaces were introduced by Čech [1]. In Čech approach, the operator satisfies idempotent condition among Kuratowski axioms. This condition need not hold for every subset A of X. When this condition is also true, the operator becomes the topological closure operator. Thus the concept of closure space is the generalisation of a topological space. In 1999, Molodtsov [2] introduced the notion of soft set to deal with problems of incomplete information. Thivagar et al. [3, 4] introduced nano topological space with respect to a subset X of an universe, which is defined in terms of lower and upper approximations and boundary region. The nano topological space is discussed in approximation space (U, R), which is constituted by universe U and R be any binary relation is an after composed set and fore composed set of topologies induced by nano topological space. The concept of soft set was first introduced by Molodtsov [2] in 1999 as a general mathematical tool for studying uncertain objects. In this paper, we define and compute Čech rough totally continuous function on Čech rough closure spaces and nano soft totally continuous function on nano soft topology. Furthermore, we define properties of Čech rough totally continuous functions and derive the basic algorithm for nano soft topology with computational model and discuss properties of nano soft totally continuous functions.

M. L. Thivagar · B. Tamilarasan (✉) · G. K. Antony
School of Mathematics, Madurai Kamaraj University, Madurai 625021, Tamil Nadu, India

P. Srivastava et al. (eds.), *Mathematical, Computational Intelligence and Engineering Approaches for Tourism, Agriculture and Healthcare*, Lecture Notes in Networks and Systems 214, https://doi.org/10.1007/978-981-16-3807-7_10

2 Preliminaries

In this section, we shall recall a few definitions that will help us to expand our topic further and arrive at new results.

Definition 2.1 [1, 5] A function $c : P(X) \rightarrow P(X)$, where $P(X)$ is a power set of X, is called a Čech closure operator for X provided the following conditions are satisfied:

1. $c(\phi) = \phi$
2. $A \subset c(A)$ for each $A \subset X$
3. $c(A \cup B) = c(A) \cup c(B)$ for each $A, B \subset X$.

Then the pair (X, c), where X is a non-empty set and c is a Čech closure operator for X, is called a Čech closure space. If (X, c) is a Čech closure space and $A \subset X$, then $c(A)$ is called the closure of A in (A, C). The Čech closure space (X, c) is said to be Kuratowski (topological) space, if $c(c(A)) = c(A)$ for each $A \subset X$.

Definition 2.2 [6, 7] Let U be a nonempty set of objects called the universe and R be an equivalence relation on U named as the indiscernibility relation. Elements belonging to the same equivalence class are said to be indiscernible with one another. The pair (U, R) is said to be the approximation space. Let $X \subseteq U$.

1. The lower approximation of X with respect to R is the set of all objects, which can be for certain classified as X with respect to R and is denoted by $L_R(X)$. That is, $L_R(X) = \cup_{x \in U}\{R(x) : R(x) \subseteq X\}$, where $R(x)$ denotes the equivalence class determined by x.
2. The upper approximation of X with respect to R is the set of all objects, which can be possibly classified as X with respect to R and is denoted by $U_R(X)$. That is, $U_R(X) = \cup_{x \in U}\{R(x) : R(x) \cap X \neq \phi\}$.
3. The boundary region of X with respect to R is the set of all objects, which can be classified as X nor as not X with respect to R and is denoted by $B_R(X)$. That is, $B_R(X) = U_R(X) - L_R(X)$.

Definition 2.3 [6] If (U, R) is an approximation space and $X \subseteq U$ is said to be a rough set (inexact) with respect to R if $B_R(X) = U_R(X) - L_R(X) \neq 0$, that is $U_R(X) \neq L_R(X)$, otherwise the set X is said to be crisp (exact) with respect to R.

Definition 2.4 [6] If (U, R) is an approximation space and $X, Y \subseteq U$. Then the set X is said to be rough subset of Y with respect to R if $L_R(X) \subseteq L_R(Y)$ and $U_R(X) \subseteq U_R(Y)$. Note that every subset X of a rough set Y is a rough subset of Y.

Definition 2.5 [8, 9] Let $P(X)$ be the power set of a rough set X in the approximation space (U, R). A function $C : P(X) \rightarrow P(X)$ is called a Čech rough closure (simply rough closure) operator for X if it satisfies the following conditions:

1. $C(\phi) = \phi$

2. $A \subset C(A)$ for each $A \subset X$
3. $C(A \cup B) = C(A) \cup C(B)$ for each $A, B \subset X$.

Then the set X together with the Čech rough closure operator C is called a Čech rough closure space (simply rough closure space) and it is denoted by (X, C). A subset A of X is said to be rough closed in (X, C) if $C(A) = A$ and is said to be rough open if its complement is rough closed, that is, $C(X - A) = X - A$. A subset A of X is said to be clopen in (X, C) if A is both rough open and rough closed in (X, C). Let (X, C) be a rough closure space, then its associated rough topology on X is denoted by τ_C and is defined by $\tau_C = \{G \subset X : C(X - G) = X - G\}$.

Remark 2.6 [8] Let (X, C) be a rough closure space and $A, B \subset X$. Then the following are true:

1. If $A \subset B$, then $C(A) \subset C(B)$.
2. $C(A \cap B) \subset C(A) \cap C(B)$.

Definition 2.7 [8] Let (X, C_1) and (Y, C_2) be two rough closure spaces on rough sets X and Y, respectively, and a function $f : X \to Y$ is said to be rough continuous at a point $x \in X$ if $f(x) \in C_2(f(A))$ for all $A \subseteq X$ and $x \in C_1(A)$ and rough continuous on X if it is rough continuous at each point of X or equivalently if $f(C_1(A)) \subseteq C_2(f(A))$ for each $A \subseteq X$.

Theorem 2.8 [8, 10] *Let (X, C_1) and (Y, C_2) be two rough closure spaces on rough sets X and Y respectively and a function $f : X \to Y$ is said to be rough continuous on X if and only if $C_1(f^{-1}(B) \subseteq f^{-1}(C_1(B))$ for every $B \subseteq Y$.*

Definition 2.9 [2] Let X be an initial universe and E be a set of parameters. Let $P(X)$ denote the power set of X and A be a non-empty subset of E. A soft set F_A on the universe X is defined by $F_A = \{(e, F(E))/e \in E, F(e) \in P(X)\}$, where $F : E \to P(X)$ such that $F(e) = \phi$ if $x \notin A$. Here F is called an approximation function of the soft set F_A. The value of $F(e)$ may be arbitrary. Some of them may be empty and some may have nonempty intersections. The class of all soft sets over X and the parameter set E is denoted by $S(X, E)$.

Definition 2.10 [2, 11] For two soft sets F_A and G_B over a common universe X, we say that F_A is a soft subset of G_B if.

1. $A \subseteq B$ and
2. for all $e \in A$, $F(e)$ and $G(e)$ are identical approximations.

we write $F_A \subseteq G_B$ ($\subseteq$ denotes soft inclusion). Consequently G_B is said to be a soft superset of F_A.

3 Čech Rough Totally Continuous Functions

In this section, we shall introduce the concept of Čech rough totally continuous function in topological spaces as well as establish a few results.

Definition 3.1 Let (X, C_1) and (Y, C_2) be two Čech rough closure spaces on rough sets X and Y, respectively. A function $f : X \to Y$ is called Čech rough totally continuous if the inverse image of every rough open set over Y is rough clopen set over X.

Theorem 3.2 *A rough set function $f : (X, C_1) \to (Y, C_2)$ is Čech rough totally continuous if and only if the inverse of every rough closed set over (Y, C_2) is rough clopen over (X, C_1).*

Proof Let $C_2(B)$ be any rough closed set over Y. Then $Y - C_2(B)$ is rough open over Y. We know by definition $f^{-1}(Y - C_2(B))$ is rough clopen over X. But $f^{-1}(Y - C_2(B)) = f^{-1}(Y) - f^{-1}\ (C_2(B)) = X - f^{-1}(C_2(B))$, which is rough clopen over X. This implies that $f^{-1}(C_2(B))$ is rough clopen over X. Conversely suppose $C_1(B))$ is rough open over Y. Then $Y - C_1(B)$ is closed over Y. By hypothesis $f^{-1}(Y - C_1(B))$ is rough clopen over X.

But $f^{-1}(Y - C_1(B)) = f^{-1}(Y) - f^{-1}\ (C_1(B)) = X - f^{-1}(C_1(B))$, which is rough clopen over X. This implies $f^{-1}(C_1(B))$ is rough clopen over X. Hence, the inverse image of every rough open set over Y is rough clopen over X. Therefore, the function f is Čech rough totally continuous.

Theorem 3.3 *Every Čech rough totally continuous function is Čech rough continuous.*

Proof Suppose $f : (X, C_1) \to (Y, C_2)$ is Čech rough totally continuous function and $C_2(B)$ is any rough open set over Y. Since f is Čech rough totally continuous function $f^{-1}(C_2(B))$ is rough clopen over X. This implies that $f^{-1}(C_2(B))$ is rough open over X. Thus the inverse image of a rough open set over Y is rough open over X. Thus f is Čech rough continuous function.

The converse of the above theorem need not be true as shown by the following example.

Example 3.4 Let $U = \{a, b, c, d\}$ with $U|R_1 = \{\{a\}, \{b\}, \{c, d\}\}$ and $X = \{a, b, d\}$ Then $U_R(X) = \{a, b, c, d\}$, $L_R(X) = \{a, b\}$ and $B_R(X) = \{c, d\} \neq \phi$. Clearly $X = \{a, b, d\} \subseteq U$ is a rough set on U with respect to R_1.

Define a map $C_1 : P(X) \to P(X)$ by $C_1\{\phi\} = \phi$, $C_1\{X\} = X$, $C_1\{a\} = X$, $C_1\{b\} = X$, $C_1\{d\} = X$, $C_1\{a, b\} = X$, $C_1\{b, d\} = \{b, d\}$, $C_1\{a, d\} = X$. Then C_1 satisfies the conditions $C_1(\phi) = \phi$, $A \subset C_1(A)$ for each $A \subset X$ and $C_1(A \cup B) = C_1(A) \cup C_1(B)$ for each $A, B \subset X$. Therefore, C_1 is Čech rough operator and (X, C_1) is a Čech rough closure space. Hence ϕ, X, $\{a\}$ are rough open sets and X, ϕ, $\{b, d\}$ are rough closed sets.

Let $V = \{1, 2, 3, 4\}$ with $V|R_2 = \{\{1, 2\}, \{3\}, \{4\}\}$ and $Y = \{2, 3, 4\}$. Then $U_R(Y) = \{1, 2, 3, 4\}$, $L_R(Y) = \{3, 4\}$ and $B_R(Y) = \{1, 2\} \neq \phi$, which implies that Y is a rough set on V with respect to R_2.

Define a map $C_2 : P(Y) \to P(Y)$ by $C_2\{\phi\} = \phi$, $C_2\{Y\} = Y$, $C_2\{2\} = Y$, $C_2\{3\} = Y$, $C_2\{4\} = Y$, $C_2\{2, 3\} = Y$, $C_2\{2, 4\} = \{2, 4\}$, $C_1\{3, 4\} = Y$. Then C_2 satisfies the conditions $C_2(\phi) = \phi$, $A \subset C_2(A)$ for each $A \subset Y$ and $C_2(A \cup B) = C_2(A) \cup C_2(B)$ for each $A, B \subset Y$. Therefore, C_2 is Čech rough operator and (Y, C_2) is a Čech rough closure space. Hence ϕ, Y, $\{3\}$ are rough open sets and Y, ϕ, $\{2, 4\}$ are rough closed sets.

A map $f : X \to Y$ defined by $f(a) = 3$, $f(b) = 2$ and $f(d) = 4$. Clearly f is Čech rough continuous. Since the inverse image of a open set $\{3\}$ is $\{a\}$ not a clopen set. Hence, f is not a Čech rough totally continuous function.

Theorem 3.5 *If $f : (X, C_1) \to (Y, C_2)$ is Čech rough continuous and (X, C_1) is Čech rough locally indiscrete, then f is Čech rough totally continuous.*

Proof Suppose f is Čech rough continuous and $C_2(B)$ is rough open over Y. Since f is rough continuous and (X, C_1) is Čech rough locally indiscrete, $f^{-1}(C_2(B))$ is rough open and rough closed over X. Hence, $f^{-1}(C_2(B))$ is rough clopen over X. Thus f is Čeach rough totally continuous.

Theorem 3.6 *The composition of two Čech rough totally continuous functions is Čech rough totally continuous.*

Proof Let $f : (X, C_1) \to (Y, C_2)$ and $G : (Y, C_2) \to (Z, C_3)$ be any two Čech rough totally continuous functions. Now, we know that $(f \circ g)(x) = g(f(x))$. We have to prove that $(f \circ g)(x)$ is Čech rough totally continuous. Let $C_3(B)$ be a open set in Z, then $(f \circ g)^{-1}(C_3(B)) = f^{-1}\big(g^{-1}(C_3(B))\big)$. Since, g is Čech rough totally continuous $g^{-1}(C_3(B))$ is a clopen set in Y. Moreover, $g^{-1}(C_3(B))$ is open in Y and f is Čech rough totally continuous $f^{-1}\big(g^{-1}(C_3(B))\big)$ is clopen in X. Hence $(f \circ g)$ is Čech rough totally continuous.

Theorem 3.7 *The composition of Čech rough totally continuous function and Čech rough continuous function is Čech rough totally continuous.*

Proof *Case 1.* Let $f : (X, C_1) \to (Y, C_2)$ be Čech rough totally continuous function and $G : (Y, C_2) \to (Z, C_3)$ be Čech rough continuous function. Now, we have to prove that $(f \circ g)(x)$ is Čech rough totally continuous. Let $C_3(B)$ be a open set in Z. Then, $(f \circ g)^{-1}(C_3(B)) = f^{-1}\big(g^{-1}(C_3(B))\big)$. Since, g is Čech rough continuous $g^{-1}(C_3(B))$ is a open set in Y. Moreover, f is Čech rough totally continuous which implies that $f^{-1}\big(g^{-1}(C_3(B))\big)$ is clopen in X. Hence $(f \circ g)$ is Čech rough totally continuous.

Case 2. Let $f : (X, C_1) \to (Y, C_2)$ be Čech rough continuous function and $G : (Y, C_2) \to (Z, C_3)$ be Čech rough totally continuous function. Now, we have to prove that $(f \circ g)(x)$ is Čech rough totally continuous. Let $C_3(B)$ be a open set in Z. Then, $(f \circ g)^{-1}(C_3(B)) = f^{-1}\big(g^{-1}(C_3(B))\big)$. Since, g is Čech rough totally continuous $g^{-1}(C_3(B))$ is a clopen set in Y and f is Čech rough continuous, which implies that $f^{-1}\left(g^{-1}(C_3(B))\right)$ is clopen in X. Hence $(f \circ g)$ is Čech rough totally continuous.

4 Nano Soft Totally Continuous Functions

In this section, we discuss some techniques on nano soft topological space and derive totally continuous function with their properties.

Definition 4.1 Let $\mathcal{U}$ be nonempty finite universe, F_A be a soft set over $\mathcal{U}$ and R is a soft equivalence relation on F_A. Elements belonging to the soft equivalence class of $F(a)$ denoted by $[F(a)]$ are said to be soft indiscernible with one another. The ordered pair $(\mathcal{U}, F_A)$ is said to be soft approximation space. Let $G_B \subseteq F_A$.

(i) If $L_R(G_B) = \bigcup_{a \in A} \{F(a) : [F(a)] \subseteq G_B\}$ is a soft lower approximation of F_A with respect to G_B.
(ii) If $U_R(G_B) = \bigcup_{a \in A} : \left\{F(a) : [F(a)] \cap G_B \neq \widetilde{\phi}\right\}$ is a soft upper approximation of F_A with respect to G_B.
(iii) If $B_R(G_B) = U_R(G_B) - L_R(G_B)$ is a boundary region of F_A with respect to G_B.

Definition 4.2 Let $\mathcal{U}$ be a nonempty set of objects called the universe, F_A is a soft set over $\mathcal{U}$. Then $(\mathcal{U}, F_A)$ is an ordered pair of soft approximation space and $\widetilde{\tau}_R(G_B) = \left\{\widetilde{\mathcal{U}}, \widetilde{\phi}, L_R(G_B), U_R(G_B), B_R(G_B)\right\}$), where $G_B \subseteq F_A$, and $\widetilde{\tau}_R(G_B)$ satisfies the following axiom:

(i) $\widetilde{\mathcal{U}}, \widetilde{\phi} \in \widetilde{\tau}_R(G_B)$.
(ii) The union of the elements of any sub-collection of soft sets $\widetilde{\tau}_R(G_B)$ is an $\widetilde{\tau}_R(G_B)$.
(iii) The intersection of the elements of any finite sub-collection of soft sets $\widetilde{\tau}_R(G_B)$ is an $\widetilde{\tau}_R(G_B)$.

That is, $\widetilde{\tau}_R(G_B)$ forms a soft topology on $\mathcal{U}$ having the at most five elements of soft sets and triplet ordered pair of $(\mathcal{U}, \widetilde{\tau}_R, E)$ is called a nano soft topological space over $\mathcal{U}$ with respect to G_B, then the members of $\widetilde{\tau}_R$ are said to be nano soft open sets in $\mathcal{U}$. The elements of $([\widetilde{\tau}_R(G_B)]^c)$ are called nano soft closed.

Define the notations are using the following definitions, let $\mathcal{U}, \mathcal{V}$ be nonempty universe sets and E_1, E_2 are parameters, F_A, F_B are soft sets over $\mathcal{U}, \mathcal{V}$. Then the soft approximation spaces $(\mathcal{U}, F_A), (\mathcal{V}, F_B)$ and $G_A \subseteq F_A, G_B \subseteq F_B$. The corresponding topologies are $\widetilde{\tau}_R(G_A)$ and $\widetilde{\tau}_R(G_B)$.

Definition 4.3 Let $(\mathcal{U}, \widetilde{\tau}_R(G_A), E_1)$ and $(\mathcal{V}, \widetilde{\tau}_R(G_B), E_2)$ be two nano soft topologies with respect to F_A and F_B respectively. A function $f : (\mathcal{U}, \tau_R(G_A), E_1) \rightarrow (\mathcal{V}, \tau_R(G_B), E_2)$ is called nano soft totally continuous if the inverse image of every nano soft open set over $\mathcal{V}$ is nano soft open set over $\mathcal{U}$.

Definition 4.4 Let $(\mathcal{U}, \widetilde{\tau}_R(G_A), E_1)$ and $(\mathcal{V}, \widetilde{\tau}_R(G_B), E_2)$ be two nano soft topologies with respect to F_A and F_B respectively. A function $f : (\mathcal{U}, \tau_R(G_A), E_1) \rightarrow (\mathcal{V}, \tau_R(G_B), E_2)$ is called nano soft totally continuous if the inverse image of every nano soft open set is nano soft clopen set.

Theorem 4.5 *A nano soft set function $f : (\mathcal{U}, \tau_R(G_A), E_1) \rightarrow (\mathcal{V}, \tau_R(G_B), E_2)$ is nano soft totally continuous if and only if the inverse of every nano soft closed set over $\mathcal{V}$ is nano soft clopen over $\mathcal{U}$.*

Proof Let F be any nano soft closed set over $\mathcal{V}$. Then F^c is nano soft open over $\mathcal{V}$. Let us assume that f is nano soft totally continuous function, by definition $f^{-1}(F^c)$ is nano soft clopen over $\mathcal{U}$. But $f^{-1}(F^c) = f^{-1}(G_B) - f^{-1}(F) = G_A - f^{-1}(F)$, which is nano soft clopen over $\mathcal{U}$. This implies that $f^{-1}(F)$ is nano soft clopen over $\mathcal{U}$. Conversely, suppose G is nano soft open over $\mathcal{V}$. Then G^c is nano soft closed over $\mathcal{V}$. By hypothesis $f^{-1}(G^c)$ is nano soft clopen over $\mathcal{U}$. But $f^{-1}(G^c) = f^{-1}(G_B) - f^{-1}(G) = G_A - f^{-1}(G)$, which is nano soft clopen over $\mathcal{U}$. This implies $f^{-1}(G)$ is nano soft clopen over $\mathcal{U}$. Hence, the inverse image of every nano soft open set over $\mathcal{V}$ is nano soft clopen over $\mathcal{U}$. Therefore, the function f is nano soft totally continuous.

Theorem 4.6 *Every nano soft totally continuous function is nano soft continuous.*

Proof Suppose $f : (\mathcal{U}, \tau_R(G_A), E_1) \rightarrow (\mathcal{V}, \tau_R(G_B), E_2)$ is nano soft totally continuous function and G is any nano soft open set in $\mathcal{V}$. Since f is nano soft totally continuous function $f^{-1}(G)$ is nano soft clopen over $\mathcal{U}$. This implies that $f^{-1}(G)$ is nano soft open over $\mathcal{U}$. Thus the inverse image of a nano soft open set over $\mathcal{V}$ is nano soft open over $\mathcal{U}$. Thus f is nano soft continuous function.

Remark 4.7 Every nano soft continuous function need not be a nano soft totally continuous function.

Example 4.8 Let $\mathcal{U} = \{h_1, h_2, h_3, h_4, h_5\}$ be the universe, $E_1 = \{a_1, a_2, a_3, a_4, a_5\}$ and $A = \{a_1, a_2, a_3, a_4\}$ be a set of parameters.

Let $F_A = \{(a_1, \{h_1, h_2\}), (a_2, \{h_2, h_3\}), (a_3, \{h_2, h_3\}), (a_4, \{h_4, h_5\})\}$and $G_A \subseteq F_A$, where $G_A = \{(a_1, \{h_1, h_2\}), (a_2, \{h_3\}), (a_3, \{h_2, h_3\})\}$ such that $F(a_1) = \{h_1, h_2\}, F(a_2) = \{h_2, h_3\}, F(a_3) = \{h_2, h_3\}, F(a_4) = \{h_4, h_5\}$ is a soft set over $\mathcal{U}$ and $R = \{F(a_1) \times F(a_1), F(a_2) \times F(a_2), F(a_3) \times F(a_3), F(a_4) \times F(a_4), F(a_1) \times F(a_2), F(a_2) \times F(a_1)\}$ is a soft equivalence relation. Then $[F(a_1)] = \{F(a_1), F(a_2)\}$, $[F(a_2)] = \{F(a_1), F(a_2)\}$, $[F(a_3)] = \{F(a_3)\}$, $[F(a_4)] = \{F(a_4)\}$ We have $L_R(G_A) = \{(a_3, \{h_2, h_3\})\}$ and $U_R(G_A) = \{(a_1, \{h_1, h_2\}), (a_2, \{h_2, h_3\}), (a_3, \{h_2, h_3\})\}$ and $B_R(G_A) = \{(a_1, \{h_1, h_2\}), (a_2, \{h_2, h_3\})\}$. Hence $\widetilde{\tau}_R(G_A) = \{\widetilde{\mathcal{U}}, \widetilde{\phi}, \{(a_3, \{h_2, h_3\})\}, \{(a_1, \{h_1, h_2\}), (a_2, \{h_2, h_3\}), (a_3, \{h_2, h_3\}), \{(a_1, \{h_1, h_2\}), (a_2, \{h_2, h_3\})\}\}$ is a nano soft topological space.

Let $\mathcal{V} = \{h_{11}, h_{12}, h_{13}, h_{14}, h_{15}\}$ be the universe, $E_2 = \{b_1, b_2, b_3, b_4, b_5\}$ and $B = \{b_1, b_2, b_4, b_5\}$ be a set of parameters.

Let $F_B = \{(b_1, \{h_{11}, h_{12}\}), (b_2, \{h_{12}, h_{13}\}), (b_4, \{h_{12}, h_{13}\}), (b_5, \{h_{14}, h_{15}\})\}$ and $G_B \subseteq F_A$, where $G_B = \{(b_1, \{h_{11}, h_{12}\}), (b_2, \{h_{13}\}), (b_4, \{h_{12}, h_{13}\})\}$, $R = \{F(b_1) \times F(b_1), F(b_2) \times F(b_2), F(b_4) \times F(b_4), F(b_5) \times F(b_5), F(b_1) \times F(b_2), F(b_2) \times F(b_1)\}$ is a soft equivalence relation. We have $L_R(G_B) = \{(b_4, \{h_{12}, h_{13}\})\}$ $U_R(G_B) = \{(b_1, \{h_{11}, h_{12}\}), (b_2, \{h_{12}, h_{13}\}), (b_4, \{h_{12}, h_{13}\})\}$ and $B_R(G_B) = \{(b_1, \{h_{11}, h_{12}\}), (b_2, \{h_{12}, h_{13}\})\}$. Hence $\widetilde{\tau}_R(G_B) = \{\widetilde{\mathcal{U}}, \widetilde{\phi}, \{(b_4, \{h_{12}, h_{13}\})\}, \{(b_1, \{h_{11}, h_{12}\}), (b_2, \{h_{12}, h_{13}\}), (b_4, \{h_{12}, h_{13}\})\}, \{(b_1, \{h_{11}, h_{12}\}), (b_2, \{h_{12}, h_{13}\})\}\}$.

A map $f : (\mathcal{U}, \tau_R(G_A), E_1) \to (\mathcal{V}, \tau_R(G_B), E_2)$ defined by $f(a_1, \{h_1, h_2\}) = (b_4, \{h_{12}, h_{13}\})$, $f(a_2, \{h_3\}) = (b_2, \{h_{13}\})$, $f(a_3, \{h_2, h_3\}) = (b_1, \{h_{11}, h_{12}\})$.

Then f is nano soft continuous. Since the inverse image of f with respect to $\mathcal{V}$, only $\widetilde{\mathcal{U}}$ and $\widetilde{\phi}$ are the clopen sets. Hence f is not a nano soft totally continuous function.

Remark 4.9 The composition of two nano soft totally continuous functions is nano soft totally continuous.

Remark 4.10 The composition of nano soft totally continuous function and nano soft continuous function is nano soft totally continuous.

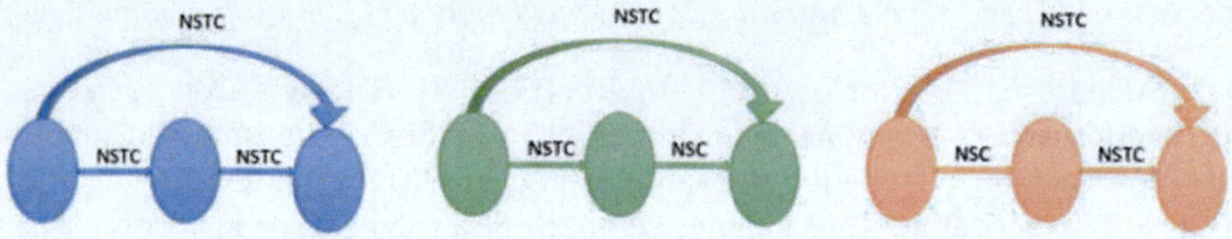

NSTC – Nano Soft Totally Continuous Function, NSC – Nano Soft Continuous Function

5 Conclusion

In this paper, we developed the notion of totally continuous function in Čech rough closure space and nano soft topological space. Also, the relationships between these new classes of Čech rough totally continuous function and nano soft totally continuous function with the existing composition function through computing technique are studied. We believe that these results will enhance the study of developments of information system through nano soft topological spaces and will help researchers to establish real-life applications through nano topology.

Acknowledgements The second author acknowledges the university grants commission for providing financial support under—UGC—BSR Meritorious scheme.

References

1. Čech, E.: Topological Spaces. Interscience Publishers, John Wiley and Sons, New York, (1966)
2. Molodtsov, D.A.: Soft set theory—first result. Comput. Math. Appl. **37**, 19–31 (1999)
3. Lellis Thivagar, M., Richard, C.: On nano forms of weakly open sets. Int. J. Math. Stat. Inv. **1**(1), 31–37 (2013)
4. Lellis Thivagar, M., Sutha Devi, V.: Computing technique for recruitment process via nano topology. Sohag J. Math. **3**(1), 37–45 (2016)
5. Čech, E.: Topological spaces, Topological Papers of Eduard Čech, pp. 436–472. Academia, Praque (1968)
6. Pawlak, Z.: Rough sets. Int. J. Comput. Inform. Sci. **11**, 341–356 (1982)
7. Lellis Thivagar, M., Saeid Jafari, Sutha Devi, V., Antonysamy, V.: A novel approach to nano topology via Neutrosophic sets. Neutrosophic Sets Syst. (University of New Mexico) **20**, 86–94 (2018)
8. Lellis Thivagar, M., Antonysamy, V., Arockia Dasan, M.: Revelation of nano topology in Čech rough closure spaces. Int. J. Math. Appl. **6**(1-A), 121–131 (2018)
9. Feng, F., Liu, X., Leoreanu-Fotea, V., Jun, Y.B.: Soft sets and soft rough sets. Inf. Sci. **181**, 1125–1137 (2011)
10. Lellis Thivagar, M., Sutha Devi, V.: Nano topology induced by lattices. Bol. Soc. Paran. Mat. **38**(5), 197–204 (2020)
11. Shabir, M., Naz, M.: On soft topological spaces. Comput. Math. Appl. **61**, 1786–1799 (2011)

Error Estimates Using Higher Modulus of Smoothness in L_p Spaces

Anshul Srivastava and ShilpiBisht

1 Introduction

Durrmeyer [5] introduced integral modification of Bernstein polynomials to approximate Lebesgue integrable functions on $[0, \infty)$. Later, Derriennic [2] worked on Bernstein–Durrmeyer operators.

In this paper, we combine weights of Szasz and Baskakov operators for approximating Lebesgue integrable functions on $[0, \infty)$. Our operator is hybrid of above two operators and is as follows:

$$M_{(n)}(f; x) = \sum_{i=0}^{\infty} p_{(n,i)}(x, c) \int_{0}^{\infty} q_{(n,i)}(t, c) f(t) dt \tag{1.1}$$

where $p_{(n,i)}(x, c) = (-1)^i x^i \frac{(i+1)}{(i+1)!} \varphi^{(i)}_{(n,c)}(x)$

and $q_{(n,i)}(t, c) = (-1)^{i+1} (t)^i \frac{(i+1)}{(i+1)!} \varphi^{(i+1)}_{(n,c)}(t)$.

Gupta [8] proved that better results are obtained for these operators than symmetric operators.

The following properties holds for $n \in N$, $\{\varphi_{(n,c)}(x)\}$ is a sequence of functions defined on [0,b], b > 0:

(i) For $c > 0, \varphi_{(n,c)}(x) = \frac{1}{(1+cx)^{n/c}}$, our operator $M_{(n)}(f; x)$ reduces to Baskakov–Durrmeyer operators.

A. Srivastava (✉)
Faculty of Applied Sciences, Dr. APJ Abdul Kalam Technical University, Lucknow, Uttar Pradesh, India
e-mail: anshul_sriv@rediffmail.com

ShilpiBisht
Department of Mathematics, Birla Institute of Applied Sciences, Bhimtal, Uttarakhand, India

P. Srivastava et al. (eds.), *Mathematical, Computational Intelligence and Engineering Approaches for Tourism, Agriculture and Healthcare*, Lecture Notes in Networks and Systems 214, https://doi.org/10.1007/978-981-16-3807-7_11

(ii) For $c = 0$, $\varphi_{(n,c)}(x) = \frac{1}{e^{nx}}$, $M_{(n)}(f; x)$ becomes Szasz–Durrmeyer operator.
(iii) For $c = -1$, $\varphi_{(n,c)}(x) = (1-x)^n$ for interval [0,1].
(iii) $\varphi_{(n,c)}(x) \in C^{\infty}[a, b]$.
(iv) $\varphi_{(n,c)}(0) = 1$.
(v) $\{\varphi_{(n,c)}(x)\}$ is a monotonic sequence.
(vi) For $n > max\{0, -c\}$, we have $\varphi^{(i+1)}_{(n,c)} = (-1)\varphi^{(i)}_{(n+c,c)}$.

$M_{(n)}(f; x)$ is given by

$$M_{(n)}(f; x) = \int_0^{\infty} A_{(n,c)}(x, t) f(t) dt$$

where kernel $A_{(n,c)}(x, t) = \sum_{i=0}^{\infty} p_{(n,i)}(x, c) q_{(n,i)}(t, c)$.

Howsoever smooth the function may be, the order of approximation for these operators is at its best $O(n^{-1})$.

For improving the rate of convergence of $M_{(n)}(f; x)$, we will consider a linear combination $M_{(n)}(f; k; x)$ of the operators $M_{(n)}(f; x)$ as

$$M_{(n)}(f; k; x) = \sum_{j=0}^{\infty} C(j, k) M_{(d_j n)}(f; x)$$

where

$$C(j, k) = \prod_{i=0, i \neq j}^{k} \frac{d_j}{d_j - d_i}, \; k \neq 0 \text{ and } C(0, 0) = 1. \tag{1.2}$$

Here, $d_0, d_1, d_2, \ldots\ldots\ldots, d_k$ are $(k+1)$ arbitrarily fixed distinct positive integers.

Let $0 < a < a_1 < a_2 < a_3 < b_3 < b_2 < b_1 < b < \infty$, $I = [a, b]$, $I_i = [a_i, b_i]$, $i = 1, 2, 3$ and $f \in L_p(I)$, $1 \leq p < \infty$.

The mth order integral modulus of smoothness of f is given by

$$\omega_m(f, \tau, p, I) = \sup_{0 < \delta \leq \tau} \|\Delta_\delta^m f(t)\|_{L_p[a, b - m\delta]}.$$

Here, Δ_δ^m is the forward difference operator with step length δ and $0 < \tau \leq (b-a)/m$.

Let $AC(I)$ and $BV(I)$ be the classes of absolutely continuous functions and functions of bounded variations, respectively, on $[a, b]$. Further, C is a positive constant not necessarily the same in different cases and $\chi(t)$ is a characterstic function of $[a_1, b_1]$.

2 Some Auxiliary Results

We will be using the following results for proving our main theorem(s):

Result 2.1. ([3]) For $1 \leq p < \infty$ and $\in L_p[0, \infty)$, we have

$$L_p^{(2k+2)}[a_1, b_1] = \left\{f \in L_p[0, \infty) : f^{(2k+1)} \in AC[a_1, b_1] \; and \; f^{(2k+2)} \in L_p[a_1, b_1]\right\}$$

Result 2.2. ([7]) For $m \in N^0$, if we define mth order moment for the operator $M_{(n)}(f; x)$ by

$$T_{(n,m)}(x, c) = \sum_{i=0}^{\infty} p_{(n,i)}(x, c) \int_0^{\infty} q_{(n,i)}(t, c)(t - x)^m dt$$

or

$$T_{(n,m)}(x, c) = \int_0^{\infty} A_{(n,c)}(x, t)(t - x)^m dt.$$

Then,

(i) $T_{(n,0)}(x, c) = 1$.
(ii) $T_{(n,1)}(x, c) = (1 + cx)/(n - c)$.
(iii) $T_{(n,2)}(x, c) = \frac{2cx^2(n+c)+2(nx+2cx+1)}{n^2-3cn+2c^2}$.

$$\begin{aligned}(n - cm - c)&T_{(n,m+1)}(x, c)\\ &= \left(x + cx^2\right)\left[T'_{(n,m)}(x, c) + 2mT_{(n,m-1)}(x, c)\right]\\ &\quad + [1 + m + cx + 2cxm]T_{(n,m)}(x, c).\end{aligned}$$

It may be remarked that $T_{(n,m)}(x, c)$ is a polynomial in x of degree m, $m \neq 1$ and $\forall x \in [0, \infty)$ so we have

$$T_{(n,m)}(x, c) = O\left(n^{-[(m+1)/2]}\right).$$

Result 2.3. ([1]) For $x \in [0, \infty)$, $r > 0$, $n \to \infty$, we have

$$M_{(n)}\left(|t - x|^r, x\right) = O\left(n^{-(r/2)}\right).$$

Result 2.4. ([4]) For the function $\widehat{T}_{(n,m)}(t)$ and $n > (cm + 2c)$, there holds the recurrence relation

$$(n - cm - 2c)\hat{T}_{(n,m+1)}(x, c) = \big(x + cx^2\big)\hat{T}'_{(n,m)}(x, c) + 2m\hat{T}_{(n,m-1)}(x, c)$$
$$+ [(1 + 2cx)(m + 1) + cx]\hat{T}_{(n,m)}(x, c).$$

Here, derivative is with respect to variable x.

Result 2.5. ([11]) For $1 < p < \infty$, $f \in L_p(I)$ then $h_f \in L_p(I)$, and we have

$$\|h_f\|_{L_p(I)} \le 2^{1/p}\left(1 + \frac{1}{(p-1)}\right)\|f\|_{L_p(I)}$$

where $h_f(x)$ is the Hardy–Littlewood majorant [10] of the function f and is given by

$$h_f(x) = \sup_{\zeta \neq x} \frac{1}{(\zeta - x)} \int_x^{\zeta} f(t)dt, a \le \zeta \le b.$$

Result 2.6. ([6]) For $1 \le p < \infty$, $f \in L_p(I)$. Let $f^{(k)} \in AC(I)$ and $f^{(k)} \in L_p(I)$, then for the bounds of intermediate derivatives of f in terms of highest order derivatives and L_p norm are given by

$$\|f^{(j)}\|_{L_p(I)} \le K_j\Big(\|f^{(k+1)}\|_{L_p(I)} + \|f\|_{L_p(I)}\Big).$$

Here K_j, $j = 1, 2, , \ldots\ldots\ldots, k$ are constants independent of f.

Result 2.7. ([12]) Let $f \in BV[a_1, b_1]$; the following inequality holds:

$$\|M_{(n)}\left(\chi(t)\int_x^t (t - w)^{2k+1} df(w); x\right)\|_{L_1[a_2,b_2]} \le C\frac{1}{n^{(k+1)}}\|f\|_{BV[a_1,b_1]}.$$

3 Error Estimation

We first define the Steklov mean, linear approximating method for estimating error. Let $f \in L_p(0, \infty]$, $1 \le p < \infty$, then for sufficiently small $\eta > 0$ the Steklov mean $f_{(\eta,m)}$ of mth order corresponding to f is defined by

$$f_{(\eta,m)} = \frac{1}{\eta^m}\int_{-\eta/2}^{\eta/2} \cdots\cdots\cdots \int_{-\eta/2}^{\eta/2} \{f(t) + (-1)^{m-1}\Delta_u^m f(t)\}dt_1dt_2 \ldots\ldots\ldots dt_m$$

where $u = \sum_{i=1}^{m} t_i$, $t \in [a_1, b_1]$ and Δ_u^m is mth forward difference.

The following properties hold for $f_{(\eta,m)}$:

(i) $f_{(\eta,m)}^{(m-1)} \in AC[a_1, b_1]$.
(ii) mth derivatives of $f_{(\eta,m)}$ exist a.e., and belong to $L_p(I)$.
(iii) $\|f_{(\eta,m)}^{(r)}\|_{L_p[a_2,b_2]} \leq K_1 \frac{1}{\eta^r} \omega_r(f, \eta, p, [a_1, b_1]), r = 1, 2, \ldots\ldots\ldots, m.$
(iv) $\|f - f_{(\eta,m)}\|_{L_p[a_2,b_2]} \leq K_2 \omega_m(f, \eta, p, [a_1, b_1])$.
(v) $\|f_{(\eta,m)}\|_{L_p[a_2,b_2]} \leq K_3 \|f\|_{L_p[a_1,b_1]}$.
(vi) $\|f_{(\eta,m)}^{(m)}\|_{L_p[a_2,b_2]} \leq K_4 \frac{1}{\eta^m} \|f\|_{L_p[a_1,b_1]}$

where K_1, K_2, K_3, K_4 are constants and are independent of f and η.

Approximation in the smooth subspace $L_p^{(2k+2)}[a_1, b_1]$ of $L_p(0, \infty]$ is studied in the next theorem.

Theorem 3.1. For $p > 1$, $f \in L_p^{(2k+2)}[a_1, b_1]$, we have sufficiently large we have

$$\left\| M_{(n)}(f, k, \cdot) - f(\cdot) \right\|_{L_p[a_2,b_2]} \leq C_1 \frac{1}{n^{(k+1)}} \left\{ \left\| f^{(2k+2)} \right\|_{L_p[a_1,b_1]} + \|f\|_{L_p(0,\infty]} \right\} \tag{3.1}$$

where C_1 is a certain constant independent of f and n.

Proof Let $p > 1$, then for all $t \in [0, \infty)$ and $x \in [a_2, b_2]$ we have

$$\begin{aligned} f(t) - f(x) = &\sum_{j=1}^{2k+1} \frac{f^{(j)}(x)}{j!}(t - x)^j + \frac{1}{(2k+1)!} \int_x^t \chi(t)(t - v)^{2k+1} f^{(2k+2)}(v) dv \\ &+ G(t, x)(1 - \chi(t)) \end{aligned} \tag{3.2}$$

where $G(t, x) = f(t) - \sum_{j=0}^{2k+1} \frac{f^{(j)}(x)}{j!}(t - x)^j \ \forall t \in [0, \infty)$ and $x \in [a_2, b_2]$.

Now operating $M_{(n)}(\cdot, k, x)$ on both sides of (3.2), we have

$$\begin{aligned} M_{(n)}(f, k, x) - f(x) = &\sum_{j=1}^{2k+1} \frac{f^{(j)}(x)}{j!} M_{(n)}\big((t - x)^j, k, x\big) \\ &+ \frac{1}{(2k+1)!} M_{(n)} \left(\int_x^t \chi(t)(t - v)^{2k+1} f^{(2k+2)}(v) dv, k, x \right) \\ &+ M_{(n)}(G(t, x)(1 - \chi(t)), k, x) \\ &= H_1 + H_2 + H_3 (\text{say}). \end{aligned}$$

By [9], we have $Q(r, k, x)$ certain polynomials in x of degree r for every $r \in N$.

Using the above fact, we get

$$\|H_1\|_{L_p[a_2,b_2]} \le C\frac{1}{n^{(k+1)}}\Big(\|f^{(2k+2)}\|_{L_p[a_2,b_2]} + \|f\|_{L_p[a_2,b_2]}\Big). \tag{3.3}$$

For estimating H_2, let $h_{f^{(2k+2)}}$ be the Hardy–Littlewood majorant of $f^{(2k+2)}$ on $[a_1, b_1]$.

By Hölder's inequality and Result 2.3,

$$\begin{aligned}
|I_1| &= \left| M_{(n)}\left(\int_x^t \chi(t)(t-v)^{2k+1} f^{(2k+2)}(v)dv; x\right)\right| \\
&\le M_{(n)}\left(\chi(t)|t-x|^{2k+1}\left|\int_x^t f^{(2k+2)}(v)dv\right|; x\right) \\
&\le M_{(n)}\Big(\chi(t)|t-x|^{2k+2}\Big|h_{|f^{(2k+2)}|}(t)\Big|; x\Big) \\
&\le \Big\{M_{(n)}(\chi(t)|t-x|^{(2k+2)q}; x)\Big\}^{1/q}\Big\{M_{(n)}\Big(\chi(t)\Big|h_{|f^{(2k+2)}|}(t)\Big|^p; x\Big)\Big\}^{1/p}.
\end{aligned}$$

Using Fubini's theorem and Result 2.5,

$$\begin{aligned}
\|I_1\|^p_{L_p[a_2,b_2]} &\le C\frac{1}{n^{(k+1)p}}\int_{a_2}^{b_2}\int_{a_1}^{b_1} A_{(n,c)}(x,t)\Big|h_{|f^{(2k+2)}|}(t)\Big|^p dt dx \\
&\le C\frac{1}{n^{(k+1)p}}\int_{a_1}^{b_1}\left[\int_{a_2}^{b_2} A_{(n,c)}(x,t)dx\right]\Big|h_{|f^{(2k+2)}|}(t)\Big|^p dt.
\end{aligned}$$

In view of $\hat{T}_{(n,0)}(t) = \frac{n}{(n-c)}$,

$$\begin{aligned}
\|I_1\|^p_{L_p[a_2,b_2]} &\le C\frac{1}{n^{(k+1)p}}\frac{n}{(n-c)}\int_{a_1}^{b_1}\Big|h_{|f^{(2k+2)}|}(t)\Big|^p dt \\
&\le C\frac{1}{n^{(k+1)p}}\Big\|h_{|f^{(2k+2)}|}\Big\|^p_{L_p[a_1,b_1]} \quad \text{for a large } n \\
&\le C\frac{1}{n^{(k+1)p}}\|f^{(2k+2)}\|^p_{L_p[a_1,b_1]}.
\end{aligned}$$

Consequently,

$$\|I_1\|_{L_p[a_2,b_2]} \le C\frac{1}{n^{(k+1)}}\|f^{(2k+2)}\|_{L_p[a_1,b_1]}.$$

Therefore

$$\|H_2\|_{L_p[a_2,b_2]} \leq C\frac{1}{n^{(k+1)}}\|f^{(2k+2)}\|_{L_p[a_1,b_1]}. \tag{3.4}$$

Let $t \in [0,\infty)\backslash[a_1,b_1]$, $x \in [a_2,b_2]$, then we have for $\delta > 0$, $|t-x| \geq \delta$. For estimating H_3, we will consider

$$\begin{aligned}
P &= \left|M_{(n)}(G(t,x)(1-\chi(t));x)\right| \\
&= \left|M_{(n)}\left(\left(f(t)-\sum_{j=0}^{2k+1}\frac{f^{(j)}(x)}{j!}(t-x)^j\right)(1-\chi(t));x\right)\right| \\
&\leq \frac{1}{\delta^{(2k+2)}}\left\{M_{(n)}\left(|f(t)|(t-x)^{2k+2};x\right)\right\} \\
&\quad + \frac{1}{\delta^{(2k+2)}}\sum_{j=0}^{2k+1}\frac{\left|f^{(j)}(x)\right|}{j!}\left\{M_{(n)}\left(|t-x|^{2k+j+2};x\right)\right\} \\
&= I_2 + I_3 \text{ (say)}.
\end{aligned}$$

Using Hölder's inequality, we have

$$\begin{aligned}
|I_2| &\leq \frac{1}{\delta^{(2k+2)}}\left\{M_{(n)}\left(|f(t)|^p;x\right)\right\}^{1/p}\left\{M_{(n)}\left(|t-x|^{2kq+2q};x\right)\right\}^{1/q} \\
&\leq C\frac{1}{n^{(k+1)}}\left\{\int_{a_1}^{b_1} A_{(n,c)}(x,t)|f(t)|^p\right\}^{1/p}.
\end{aligned}$$

From Fubini's theorem, we have

$$\begin{aligned}
\|I_2\|^p_{L_p[a_2,b_2]} &\leq C\frac{1}{n^{(k+1)p}}\int_{a_2}^{b_2}\int_{a_1}^{b_1} A_{(n,c)}(x,t)|f(t)|^p\,dt\,dx \\
&\leq C\frac{1}{n^{(k+1)p}}\int_{a_1}^{b_1}\left(\int_{a_2}^{b_2} A_{(n,c)}(x,t)\,dx\right)|f(t)|^p\,dt.
\end{aligned}$$

Now following on the same lines as for calculating I_1, we have

$$\|I_2\|_{L_p[a_2,b_2]} \leq C\frac{1}{n^{(k+1)}}\|f\|_{L_p[0,\infty)}.$$

Using Results 2.3 and 2.6,

$$\|I_3\|_{L_p[a_2,b_2]} \le C\frac{1}{n^{(k+1)}}\left(\|f^{(2k+2)}\|_{L_p[a_2,b_2]} + \|f\|_{L_p[a_2,b_2]}\right)$$

so we have

$$\|H_3\|_{L_p[a_2,b_2]} \le C\frac{1}{n^{(k+1)}}\left(\|f^{(2k+2)}\|_{L_p[a_2,b_2]} + \|f\|_{L_p[0,\infty)}\right) \tag{3.5}$$

Combining (3.3), (3.4) and (3.5),

$$\left\|M_{(n)}(f,k,\cdot) - f(\cdot)\right\|_{L_p[a_2,b_2]} = C\frac{1}{n^{(k+1)}}\left(\left\|f^{(2k+2)}\right\|_{L_p[a_2,b_2]} + \|f\|_{L_p[a_2,b_2]}\right)$$
$$+ C\frac{1}{n^{(k+1)}}\left\|f^{(2k+2)}\right\|_{L_p[a_1,b_1]} + C\frac{1}{n^{(k+1)}}\left(\left\|f^{(2k+2)}\right\|_{L_p[a_2,b_2]} + \|f\|_{L_p[0,\infty)}\right).$$

Hence the theorem.

Theorem 3.2. For sufficiently large n, $f \in L_1[0,\infty)$, $f^{(2k+1)} \in BV[a_1,b_1]$ and $f^{(2k+1)}$ is continuous almost everywhere on $[a_1,b_1]$, and f has derivatives up to order $(2k+1)$ on $[a_1,b_1]$ then there holds

$$\left\|M_{(n)}(f,k,\cdot) - f(\cdot)\right\|_{L_1[a_2,b_2]}$$
$$\le C_2\frac{1}{n^{(k+1)}}\left(\left\|f^{(2k+1)}\right\|_{BV[a_1,b_1]} + \left\|f^{(2k+1)}\right\|_{L_1[a_2,b_2]} + \|f\|_{L_1[0,\infty)}\right) \tag{3.6}$$

where C_2 is a certain constant independent of f and n.

Proof For $x \in [a_2,b_2]$ and $t \in [0,\infty)$, we have

$$f(t) - f(x) = \sum_{j=1}^{2k+1}\frac{f^{(j)}(x)}{j!}(t-x)^j + \frac{1}{(2k+1)!}\int_x^t \chi(t)(t-v)^{2k+1}df^{(2k+1)}(v)$$
$$+ G(t,x)(1-\chi(t)) \tag{3.7}$$

where $G(t,x) = f(t) - \sum_{j=0}^{2k+1}\frac{f^{(j)}(x)}{j!}(t-x)^j$.

Operating $M_{(n)}(\cdot,k,x)$ on both sides of (3.7), we have

$$M_{(n)}(f,k,x) - f(x) = \sum_{j=1}^{2k+1}\frac{f^{(j)}(x)}{j!}M_{(n)}\left((t-x)^j,k,x\right)$$
$$+ \frac{1}{(2k+1)!}M_{(n)}\left(\int_x^t \chi(t)(t-v)^{2k+1}df^{(2k+1)}(v),k,x\right)$$
$$+ M_{(n)}(G(t,x)(1-\chi(t)),k,x)$$

$$= H_4 + H_5 + H_6 \text{ (say)}.$$

Proceeding as in Theorem 3.1, we have

$$\|H_4\|_{L_1[a_2,b_2]} \leq C\frac{1}{n^{(k+1)}}\left(\|f^{(2k+1)}\|_{L_1[a_2,b_2]} + \|f\|_{L_1[a_2,b_2]}\right)$$

and

$$\|H_5\|_{L_1[a_2,b_2]} \leq C\frac{1}{n^{(k+1)}}\|f^{(2k+1)}\|_{BV[a_1,b_1]}.$$

For, estimating H_6, let $t \in [0,\infty)\backslash[a_1,b_1]$, $x \in [a_2,b_2]$ then we have for

$$\delta > 0, |t-x| \geq \delta$$

$$\|H_6\|_{L_1[a_2,b_2]} \leq C\int_{a_2}^{b_2}\int_0^{\infty} A_{(n,c)}(x,t)|f(t)(1-\chi(t))|dtdx$$

$$+\sum_{i=0}^{2k+1}\frac{1}{i!}\int_{a_2}^{b_2}\int_0^{\infty} A_{(n,c)}(x,t)\left|f^{(i)}(x)\right||t-x|^i(1-\chi(t))dtdx$$

$$= Q_1 + Q_2 \text{(say)}.$$

We can find positive constant M and C' for sufficiently large n such that $\frac{(t-x)^{2(k+1)}}{t^{2(k+1)}+1} > C'$ for every $t \geq M$, $x \in [a_2,b_2]$.

For estimating Q_1, we have from Fubini's theorem

$$Q_1 = \int_0^M\int_{a_2}^{b_2} A_{(n,c)}(x,t)|f(t)(1-\chi(t))|dxdt$$

$$+\int_M^{\infty}\int_{a_2}^{b_2} A_{(n,c)}(x,t)|f(t)(1-\chi(t))|dxdt$$

$$= Q_3 + Q_4 \text{(say)}.$$

Now

$$Q_3 \leq \frac{1}{\delta^{(2k+2)}}\int_0^M\int_{a_2}^{b_2} A_{(n,c)}(x,t)|f(t)|(t-x)^{2(k+1)}dxdt$$

$$\leq C\frac{1}{n^{(k+1)}}\int_0^M |f(t)|dt$$

$$Q_4 \leq \frac{1}{C'}\int_M^\infty\int_{a_2}^{b_2} A_{(n,c)}(x,t)\frac{(t-x)^{2(k+1)}}{t^{2(k+1)}+1}|f(t)|dxdt$$

$$\leq C\frac{1}{n^{(k+1)}}\int_M^\infty |f(t)|dt$$

from $Q_1 = Q_3 + Q_4$

$$Q_1 \leq C\frac{1}{n^{(k+1)}}\int_0^M |f(t)|dt + C\frac{1}{n^{(k+1)}}\int_M^\infty |f(t)|dt$$

so $Q_1 \leq C\frac{1}{n^{(k+1)}}\int_0^\infty |f(t)|dt$.

This implies

$$Q_1 \leq C\frac{1}{n^{(k+1)}}\|f\|_{L_1[0,\infty)}.$$

Similarly, we can obtain

$$Q_2 \leq C\frac{1}{n^{(k+1)}}(\|f^{(2k+1)}\|_{L_1[a_2,b_2]} + \|f\|_{L_1[a_2,b_2]}).$$

Hence

$$\left\|M_{(n)}(G(t,x)(1-\chi(t));x)\right\|_{L_1[a_2,b_2]}$$
$$\leq C\frac{1}{n^{(k+1)}}\left(\left\|f^{(2k+1)}\right\|_{L_1[a_2,b_2]} + \|f\|_{L_1[0,\infty)}\right).$$

So

$$\|H_6\|_{L_1[a_2,b_2]} \leq C\frac{1}{n^{(k+1)}}(\|f^{(2k+1)}\|_{L_1[a_2,b_2]} + \|f\|_{L_1[0,\infty)}).$$

Combining the estimates $H_4, H_5, H_6,$ we have the theorem.

Theorem 3.3. For very large n, $f \in L_p[0,\infty)$, $p \geq 1$,

$$\left\|M_{(n)}(f,k,\cdot) - f\right\|_{L_p[a_2,b_2]} \leq C_k\left(\omega_{2k+2}(f, n^{-1/2}, p, [a_1,b_1]) + \frac{1}{n^{(k+1)}}\|f\|_{L_p[0,\infty)}\right).$$

Proof Here,

$$\begin{aligned}
&\left\| M_{(n)}(f,k,\cdot) - f \right\|_{L_p[a_2,b_2]} \\
&\le \left\| M_{(n)}\big(f - f_{(\eta,2k+2)}, k, \cdot\big) \right\|_{L_p[a_2,b_2]} \\
&\quad + \left\| M_{(n)}\big(f_{(\eta,2k+2)}, k, \cdot\big) - f_{(\eta,2k+2)} \right\|_{L_p[a_2,b_2]} \\
&\quad + \left\| f_{(\eta,2k+2)} - f \right\|_{L_p[a_2,b_2]} \\
&= W_1 + W_2 + W_3 \text{ (say)}.
\end{aligned}$$

Let $\chi\,(t)$ be the characterstic function of $[a_3, b_3]$ such that we have

$$\begin{aligned}
&M_{(n)}\big((f - f_{(\eta,2k+2)})(t); x\big) = M_{(n)}\big(\chi(t)(f - f_{(\eta,2k+2)})(t); x\big) \\
&\quad + M_{(n)}\big((1-\chi(t))(f - f_{(\eta,2k+2)})(t); x\big) \\
&= W_4 + W_5 \text{(say)}.
\end{aligned}$$

We have

$$\int_{a_2}^{b_2} |W_4|^p dx \le \int_{a_2}^{b_2}\int_{a_3}^{b_3} A_{(n,c)}(x,t)\left|(f - f_{(\eta,2k+2)})(t)\right|^p dt dx.$$

By Fubini's theorem, we get

$$\|W_4\|_{L_p[a_2,b_2]} \le 2\|f - f_{(\eta,2k+2)}\|_{L_p[a_2,b_2]}.$$

Similarly, we have for $p \ge 1$

$$\|W_5\|_{L_p[a_2,b_2]} \le C\frac{1}{n^{(k+1)}}\|f - f_{(\eta,2k+2)}\|_{L_p[0,\infty)}$$

$$W_1 \le C\left(\omega_{2k+2}(f,\eta,p,[a_1,b_1]) + \frac{1}{n^{(k+1)}}\|f\|_{L_p[0,\infty)}\right).$$

Since $\|f^{(2k+1)}_{(\eta,2k+2)}\|_{BV[a_3,b_3]} = \|f^{(2k+1)}_{(\eta,2k+2)}\|_{L_1[a_3,b_3]}$,
Using Theorems 3.1, 3.2 and properties of Steklov mean, we have

$$\begin{aligned}
W_2 &\le C\frac{1}{n^{(k+1)}}\left(\left\| f_{(\eta,2k+2)}\right\|_{L_p[0,\infty)} + \left\| f^{(2k+2)}_{(\eta,2k)}\right\|_{L_p[a_3,b_3]}\right) \\
&\le C\frac{1}{n^k}\left(\|f\|_{L_p[a_1,b_1]} + \frac{1}{\eta^{2k}}\omega_{2k}(f,\eta,p,[a_1,b_1])\right)
\end{aligned}$$

and

$$W_3 \leq C\omega_{2k+2}(f, \eta, p, [a_1, b_1]).$$

For $\eta = n^{-1/2}$ combining estimates W_1, W_2, W_3, we have the theorem.

4 Conclusion

Rate of convergence has been improved by taking linear combinations of our operators using higher modulus of smoothness. It has been seen that hybrid operators give better results than any other symmetric operators.

References

1. Agarwal, P.N., Gairola, A.R.: On L_p - inverse theorem for a linear combination of Szasz-beta operators. Thai J. Math. **9**, 11–20 (2011)
2. Derriennic, M.M.: Sur'l approximation de functions integrablesur [0,1] par des polynomes de Bernstein modifies. J. Approx. Theory **31**, 325–343 (1981)
3. Ditzian, Z., Ivanov, K.: Bernstein type operators and their derivatives. J. Approx. Theory **56**, 72–90 (1989)
4. Ditzian, Z., Totik, V.: Moduli of Smoothness. Springer Series in Computational Mathematics, vol. 9. Springer, Berlin (1987)
5. Durrmeyer, J.L.: uneformule'd inversion de la transformee de Laplace: applications a la Theorie des Moments, These de 3e cycle, Faculte des Science de l'universite de Paris
6. Goldberg, S., Meir, A.: Minimum moduli of ordinary differential operators. Proc. Lond. Math. Soc. **23**, 1–15 (1971)
7. Gupta, V., Mohapatra, R.: On the rate of convergence for certain summation-integral type operators. Math. Ineq. Appl. **9**(3), 465–472 (2006)
8. Gupta, V.: A note on modified Baskakov type operators. Approx. Theory Appl. **10**(3), 74–78 (1994)
9. Gupta, V., Mohapatra, R.N., Finta, Z.: On certain family of mixed summation–integral type operators. Math. Comput. Modelling **42**, 181–191 (2005)
10. Hardy, G.H., Littlewood, J.E., Poyla, G.: Hölder's Inequality and its Extensions, pp. 21–26 Cambridge University Press (1988)
11. Hewitt, E., Stromberg, k.: Real and Abstract Analysis. McGraw Hill, New York (1965)
12. Maheshwari, P.: Some direct estimates for Szasz-beta operators in L_p norm. Int. J. Pure Appl. Math. Sci. **3**(1), 1–11 (1995)

Space-Time Admitting W_6-Curvature Tensor

S. P. Maurya, S. K. Pandey, and R. N. Singh

2000 Mathematics Subject Classification 53C25 · 53C50 · Secondary 53C80 · 53B20

1 Introduction

A Lorentzian manifold consists of three types of vectors such as timelike (i.e., $g(v, v) < 0$), spacelike (i.e., $g(v, v) > 0$), and null vector (i.e., $g(v, v) = 0$ for $v \neq 0$). In general, a Lorentzian manifold (M, g) may not have a globally timelike vector field. If (M, g) admits a globally timelike vector field, it is called time orientable Lorentzian manifold, physically known as space-time. The foundation of general relativity is based on a four-dimensional space-time manifold which is the stage of present modeling of the physical world a torsionless, time-oriented Lorentzian manifold (M, g).

An n-dimensoinal generalized Robertson–Walker (GRW) space-time with $n \geq 3$ is a Lorentzian manifold which is a warped product of an open interval I of $\Re$ and an $(n-1)$-dimensional Riemannian manifold [8–10]. These Lorentzian manifold broadly extends the classical Robertson–Walker (RW) space-time. RW space-time

The research of Mr. Satya Prakash Maurya is supported by Council of Scientific and Industrial Research, India under grant no. $09/405(0001)/2019 - EMR - I$.

S. P. Maurya (✉) · S. K. Pandey · R. N. Singh
Department of Mathematical Sciences, A.P.S. University, Rewa 486003, MP, India
e-mail: rnsinghmp@rediffmail.com

P. Srivastava et al. (eds.), *Mathematical, Computational Intelligence and Engineering Approaches for Tourism, Agriculture and Healthcare*, Lecture Notes in Networks and Systems 214, https://doi.org/10.1007/978-981-16-3807-7_12

is regarded as cosmological models since it is spatially homogenous and spatially isotropic, whereas GRW space-time serve as inhomogeneous extension of RW space-times that admit an isotropic radiation [14]. A Lorentzian manifold is a perfect fluid space-time, if its Ricci tensor S has the form

$$S(X, Y) = \alpha g(X, Y) + \beta A(X)A(Y), \tag{1}$$

where α and β are scalars, A is a non-zero one-form such that $g(X, v) = A(X)$ for all v and v is the velocity vector field such that $g(v, v) = -1$. Perfect- fluid space-times in a language of differential geometry are called quasi-Einstein spaces, where A is metrically equivalent to a unit spacelike vector field.

The Einstein's field equation with cosmological constant is given by [12]

$$S(X, Y) - \frac{r}{2}g(X, Y) + \lambda g(X, Y) = kT(X, Y), \tag{2}$$

where S and r denotes the Ricci tensor and scalar curvature, respectively, λ is the cosmological constant, $T(X, Y)$ is the energy-momentum tensor and $k \neq 0$.

Einstein's field equation without cosmological constant is given by [12]

$$S(X, Y) - \frac{r}{2}g(X, Y) = kT(X, Y). \tag{3}$$

The Einstein's field Eqs. (2) and (3) imply that the energy-momentum tensor is conservative. This requirement is satisfied if the energy-momentum tensor is covariant constant [1]. Chaki and Ray [1] showed that a general relativistic space-time with covariant constant energy-momentum tensor is Ricci symmetric, that is, $\nabla S = 0$, where S is Ricci tensor of the space-time.

A symmetric (0, 2) type tensor field T on a semi-Riemannian manifold (M^n, g) is said to be a Codazzi tensor if it satisfies the Codazzi equation

$$(\nabla_X T)(Y, Z) = (\nabla_Y T)(X, Z), \tag{4}$$

for arbitrary vector fields X, Y, and Z. The geometrical and topological consequences of the existence of a non-trivial Codazzi tensor on a Riemannian manifold have been studied by Derdzinski and Shen [5].

A semi-Riemannian manifold is semi-symmetric [15] if its curvature tensor R satisfies

$$R(\zeta, \xi).R = 0, \tag{5}$$

where $R(\zeta, \xi)$ acts as a derivation on R. Also, a semi-Riemannian manifold is Ricci semi-symmetric [11] if its Ricci tensor S satisfies

$$R(\zeta, \xi).S = 0, \tag{6}$$

where $R(\zeta, \xi)$ acts as a derivation on S. It is known that every semi-symmetric manifold is Ricci semi-symmetric. The converse does not hold in general.

Pokhariyal and Mishra [13] introduced some curvature tensors similar to projective curvature tensor on an n-dimensional Riemannian manifold whose physical significance and geometric properties have been studied by several authors [2–4]. Our work is confined to study geometric behavior of W_6-curvature tensor on relativistic space-time. The W_6-curvature tensor on an n-dimensional Riemannian manifold is defined as

$$W_6(U, X, Y, Z) = R(U, X, Y, Z) + \frac{1}{n-1}[g(U, X)S(Y, Z) - g(U, Z)S(X, Y)]. \tag{7}$$

where $R(U, X, Y, Z) = g(R(U, X)Y, Z)$ is the Riemannian curvature tensor and $S(X, Y)$ is Ricci tensor.

2 W_6-Curvature Semi-symmetric Space-Time

Definition A four-dimensional relativistic space-time M is said to be W_6-curvature semi-symmetric space-time if

$$R(\zeta, \xi).W_6 = 0, \tag{8}$$

where $R(\zeta, \xi)$ acts as a derivation on W_6.
From Eq. (7), we have

$$\begin{aligned} R(\zeta, \xi).W_6(U, X, Y, Z) &= R(\zeta, \xi).R(U, X, Y, Z) \\ &+ \frac{1}{n-1}[g(U, X)R(\zeta, \xi).S(Y, Z) - g(U, Z)R(\zeta, \xi).S(X, Y)], \end{aligned} \tag{9}$$

which on contraction gives

$$R(\zeta, \xi).W_6^*(U, Z) = \frac{2}{3}R(\zeta, \xi).S(U, Z), \tag{10}$$

where $W_6^*(U, Z) = W_6(U, X, X, Z)$. Thus, we can state as follows:

Theorem 2.1 *A four-dimentional relativistic space-time M is Ricci semi-symmetric if and only if $W_6^*(U, Z) = W_6(U, X, X, Z)$ is semi-symmetric.*

On the basis of the above result, we have the following corollary as

Corollary 2.2 *A four-dimentional relativistic space-time M is Ricci semi-symmetric if W_6-curvature tensor is semi-symmetric.*

A space-time is said to be conformally semi-symmetric if the conformal curvature tensor C is semi-symmetric. Thus, we can state as follows:

Theorem 2.3 *A W_6-semi-symmetric space-time M is conformally semi-symmetric if and only if it is semi-symmetric.*

From Eq. (2), we have

$$R(\zeta, \xi).S(X, Y) = kR(\zeta, \xi).T(X, Y) \tag{11}$$

i.e. M is Ricci semi-symmetric if and only if the energy-momentum tensor is semi-symmetric. Thus, by virtue of Theorem 2.1 and Eq. (11), we can state as follows:

Theorem 2.4 *A four-dimentional relativistic space-time M obeying Einstein's field equation with cosmological constant is W_6-semi-symmetric if and only if its energy-momentum tensor T is semi-symmetric.*

It is clear that $\nabla W_6(U, X, X, Z) = 0$ implies $R(\zeta, \xi).W_6(U, X, X, Z) = 0$. Thus, we can state as follows:

Corollary 2.5 *A four-dimentional relativistic space-time M with covariantly constant W_6-curvature tensor is conformally semi-symmetric.*

Corollary 2.6 *In a four-dimentional relativistic space-time M with covariantly constant W_6-curvature tensor, the energy-momentum tensor T is semi-symmetric.*

Definition A Ricci tensor S is called recurrent if

$$(\nabla_X S)(Y, Z) = A(X)S(Y, Z), \tag{12}$$

where A is associated recurrence one-form. Assume that the Ricci tensor is recurrent, then

$$\begin{aligned} R(\zeta, \xi).S(Y, Z) &= [\nabla_\zeta \nabla_\xi - \nabla_\xi \nabla_\zeta]S(Y, Z) \\ &= \nabla_\zeta(\nabla_\xi S)(Y, Z) - \nabla_\xi(\nabla_\zeta S)(Y, Z) \\ &= \nabla_\zeta[A(\xi)S(Y, Z)] - \nabla_\xi[A(\zeta)S(Y, Z)], \end{aligned}$$

which gives

$$R(\zeta, \xi).S(Y, Z) = [(\nabla_\zeta A)(\xi) - (\nabla_\xi A)(\zeta)]S(Y, Z). \tag{13}$$

Thus, in view of Eqs. (11), (13), and theorem (11), we can state as follows:

Corollary 2.7 *The following statements are equivalent on a four-dimentional relativistic space-time M obeying Einstein's field equation with cosmological constant*
(1) *The Ricci tensor S is recurrent with closed recurrence one-form,*
(2) *The energy-momentum tensor T is semi-symmetric,*
(3) $W_6^*(U, Z) = W_6(U, X, X, Z)$ *is semi-symmetric.*

Definition A 4-dimentional relativistic space-time M is called W_6-symmetric if

$$(\nabla_V W_6)(U, X, Y, Z,) = 0.$$

Applying the covariant derivative on both sides of equation (7), we obtain

$$\begin{aligned}(\nabla_V W_6)(U, X, Y, Z) &= (\nabla_V R)(U, X, Y, Z) \\ &+ \frac{1}{3}[g(U, X)(\nabla_V S)(Y, Z) - g(U, Z)(\nabla_V S)(X, Y)].\end{aligned} \tag{14}$$

If M is a W_6-symmetric space-time, then

$$(\nabla_V R)(U, X, Y, Z) = \frac{1}{3}[g(U, Z)(\nabla_V S)(X, Y) - g(U, X)(\nabla_V S)(Y, Z)]. \tag{15}$$

Putting $X = Y = e_i$, and taking summation over i, $1 \leq i \leq 4$, we obtain

$$(\nabla_V S)(U, Z) = \frac{1}{3}[g(U, Z)\nabla_V r - (\nabla_V S)(U, Z)],$$

and hence

$$(\nabla_V S)(U, Z) = \frac{1}{4}g(U, Z)\nabla_V r.$$

Thus, we can state as follows:

Theorem 2.8 *A four-dimentional relativistic space-time M with W_6-symmetric is Ricci symmetric if and only if scalar curvaturer is constant.*

3 Space-Time with Conservative W_6-Curvature Tensor

Definition A four-dimentional relativistic space-time is said to be W_6-conservative if $(div W_6)(U, X, Y, Z) = 0$, where "$div$" denotes the divergence.

From Eq. (7), we have

$$W_6(U, X)Y = R(U, X)Y + \frac{1}{n-1}[g(U, X)Q(Y) - S(X, Y)U], \tag{16}$$

where $Q(X)$ is Ricci operator.

The divergence of W_6-curvature tensor is given by

$$(div W_6)(U, X)Y = g((\nabla_{e_i} W_6)(U, X)Y, e_i)$$

i.e.

$$(divW_6)(U,X)Y = g((\nabla_{e_i}R)(U,X)Y,e_i) + \frac{1}{n-1}[g(U,X)g((\nabla_{e_i}Q)Y,e_i) - (\nabla_{e_i}S)(X,Y)g(U,e_i)]$$

i.e.

$$(divW_6)(U,X)Y = (\nabla_U S)(X,Y) - (\nabla_X S)(U,Y) + \frac{1}{3}\left[g(U,X)\frac{1}{2}dr(Y) - (\nabla_U S)(X,Y)\right],$$

which on contraction gives

$$(divW_6)(U,X)Y = \frac{2}{3}(\nabla_U S)(X,Y) - (\nabla_X S)(U,Y) + \frac{1}{6}dr(Y). \tag{17}$$

Putting $X = Y = e_i$, and taking summation over i, $1 \leq i \leq 4$, we obtain

$$(divW_6)(U,X)Y = \frac{1}{3}dr(U). \tag{18}$$

If W_6-curvature tensor is conservative then in virtue of equation (18), we obtain scalar curvature r is constant which implies that the Ricci tensor is of Codazzi type. Thus, we can state as follows:

Theorem 3.1 *A four-dimentional relativistic space-time M admits a conservative W_6-curvature tensor if and only if the Ricci tensor is Codazzi tensor. In both cases, the scalar curvature is constant.*

Now, Guifoyle and Nolan [6], named "Yang Pure Space" a four-dimensional Lorentzian manifold (M, g) whose metric tensor solves Yang's equation

$$(\nabla_X S)(Y,Z) - (\nabla_Y S)(X,Z) = 0.$$

Thus, we can state as follows:

Theorem 3.2 *A four-dimentional relativistic space-time M admits a conservative W_6-curvature tensor is a Yang Pure space.*

Since we known that [6], a four-dimensional relativistic perfect fluid space-time with $\sigma + \rho \neq 0$ is a Yang Pure space-time if and only if space-time is RW space-time.

Theorem 3.3 *A four-dimentional relativistic space-time M admits a conservative W_6-curvature tensor is a RW space-time.*

It is known that divergence of conformal (Weyl) curvature tensor can be written as

$$(divC)(X, Y)Z = \frac{n-3}{n-2}[(\nabla_X S)(Y, Z) - (\nabla_Y S)(X, Z) - \frac{1}{2(n-1)} g(Y, Z)dr(X) - g(X, Y)dr(Z)]. \tag{19}$$

In virtue of equations (19), we observe that $(divC)(X, Y)Z = 0$ if Ricci tensor is Codazzi tensor. Thus, $div\ W_6 = 0$. Thus, we can state as follows:

Theorem 3.4 *A four-dimentional relativistic space-time M satisfying* $divC = 0$ *if and only if* $div\ W_6 = 0$.

Since we known that [8–10], a perfect fluid space-time of $dimention \geq 3$ satisfies conservative conformal curvature tensor and the velocity vector field irrotational, then the space-time is a GRW space-time with $A(C(X, Y)Z) = 0$. Thus, we can state as follows:

Theorem 3.5 *A four-dimentional relativistic space-time M satisfying* $divC = 0$ *is a GRW space-time.*

In virtue of equations (2) and (4), thus we can state as follows:

Theorem 3.6 *A four-dimensional relativistic space-time M admits a conservative* W_6*-curvature tensor if and only if the energy-momentum tensor T is Codazzi tensor provided that the scalar curvature is constant in both the cases.*

4 W_6-Flat Space-Time

Let M be a four-dimensional space-time of general relativity, then in view of Eq. (7), we have

$$W_6(U, X, Y, Z) = R(U, X, Y, Z) + \frac{1}{3}[g(U, X)S(Y, Z) - g(U, Z)S(X, Y)]. \tag{20}$$

If $W_6(U, X, Y, Z) = 0$, then from equation (20), we have

$$R(U, X, Y, Z) = \frac{1}{3}[g(U, Z)S(X, Y) - g(U, X)S(Y, Z)] \tag{21}$$

Taking a frame field over X and Y, we have

$$S(U, Z) = \frac{1}{3}[g(U, Z)S(X, X) - g(U, X)S(X, Z)]$$

i.e.,

$$S(U, Z) = \frac{1}{3}[rg(U, Z) - S(U, Z)]$$

i.e.,

$$S(U, Z) + \frac{1}{3}S(U, Z) = \frac{r}{3}g(U, Z),$$

which gives

$$S(U, Z) = \frac{r}{4}g(U, Z). \tag{22}$$

Thus, a W_6-curvature flat space-time is an Einstein manifold. Next, using Eq. (22) in Eq. (21), we get

$$R(U, X, Y, Z) = \frac{r}{12}[g(U, Z)S(X, Y) - g(U, X)S(Y, Z)], \tag{23}$$

which shows that M is of constant curvature. Thus, we can state as follows:

Theorem 4.1 *A four-dimensional relativistic W_6-flat space-time is an Einstein space-time and it is of constant curvature.*

It is known that a Lorentzian manifold of constant curvature is a manifold of conformally flat. Thus, we have the following corollary

Corollary 4.2 *A four-dimensional relativistic W_6-flat space-time is a conformally flat space-time.*

In 1980, Kramer et al. [7] have been proved that a space is of O-type if the conformal curvature tensor vanishes on it. Thus, we have a theorem as follows:

Theorem 4.3 *A four-dimensional relativistic W_6-flat space-time is of O-type.*

Now, we consider a perfect fluid space-time with flat W_6-curvature tensor having Einstein's field equation in the presence of cosmological constant. Let L_ξ be the Lie derivative operator along the vector field ξ generating the symmetry. The matter collineation defined by $(L_\xi T)(U, V) = 0$ represents the symmetry of energy-momentum tensor T.

In view of Eqs. (22), (2) takes the form

$$(\lambda - \frac{r}{4})g(X, Y) = kT(X, Y). \tag{24}$$

If ξ be a Killing vector field on the space-time with W_6-flat curvature tensor, then

$$(L_\xi g)(X, Y) = 0. \tag{25}$$

Taking the Lie derivative of equation (24) along ξ, we obtain

$$(\lambda - \frac{r}{4})(L_\xi g)(X, Y) = k(L_\xi T)(X, Y). \tag{26}$$

In virtue of equation (25), Eq. (26) shows that $(L_\xi T)(X, Y) = 0$, which shows that the space-time admits matter collineation.

Conversely, If $(L_\xi T)(X, Y) = 0$, it follows that from Eq. (26), that ξ is Killing vector field. Hence, we can state as follows:

Theorem 4.4 *If a four-dimensional relativistic space-time having Einstein's field equation in the presence of cosmological constant W_6-flat curvature tensor, then the space-time satisfies matter collineation along a vector field ξ if and only if ξ is a Killing vector field.*

Next, Let us assume that the vector field ξ is a conformal Killing vector field, then we obatin

$$(L_\xi g)(X, Y) = 2\phi g(X, Y), \tag{27}$$

where ϕ is scalar, which in view of Eq. (26), gives

$$(\lambda - \frac{r}{4})2\phi g(X, Y) = k(L_\xi T)(X, Y). \tag{28}$$

Using Eq. (24) in Eq. (28), we obtain

$$(L_\xi T)(X, Y) = 2\phi T(X, Y). \tag{29}$$

From above equation, we see that the energy-momentum tensor has Lie inheritance property along ξ.

Conversely, if Eq. (29) holds, then it follows that Eq. (27) holds, i.e., the vector field ξ is a conformal Killing vector field. Thus, we have a theorem as follows:

Theorem 4.5 *In a four-dimensional relativistic space-time having Einstein's field equation in the presence of cosmological constant W_6-flat curvature tensor, a vector field ξ is conformal Killing vector field if and only if the energy-momentum tensor T has a symmetry inheritance property along ξ.*

Taking covariant derivative of equation (24), we obtain

$$(\nabla_Z T)(X, Y) = \frac{dr(Z)}{4k} g(X, Y). \tag{30}$$

Since r is constant in a W_6-curvature flat space-time, we have

$$dr(Z) = 0, \forall Z. \tag{31}$$

Using Eq. (31) in Eq. (30), we obtain

$$(\nabla_Z T)(X, Y) = 0. \tag{32}$$

Thus, we have the result as follows:

Theorem 4.6 *In a four-dimensional relativistic W_6-flat space-time having Einstein's field equation in presence of cosmological constant, the energy-momentum tensor T is covariant constant.*

Again, we consider a perfect fluid space-time with W_6-flat curvature tensor having Einstein's field equation in the absence of cosmological constant. The energy-momentum tensor T of a perfect fluid is given by [12]

$$T(X, Y) = (\sigma + \rho)A(X)A(Y) + \rho g(X, Y), \tag{33}$$

where ρ is the isotropic pressure, σ is the energy density and A is the a non-zero one-form such that $g(X, v) = A(X)$, for all X, v being the velocity vector field of the flow, i.e., $g(v, v) = -1$.

Using Eqs. (22) and (33) in Eq. (3), we get

$$-\left(\frac{r}{4} + k\rho\right)g(X, Y) = k(\rho + \sigma)A(X)A(Y), \tag{34}$$

which on contraction gives

$$r = k(\sigma - 3\rho). \tag{35}$$

Now, taking $X = Y = v$ in Eq. (34) and using $g(v, v) = -1$, we obtain

$$r = 4k\sigma. \tag{36}$$

From Eqs. (35) and (36), we have

$$\sigma + \rho = 0. \tag{37}$$

which gives that the perfect fluid behave as a cosmological constant. Thus, in view of Eqs. (37), (33) reduces to

$$T(X, Y) = \rho g(X, Y). \tag{38}$$

For a W_6-flat space-time, the scalar curvature is constant. Thus, σ is constant. Consequently, ρ is constant. Therefore, the covariant derivative of equation (38), we obtain

$$(\nabla_U T)(X, Y) = 0, \tag{39}$$

which shows that the energy-momentum tensor is covariantly constant. Thus, we have theorem as follows:

Theorem 4.7 *In a four-dimensional relativistic perfect fluid W_6-flat space-time following Einstein's field equation in the absence of cosmological constant, $\sigma + \rho = 0$ and the isotropic pressure and energy density are constants. Moreover, energy-momentum tensor is covariantly constant.*

Since the energy-momentum tensor is divergence free in a space-time, therefore, from Eq. (38), ρ is constant and consequently, from equation (37), σ is also constant. This implies

$$U\sigma = -(\sigma + \rho) div V \tag{40}$$

and

$$(\sigma + \rho)\nabla_V V = -grad\rho - (V_\rho)V. \tag{41}$$

Equations (40) and (41) gives as $div V = 0$ (expansion scalar vanishes) and $\nabla_V V = 0$ (acceleration vector vanishes) as both pressure ρ energy density σ are constant. Thus, we have

Theorem 4.8 *In a four-dimensional relativistic perfect fluid space-time with flat W_6-curvature tensor having Einstein's field equation in the absence of cosmological constant, $div V = 0$, $\nabla_V V = 0$ and $\sigma + \rho = 0$.*

Taking the frame field after contraction over X and Y of Eq. (3),we obtain

$$r = -kt, \tag{42}$$

where t is $tr(T)$.
Therefore, Eq. (3) can be written as

$$S(X, Y) = k[T(X, Y) - \frac{t}{2}g(X, Y)]. \tag{43}$$

Einstein's field equation in the absence of cosmological constant for a purely electromagnetic distribution takes the form [12]

$$S(X, Y) = kT(X, Y). \tag{44}$$

From Eqs. (43) and (44), we obtained $t = 0$. Thus, from Eq. (42), we get $r = 0$. Therefore, from Eq. (23), we obtain $R(U, X, Y, Z) = 0$, which shows that the space is flat. Thus, we arrive at

Theorem 4.9 *A a four-dimensional relativistic W_6-flat space-time having Einstein's field equation in the absence of cosmological constant for a purely electromagnetic distribution is an Euclidean space.*

References

1. Chaki, M.C., Roy, S.: Space-time admitting some geometric structure. Int. J. Theory Phys. **35**, 1027–1032 (2016)
2. De, U.C., Mallick, S.: Space-time admitting M-projective curvature tensor. Bulg. J. Phys. **39**, 331–338 (2012)
3. De, U.C., Mallick, S.: Space-time admitting W_2-curvature tensor. Int. J. Geom. Methods Mod. Phys. **11**, 04, 1450030 (2014)
4. De, U.C., Velimirovic, L.: Space-time with semi-symmetric energy-momentum tensor. Int. J. Theory Phys. **54**, 06, 1779–1783 (2015)
5. Derdzinski, A., Shen, C.L.: Proc. Lond. Math. Soc. 47 (1983)
6. Guifoyle, B.S., Nolan, B.C.: Yang's gravitational theory. Gen. Rel. Gra. **30**(3), 473–495 (1998)
7. Kramer, D., Stephani, H., Herlt, E., MacCallum, M.A.H.: Exact Solution of Einstein's Field Equations. Cambridge University Press, Cambridge (1980)
8. Mantica, C A., Molinari, L.A., De, U.C.: A Condition for a perfect-fluid space-time to be generalized Robertson-Walker space-time. J. Methods Phys. **57**, 022508 (2016)
9. Mantica, C.A., Suh, Y.J., De, U.C.: A note on generalized Robertson-Walker space-time. Int. J. Geo. Methods Mod. Phys. 298806188 (2016)
10. Mantica, C.A., Molinari, L.A.: Generalized Robertson-Walker space-time —a survey. Int. J. Geo. Methods Mod. Phys. **14**(3), 170001 (2017)
11. Mirzoyan, V.A.: Ricci semisymmetric submanifolds. Itogi Nauki i Tekhniki. Ser. Probl. Grom. **23**, 29–66 (1991)
12. O'Neill, B.: Semi-Riemannian Geometry with Applications to Relativity. Academic Press, New-York, London (1983)
13. Pokhariyal, G.P., Mishra, R.S.: Curvature tensors and their relativistic significance. Yokohama Math. J. **18**, 105–108 (1970)
14. Sanchez, M.; On the geometry of Robertson-Walker space-times: geodesics. Gen. Relativ. Gravit. **30**(6), 915–932 (1998)
15. Szabo, Z.I.: Structure theorems on Riemannian spaces satisfying $R(X, Y).R = 0$. J. Diff. Geom. **17**, 531–582 (1982)

General Characterization of at Most Twin Outer Perfect Domination Number with Colouring of Graphs

G. Mahadevan

1 Introduction

The concept of complementary perfect domination number was introduced by Paulraj Joseph et al. [7, 8, 11]. 'A set $S \subseteq V$ is called a complementary perfect dominating set, if S is a dominating set of G and the induced sub graph $\langle V - S \rangle$ has a perfect matching'. The minimum cardinality taken over all complementary perfect dominating set is called the complementary perfect domination number and is denoted by $\gamma_{cp}(G)$. Mustapha Chellali et al. in [10] first studied the concept of [1, 2] sets. A subset $S \subseteq V$ in a graph G $[j, k]$ if, $j \leq |N(V) \cap S|k$ for every vertex $v \in V - S$, $j \leq |N(v) \cap S| \leq k$ for any non-negative integer j and k. Xiaojing yang and Baoyindureng wu [14] extended the study of this parameter. A vertex set S in graph G is a [1, 2]-set if, $1 \leq |N(v) \cap S| \leq 2$ for every vertex $v \in V - S$, that is, every vertex $v \in V - S$, is adjacent to either one or two vertices in S. The minimum cardinality of a [1, 2]-set of G is denoted by $\gamma_{[1,2]}(G)$ and is called [1, 2]-domination number of G. In all the recent papers related to [1, 2] of different various domination parameters, the authors imposed some condition on the [1, 2] dominating set and nothing is mentioned about the complement. In all the above papers, discussion of the concept of perfect matching is missing. This point stimulates our brain and we impose the condition of perfect matching in the subgraph induced by V −S by maintaining [1, 2]-dominating set. Motivated by the above, in [2], Mahadevan et al., introduced the concept of at most twin outer perfect domination number of a graph. Further results are discussed in [3].

The basic Graph notations and definitions are referred from [1] and domination related definition and notations are referred from [13].

G. Mahadevan (✉)
Department of Mathematics, The Gandhigram Rural Institute-Deemed to be University
Gandhigram, Dindigul, India

P. Srivastava et al. (eds.), *Mathematical, Computational Intelligence and Engineering Approaches for Tourism, Agriculture and Healthcare*, Lecture Notes in Networks and Systems 214, https://doi.org/10.1007/978-981-16-3807-7_13

Motivation 1.1 In [12], Joseph and Arumugam obtained the relationship between chromatic number and domination number of a graph G. Based on these, later many researchers obtained excellent relationship between one colouring parameter with other domination theory parameters. Recently, investigating the relationship between chromatic number and different domination parameters is quite interesting, leading to real life situations. In [4–6, 8, 9], the authors obtained many different types of relationship for addition and equalities of colouring parameters with other domination theoretical parameters. In [2], it has been already investigated, the graphs whose sum of at most twin outer perfect domination number with chromatic number of order up to $2n - 4$. Motivated by the above, in this paper, we discover all general classes of graphs for which their sum of atop number and chromatic number is equal to 2n-5 and 2n-6 for any n > 3.

Definition 1.2 [2] A set $S \subseteq V(G)$ is called an at most twin outer perfect dominating set (atopd-set) in G, if for every node $v \in V - S, 1 \leq |N(V) \cap S| \leq 2$ *and* $\langle V - S\rangle$ has at least one perfect matching. The minimum cardinality taken over by all the at most twin outer perfect dominating set in G is called the at most twin outer perfect domination number (atopd-number) of G and is denoted by $\gamma_{atop}(G)$, and the atopd-set with minimum cardinality is also called $\gamma_{atop}-$ set.

Notations 1.3 Intensity of a node v_k is the number of edges incident to that node and is denoted by i (v_k). $K_p(s_1,s_2,\ldots\ldots s_n)$ is the graph obtained from K_p by pasting s_1 edges to any one node u_i of K_p, s_2 edges to any node u_j of K_p, for $i \neq j$, s_3 edges to any node u_k for $i \neq j \neq k, \ldots\ldots\ldots s_n$ edges to all the distinct nodes of K_p. $K_p(uP_k)$ is the graph obtained from K_p by pasting an end node of P_k to a node of K_p, u times $K_p(u((P_r,P_s)$, is the graph obtained from K_p by pasting the end nodes of P_r and P_s to the any one node u in K_p. $C_p(P_k)$ is the graph obtained from C_p by attaching an end node of P_k to a node of C_p. Wounded spider is subdivision of at most $n - 1$ edges of a star and GX H denotes the Cartesian product of two graphs.

In this section, let us now analyse the general structure classes of all graphs for which the sum of atop with chromatic number is 2n-6 for any n>3. Let G be a connected graph with $\chi(G) = n - 3$. Since $\chi(G) = n - 3$, G has maximum complete subgraph (denoted as MCS) with $n - 3$ nodes. Let V (MCS) = { $u_1, u_2, \ldots\ldots u_{n-3}$ }. Let $C = V(G) - V(MCS)$. Let the family of graphs be $\mathcal{G}_1$ ={$K_4(3P_3)$, $K_4(2P_2$, P1, 0,0), $K_3(2P_2, P_2,0)$, $K_4(P_3, P_3,0)$, $K_3(P_2, P_2, P_2)$, G_i for i = 1 to 4, P_5, Wounded Spider, $K_3(u((P_2,P_3)$, P_5, $K_3(P_3,P_2,0)$ and G_i for i = 1 to 8 as shown in Fig. 1}.

Lemma 2.1. Let G be a connected graph with $\chi(G) = n - 3$ and $\gamma_{atop} = n - 2$. If $<C> = K_3$ or $\overline{K_3}$, then $G \cong K_4(3P_3)$, $K_4(2P_2, P_1, 0,0)$, $K_3(2P_2, P_2,0)$, $K_4(P_3, P_3,0)$, $K_3(P_2, P_2, P_2)$, G_i for i = 1 to 4.

Proof Let G be a connected graph with $\chi(G) = n - 3$ and $\gamma_{atop} = n - 2$. Since $\chi(G) = n - 3$, G has MCS with $n - 3$ nodes. Let V (MCS) = { $u_1, u_2, \ldots\ldots u_{n-3}$ }. Let C = V (G) – V (MCS).

Case 1 $<C> = K_3 = (v_1, v_2, v_3)$.

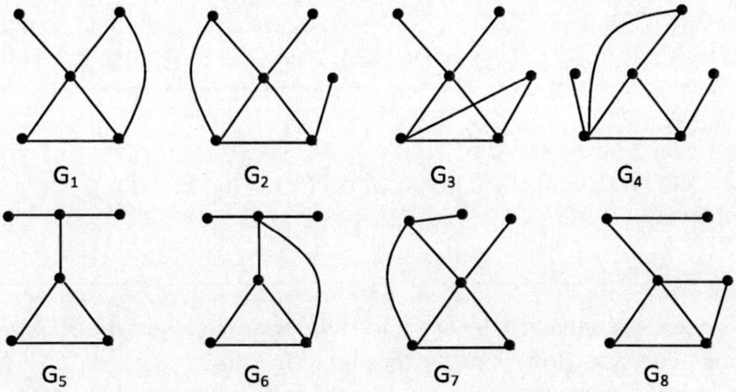

Fig. 1 Classes of graphs from G_1 to G_8

Suppose |V (MCS) |is even. Since G is connected, let the node v_1 be adjoining to at least one node $u_1 \in V(MCS)$. Then $\{v_1, u_1, u_2\}$ is an atop-set of G. Thus $\gamma_{atop} = 3$. Thus $n = 5$ so that $MCS = K_2$. In this situation, chromatic number is 2, which is a contradiction.

Subcase 1 Let all the nodes of C be adjoining to single node say u_1 in MCS. Suppose |V (MCS) |is even. In this case, $\{v_1, v_2, v_3, u_1, u_2\}$ is a γ_{atop} set of G. Thus $\gamma_{atop} = 5$. Thus $n = 7$ so that $MCS = K_4$. In this situation, $G \cong K_4 (3P_3)$. If $i(v_i) > 1$, then atop is < 5, which is a contradiction.

Suppose |V (MCS)| is odd. Then $\{v_1, v_2, v_3, u_1\}$ is a γ_{atop} set of G so that $n = 6$. Hence $MCS = K_3$. If $i(v_1) = i(v_2) = i(v_3) = 1$, then $G \cong K_3(3P_3)$. If $i(v_1) = i(v_2) = 1$ and $i(v_3) = 2$, then $G = G_1$. If we still increase, then atop < 4, which is a contradiction.

Subcase 2 Let the nodes v_1 and v_2 are adjoining to singe node say u_1, and v_3 is adjoining to u_2.

Suppose |V (MCS) |is even. In this case, $\{v_1, v_2, v_3, u_1, u_2\}$ is an atop-set of G. Thus $\gamma_{atop} = 5$. Thus $n = 7$ so that $MCS = K_4$. In this situation, if $i(v_1) = i(v_2) = i(v_3) = 1$, then $G \cong K_4(2P_2, P_1, 0,0)$. If $i(v_i) > 1$, then atop is < 5, which is a contradiction.

Suppose |V (MCS)| is odd. Then $\{v_1, v_2, v_3, u_1\}$ is a γ_{atop} set of G so that $n = 6$. Hence $MCS = K_3$. If $i(v_1) = i(v_2) = i(v_3) = 1$, then $G \cong K_3(2P_2, P_2,0)$. If $i(v_1) = 2$ and $i(v_2) = i(v_3) = 1$, then $G \cong G_2$. If $i(v_1) = i(v_2) = 1$ and $i(v_3) = 2$, then $G \cong G_3$. If we still increase, then atop < 4, which is a contradiction.

Subcase 3: Let all the nodes v_i of C are adjoining to distinct three nodes u_1, u_2, u_3 in MCS. Say v_1 is adjoining to u_1 and v_2 is adjoining to u_2 and v_3 is adjoining to u_3.

Suppose |V (MCS) |is even. In this case, $\{v_1, v_2, v_3, u_1, u_2\}$ is an atop-set of G. Thus $\gamma_{atop} = 5$. Thus $n = 7$ so that $MCS = K_4$. In this situation, $G \cong K_4 (P_3, P_3, P_3, 0)$. If $i(v_i) > 1$, then atop is 5, which is a contradiction.

Suppose |V (MCS)| is odd. Then $\{v_1, v_2, v_3, u_1\}$ is a γ_{atop} set of G so that n = 6. Hence MCS = K_3. If $i(v_1) = i(v_2) = i(v_3) = 1$, then $G \cong K_3(P_2, P_2, P_2)$. If $i(v_1) = 2$ and $i(v_3) = 1$ and $i(v_2) = 2$, then $G \cong G_4$. If we still increase, then atop < 4, which is a contradiction.

Lemma 2.2 Let G be a connected graph with $\chi(G) = n - 3$ and $\gamma_{atop} = n - 2$. If $< C > = P_3$, then $G = P_5, G_5, G_6$.

Let $<C> = P_3 = (v_1,v_2,v_3)$.

Subcase 1 Since G is connected, let any one of the nodes in $\{v_1, v_3\}$ of C is adjoining to at least one node $u_1 \in$ (MCS), say v_1 is adjoining to u_1.

Suppose |V (MCS)| is even. Then $\{v_3, u_1, u_2\}$ is a $\gamma_{atop}-$ set of G so that $\gamma_{atop} =$ 3 and hence n = 5 so that MCS = K_2. In this case, $G \cong P_5$. If we raise intensity of any one node, then chromatic number is 3, which is a contradiction.

If |V (MCS)| is odd, then $\{v_3, u_1\}$ is a γ_{atop} set of G so that $\gamma_{atop} = 2$ and hence n = 4 so that chromatic number is 1, which is a contradiction.

Sub case 2 Since G is connected, let the node v_2 of C be adjoining to at least one node $u_1 \in$ V (MCS), say u_1.

Suppose |V (MCS)| is even. Then $\{v_1, v_3, u_2\}$ is a γ_{atop} set of G so that γ_{atop} 3 and hence n = 2 so that MCS = K_2. If $i(v_1) = i(v_3) = 1$ and $i(v_2) = 3$, then $G \cong P_3(K_{1,2})$. If $i(v_1) = i(v_3) = 1$ and $i(v_2) = 4$, then $G \cong G_6$. If we raise further the intensity of the nodes, then chromatic number is 4, which is a contradiction.

If |V (MCS)| is odd, then $\{v_1, v_2, v_3, u_1\}$ is a γ_{atop} set of G so that $\gamma_{atop} = 4$ and hence n = 6 so that MCS = K_3. If $i(v_1) = i(v_3) = 1$ and $i(v_2) = 3$, then $G \cong G_5$. If $i(v_1) = i(v_3) = 1$ and $i(v_2) = 4$, then $G \cong G_6$. If we raise further the intensity of the nodes, then chromatic number is 4, which is a contradiction.

Lemma 2.3. Let G be a connected graph with $\chi(G) = n - 3$ and $\gamma_{atop} = n - 2$. If $< C > = P_2UK_1$, then, G is isomorphic to Wounded Spider, $K_3(u((P_2,P_3))$, P_5, $K_3(P_3,P_2,0)$, G_7, G_8.

Let $< S > = P_2UK_1$.

Let v_1 and v_2 be the nodes of P_2 and let the isolated node be v_3.

Since G is connected, any one of $\{v_1,v_2\}$ and v_3 in C must be adjoining to the same node say u_1 in MCS or any one of $\{v_1,v_2\}$ is adjoining to u_1 of MCS and v_3 is adjoining to u_2 of MCS.

Sub case 1 Let v_1 and v_3 be adjoining to u_1 in MCS.

Suppose |V(MCS)| is even. Then $\{v_2, v_3, u_1\}$ is a γ_{atop} set of G so that $\gamma_{atop} =$ 3 and hence n = 5 so that MCS = K_2. If $i(v_1) = 2$ and $i(v_2) = i(v_3) = 1$, then $G \cong$ Wounded Spider. If we raise intensity of any one node, then chromatic number is 3, which is a contradiction.

If |V(MCS)| is odd, then $\{v_1,v_2,v_3, u_1\}$ is an atop-set of G so that $\gamma_{atop} = 4$ and hence n = 6 and hence MCS = K_3. If $i(v_1) = 2$, $i(v_2) = i(v_3) = 1$, then $G \cong$ $K_3(u((P_2,P_3)$. If $i(v_1) = 3$, and $i(v_2) = i(v_3) = 1$, then $G \cong G_7$. If $i(v_1) > 3$, then chromatic number is 4, which is a contradiction. If $i(v_1) = i(v_2) = 2$, then atop number is less than 4, which is a contradiction. If $i(v_1) = i(v_3) = 2$ and $i(v_2) = 1$, then $G \cong G_8$.

Sub case 2 Let v_1 be adjoining to u1 in MCS and v_3 be adjoining to u_2 in MCS.

Suppose |V(MCS)| is even. Then $\{v_2, v_3, u_2\}$ is an atop-set of G so that $\gamma_{atop} = 3$ and hence n = 5 so that MCS = K_2. In this case, $G \cong P_5$. If we raise the degree, the chromatic number is 3, which is a contradiction.

Suppose |V(MCS)| is odd. Then $\{v_2, v_3, u_2, u_3\}$ is an atop-set of G so that $\gamma_{atop} = 4$ and hence n = 6 so that MCS = K_3. In this case, $G \cong K_3(P_3, P_2, 0)$. If we raise the degree, the chromatic number is 3, which is a contradiction.

Proposition 2.4 Let G be a connected graph with $\chi = n - 3$. Then $\gamma_{atop} = n - 2$, if and only if G is any one of the graph in the family $\mathcal{G}_1$

Proof Let G be a connected graph with $\chi = n - 3$. Assume Then $\gamma_{atop} = n - 2$. Since $\chi(G) = n - 3$, G has MCS with |V(MCS)| = n − 3. Let V(MCS) = $\{u_1, u_2, \ldots\ldots u_{p\text{-}3}\}$ and let $C = V - V(MCS) = \{v_1, v_2, v_3\}$. Then $<C> = K_3$ or $\overline{K_3}$ or $K_2 \cup K_1$ or P_3. The proof follows from Lemma 2.1 to 2.3. Conversely, if G is any one of the graph in the family g_1, then it can be verified that Then $\gamma_{atop} = n - 2$.

Proposition 2.5 Let G be a connected graph with $\chi = n - 1$. Then $\gamma_{atop} = n - 4$ if and only if G is isomorphic to any one graph in the Class $\mathcal{G}_2$ $\{K_5(P_2), K_6(P_2)\}$.

Proof Let G be a connected graph with $\chi = n - 1$. Assume Then $\gamma_{atop} = n - 2$. Since $\chi = n - 1$, G has a maximum complete subgraph whose cardinality is n − 1. Let V(MCS) = $\{u_1, u_2 \ldots\ldots, u_{p\text{-}1}\}$ and V −V(MCS) = $\{v_1\}$. Since G is connected, v_1 is adjoining to at least one node say $u_1 \in$ V(MCS).

Suppose |V(MCS)| is even. Then $\{v_1, u_1, u_2\}$ is an atopd-set of G. Hence $\gamma_{atop} =$ 3 which gives p = 7. Thus MCS must be K_6 and hence the graph is $K_6(P_2)$. If we increase the intensity of the pendant node, it will lead to contradiction to atop-set. Hence no other new graph exists in this case.

Suppose |V(MCS)| is odd. Then $\{v_1, u_1\}$ is an atopd-set of G. Hence $\gamma_{atop} = 2$ which gives n = 6. Thus MCS = K_5 so that $G \cong K_5(P_2)$. If we increase the intensity of the pendant node, it will lead to contradiction to atop-set. Hence no other new graph exists in this case. Converse is trivial.

Theorem 2.6 For any connected graph G with $n \geq 3$ nodes, $\gamma_{atop}(G) + (G) = 2n - 5$ if and only if $G \cong \mathcal{G}_1, \mathcal{G}_2$

Proof If G is any one of the graph stated in the theorem, it can be verified their sum of atop number and chromatic number is 2n-5 for any n > 3.

Conversely, assume that γ_{atop} (G) $+ \chi(G) = 2p - 5$. Then, the all possible structures are (i) $\gamma_{atop} = n - 2$ and $\chi = n - 3$ (ii) $\gamma_{atop} = n - 3$ and $\chi = n - 2$ (iii) $\gamma_{atop} =$

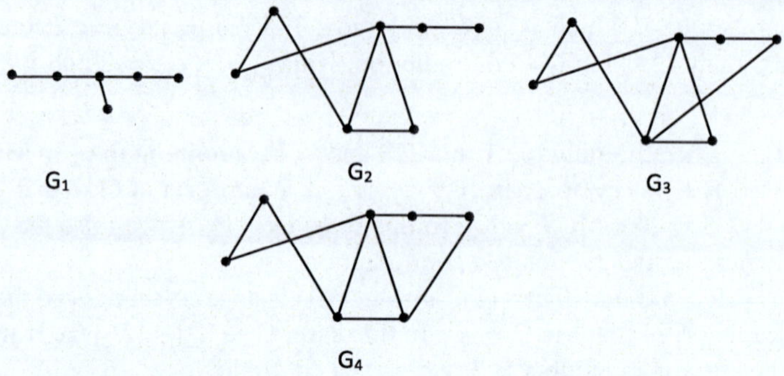

Fig. 2 Classes of graphs from G_1 to G_4

$n-4$ and $\chi = n-1$ (iv) $\gamma_{atop} = n-5$ and $\chi = n$. Consider the case for which $\gamma_{atop} = n-3$ and $\chi = n-3$. Since $\gamma_{atop} = n-3$, there exists a γ_{atop} et with n-3 nodes. In this situation, it is not possible to form perfect matching in the complement. Hence this case is excluded. In a similar manner, the case $\gamma_{atop} = n-5$ and $\chi = n$ is also excluded. For all the remaining cases, the proof follows from Propositions 2.4 and 2.5.

In this section, Let Us Now Analyse the General Structure Classes of All Graphs for Which the Sum of Atop with Chromatic Number is 2n-6 for Any N > 3.

Lemma 3.1. Let G be connected graph with $\chi(G) = n-4$, and $\gamma_{atop} = n-2$. If $< C > = K_2 \cup K_2$, or $K_3 \cup$, or $K_3(P_2)$, then G isomorphic to $C_4(P_3)$, $C_3(P_3, P_3)$, $C_3(2P_3)$ and G_i for i = 1 to 4 as in Fig. 2.

Proof **Case1** $< C > = K_2 \cup K_2 = (v_1, v_2, v_3, v_4)$. Since $\chi(G) = n-4$, G has MCS with MCS = n_4. Let V(MCS) = $\{u_1, u_2, \ldots\ldots u_{p-4}\}$ and let C = V − (MCS) = $\{v_1, v_2, v_3, v_4\}$.

Sub case 1 Suppose |V(MCS)| is even. Since G is connected, let the node v_1 and v_3 be adjoining to at least one node $u_1 \in$ V(MCS). Then $\{v_3, v_4, u_1, u_2\}$ is a γ_{atop} set of G. Thus $\gamma_{atop} = 4$. Thus n = 6 so that MCS = K_2. In this situation, chromatic number is 2, hence the graph G is isomorphic to G_1 If $i(v_1) = i(v_2) = i(v_3) = 2, i(v_4) = 1$ then G = $C_4(P_3)$.

Suppose |V(MCS)| is odd. Since G is connected, let the node v_1 and v_3 be adjoining to at least one node $u_1 \in$ V(MCS). Then $\{v_1, v_2, v_3, v_4, u_1\}$ is a atop-set of G. Thus $\gamma_{atop} = 5$. Thus n = 7 so that MCS = K_3 the graph G is isomorphic to $C_3(2P_3)$ If $i(v_1) = i(v_2) = i(v_3) = 2, i(v_4) = 1$ then $G \cong G_2$. If $i(v_1) = i(v_2) = i(v_3) = i(v_4) = 2$ then $G \cong G_3$ and G_4.

Sub Case 2 Let v_1 be adjoining to u_1 and v_4 be adjoining to any other node.

Suppose |V(MCS)| is even. Then $\{v_2, v_3, v_4, u_2\}$ is a γ_{atop} set of G. Thus $\gamma_{atop} = 4$. Thus n = 6 so that MCS = K_2. In this situation, chromatic number is 2, hence the graph G is isomorphic to P_6. If $i(v_1) = i(v_2) = i(v_3) = 2$, $i(v_4) = 1$, then $G \cong C_4(P_3)$.

Suppose |V(MCS)| is odd. Then $\{v_1, v_2, v_3, v_4, u_3\}$ is a atop-set of G. Thus $\gamma_{atop} = 5$. Thus n = 7 so that MCS = K_3 the graph G is isomorphic to $C_3(P_3, P_3)$.

Sub Case 3 Let v_1, v_2, v_3 be adjoining to u_1.

Suppose |V(MCS)| is even. Then $\{v_1, v_3, v_4, u_2\}$ is a γ_{atop} set of G. Thus $\gamma_{atop} = 4$. Thus n = 6 so that MCS = K_2. In this situation, chromatic number is 2, which is a contradiction.

Suppose |V(MCS)| is odd. Then $\{v_3, v_4, u_1\}$ is a γ_{atop} set of G. Thus $\gamma_{atop} = 3$. Thus n = 5 so that MCS = K_2. In this situation, chromatic number is 2, which is a contradiction.

Sub Case 4 Let v_1, v_3 be adjoining to u_1 and v_2 be adjoining to any other node.

Suppose |V(MCS)| is even. Then $\{v_3, v_4, u_1, u_2\}$ is a γ_{atop} set of G. Thus γ_{atop}= 4. Thus n = 6 so that MCS = K_2. In this situation, chromatic number is 2, hence the graph G is isomorphic to $C_4(P_3)$.

Suppose |V(MCS)| is odd. Then $\{v_1, v_2, v_3, v_4, u_1\}$ is a γ_{atop} set of G. Thus $\gamma_{atop} = 5$. Thus n = 7 so that MCS = K_3 then $G \cong G_2$.

Case2 $< C > = K_2 \cup K_2$ or $K_3(P_2)$.

In both the cases, by the tedious arguments, it can be verified that no new graph exists.

Lemma 3.2. Let G be connected graph with $\chi(G) = n - 4$, and $\gamma_{atop} = n - 2$. If $< C > = \overline{K_4}$, then, G is isomorphic to $K_4(K_{1,4})$, $C_3(K_{1,3})$,$K_4(K_{1,4}, K_{1,4})$, $K_4(K_{1,3}, P_2)$, $C_3(K_{1,3}, P_2)$, $K_4(P_2, P_2, P_2, P_2)$ and G_i for i = 1to 4 as in Fig. 3.

Proof Let $< C > = \overline{K_4} = (v_1, v_2, v_3, v_4)$. Since $\chi(G) = n - 4$, G has MCS with MCS = n − 4. Let V(MCS) = $\{u_1, u_2, \ldots\ldots u_{p-4}\}$ and let C = V − V(MCS) = $\{v_1, v_2, v_3, v_4\}$.

Case 1 Let all the nodes be adjoining to u_1.

Suppose |V(MCS)| is even. Then $\{v_1, v_2, v_3, v_4, u_1, u_2\}$ is a γ_{atop} set of G. Thus $\gamma_{atop} = 6$. Thus n = 8 so that MCS = K_4. Then $G \cong K_4(K_{1,4})$.

Suppose |V(MCS)| is odd. Then $\{v_1, v_2, v_3, v_4, u_1\}$ is a γ_{atop} set of G. Thus $\gamma_{atop} = 5$. Thus n = 7 so that MCS = K_3 the graph G is isomorphic to $C_3(K_{1,3})$. If $i(v_1) = 2$, $i(v_2) = i(v_3) = i(v_4) = 1$ then $G \cong G_5$.

Case 2 Let v_1, v_2 be adjoining to u_1 and v_3, v_4 be adjoining to any other node.

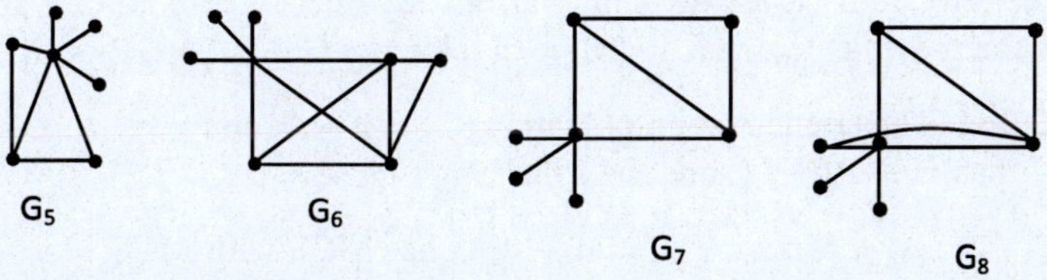

Fig. 3 Classes of graphs from G_5 to G_8

Suppose |V(MCS)| is even. Then $\{v_1, v_2, v_3, v_4, u_1, u_2\}$ is a γ_{atop} set of G. Thus $\gamma_{atop} = 6$. Thus n = 8 so that MCS = K_4. Then G ≅ $K_4(K_{1,4}, K_{1,4})$.

Suppose |V(MCS)| is odd. Then $\{v_1, v_2, v_3, v_4, u_1\}$ is a γ_{atop} set of G. Thus $\gamma_{atop} = 5$. Thus n = 7 so that MCS = K_3. This is a contradiction as it does not satisfy the γ_{atop} condition.

Case 3 Let v_1, v_2, v_3 be adjoining to v_1 and v_4 be adjoining to any other nodes.

Suppose |V(MCS)| is even. Then $\{v_1, v_2, v_3, v_4, u_1, u_2\}$ is a γ_{atop} set of G. Thus $\gamma_{atop} = 6$. Thus n = 8 so that MCS = K_4. Then G ≅ $K_4(K_{1,3}, P_2)$. If $i(v_1) = i(v_2) = i(v_3) = 1, i(v_4) = 2$ then G ≅ G_6.

Suppose |V(MCS)| is odd. Then $\{v_1, v_2, v_3, v_4, u_1\}$ is a γ_{atop} set of G. Thus $\gamma_{atop} = 5$. Thus n = 7 so that MCS = K_3. Then G ≅ $C_3(K_{1,3}, P_2)$. If $i(v_1) = i(v_2) = i(v_3) = 1$, $i(v_4) = 2$ then G ≅ G_7. If $i(v_1) = i(v_2) = 1$, $i(v_3) = i(v_4) = 2$, then G ≅ G_8.

Case 4 Let all the nodes of C be adjoining to four different nodes in MCS.

Suppose |V(MCS)| is even. Then $\{v_1, v_2, v_3, v_4, u_1, u_2\}$ is a γ_{atop} set of G. Thus $\gamma_{atop} = 6$. Thus n = 8 so that MCS = K_4. Then G ≅ $K_4(P_2, P_2, P_2, P_2)$.

No graphs exists if MCS is odd.

Lemma 3.3 Let G be connected graph with $\chi(G) = n-4$, and $\gamma_{atop} = n - 2$. If $\langle C \rangle = K_4$ or C_4 or $K_3(P_2)$, then G ≅ $C_4(P_3)$, $P_3 \times P_2$.

Proof Let $\langle C \rangle = K_4$ or C_4 or $K_3(P_2)$, then, by using various arguments for odd and even cases as per previous lemmas, it can be obtained as G ≅ $C_4(P_3)$, $P_3 \times P_2$.

Lemma 3.4 Let G be connected graph with $\chi(G) = n - 4$, and $\gamma_{atop} = n - 2$. If $\langle C \rangle = K_2 \cup \overline{K_2}$, then G is isomorphic to $P_3(K_{1,3})$, $K_3(u(2P_2, P_3))$.

Proof Let v_1 and v_2 be the nodes of K_2 and v_3 and v_4 be the isolated nodes.

Case 1 If any one node from $\{v_1,v_2\}$ and any one node from $\{v_3,v_4\}$ are adjoining to the same node u_1 of MCS.

Case 2 If any one node from $\{v_1,v_2\}$ is adjoining to u_1 in MCS and both the nodes of v_3 and v_4 are adjoining to u_2.

Case 3 If any one node from $\{v_1, v_2\}$ is adjoining to u_1 in MCS and both the nodes of v_3 and v_4 are adjoining to any two distinct nodes of MCS.

Case 4 If any one node from $\{v_1,v_2\}$ is adjoining to u_1 in MCS and node v_3 is adjoining to the same u_1 and other node v_4 is adjoining to u_2 in MCS.

In all the above four cases, by using various arguments for both odd and even cases as per the previous lemmas, it can be obtained as G ≅ $C_4(P_3)$, $P_3 \times P_2$.

Lemma 3.5 Let G be connected graph with $\chi(G) = n - 4$, and $\gamma_{atop} = n - 2$. If $\langle C \rangle = K_{1,3}$, then G ≅ $P_4(K_{1,2})$ and the following Fig. 4.

Proof $\langle C \rangle = K_{1,3}$. Since $\chi(G) = n - 4$, G has MCS with MCS = n − 4. Let V(MCS) = $\{u_1,u_2,\ldots\ldots u_{p-4}\}$ and let C = V − V(MCS) = $\{v_1, v_2, v_3, v_4\}$. Let v_1 be the root vertex and it is adjacent to u_1.

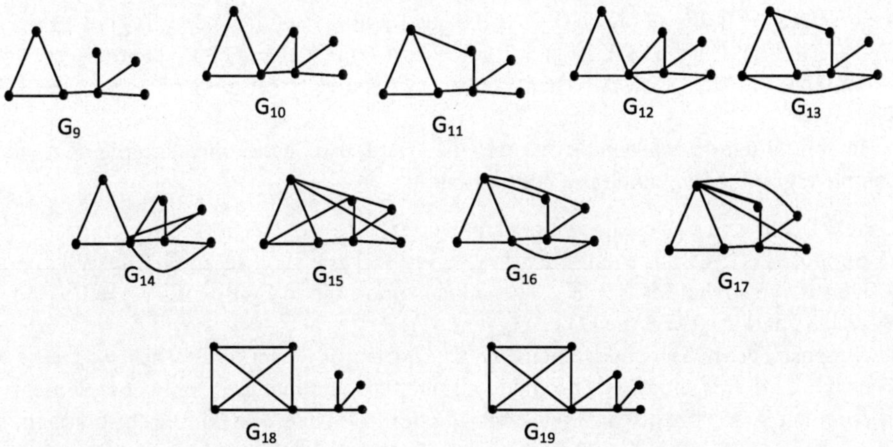

Fig. 4 Classes of graphs from G^9 to G^{19}

Suppose |V(MCS)| is odd. Then $\{v_1,v_2,v_3,v_4,u_1\}$ is an γ_{atop} set. Thus $\gamma_{atop} = 5$. Thus n = 7. Then $G \cong G_9$. If $i(v_4) = 2$, then $G \cong G_{10},G_{11}$. If $i(v_3) = i(v_4) = 2$, then $G \cong G_{12},G_{13}$. If $i(v_2) = i(v_3) = i(v_4) = 2$, then $G \cong G_{14},G_{15},G_{16}, G_{17}$.

Suppose |V(MCS)| is even. Then $\{v_1,v_2,v_3,v_4,u_1,u_2\}$ is an γ_{atop} set. Thus $\gamma_{atop} = 6$. Then n = 8. Then $G \cong G_{18},G_{19}$. If we raise the intensity of any node, atop < 6, which is contradiction. Let any one of the pendant nodes be the root vertex and it is adjacent to u_1.

Let v_4 be adjacent to u_1.

Suppose |V(MCS)| is odd. Then $\{v_2,v_3,u_1\}$ is an γ_{atop} set. Thus $\gamma_{atop} = 3$. Thus n = 5. In this case, chromatic number is 1, which is not possible.

Suppose |V(MCS)| is even. Then $\{v_2,v_3,u_1,u_2\}$ is an γ_{atop} set. Thus $\gamma_{atop} = 4$. Thus n = 6. Then MCS is K_2. In this case, the only possible graph is $P_4(K_{1,2})$.

Lemma 3.6 Let G be connected graph with $\chi(G) = n - 4$, and $\gamma_{atop} = n - 2$. If $< C > = P_4$, then $G \cong P_6, C_5(P_2), C_3(P_3,P_2)$.

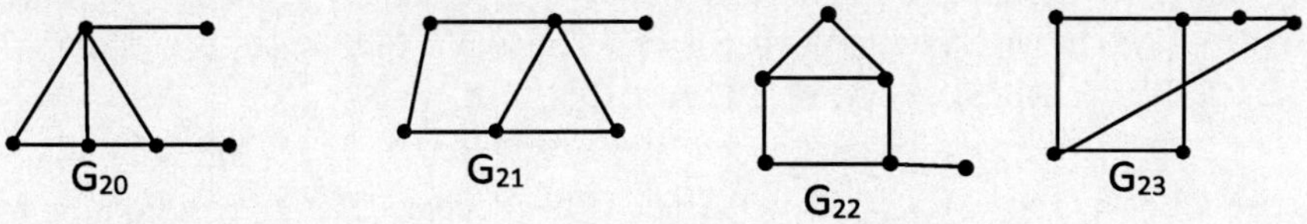

Proof Let $< C > = P_4 = (v_1,v_2,v_3,v_4)$.

Let v_1 be adjacent to u_1.

Suppose |V(MCS)| is even. Then $\{v_2, v_3, v_4, u_2\}$ is a γ_{atop} set of G. Thus $\gamma_{atop} = 4$. Thus n = 6 so that MCS = K_2. Then G is isomorphic to P_6. If the $i(v_1) = i(v_2) =$

$i(v_3) = i(v_4) = 2$, then $G \cong C_5(P_2)$. If the $i(v_1) = i(v_3) = 2, i(v_4) = 1, i(v_2) = 3$, then $G \cong C_3(P_3, P_2)$. If $i(v_1) = 2$, $i(v_2) = i(v_3) = 3$, $i(v_4) = 1$ then $G \cong G_{20}$ and G_{21}.

Suppose |V(MCS)| is odd. Then $\{v_3, v_4, u_1\}$ is a γ_{atop} set of G. Thus $\gamma_{atop} = 3$. Thus $n = 5$.

In this situation, chromatic number is 1 and that leads only complement of complete graph only, which is a contradiction.

Case 2 Let v_1 and v_4 be adjacent to same vertex.

Suppose |V(MCS)| is even. Then $\{v_2, v_3, v_4, u_2\}$ is a γ_{atop} set of G. Thus $\gamma_{atop} = 4$. Thus $n = 6$ so that $MCS = K_2$. Then G is isomorphic to $C_5(P_2)$. If $i(v_1) = i(v_3) = i(v_4) = {}_{0.2}$ and $i(v_2) = 3$ then $G = G_{22}$ and G_{23}.

Suppose |V(MCS)| is odd. Then $\{v_3, v_4, u_1\}$ is a γ_{atop} set of G. Thus $\gamma_{atop} = 3$. Thus $n = 5$. In this situation, chromatic number is 1 and that leads only complement of complete graph only, which is a contradiction. Consider the families of graphsg_3 = $\{C_4(P_3), C_3(P_3, P_3), C_3(2P_3), K_4(K_{1,4}), C_3(K_{1,3}), K_4(K_{1,4}, K_{1,4}), K_4(K_{1,3}, P_2), C_3(K_{1,3}, P_2), K_4(P_2, P_2, P_2, P_2), P_3(K_{1,3}), K_3(u(2P_2, P_3))$ and G_i for $i = 1$ to 23 as stated in Lemma 3.1 and Lemma 3.6$\}$.

Proposition 3.7 Let G be a connected graph with $\chi = n - 4$. Then $\gamma_{atop} = n - 2$ if and only if G is any one of the graph in the family $\mathcal{G}_3$.

Proof Let G be a connected graph with $\chi = n - 4$. Assume $\gamma_{atop} = n - 2$. Since $\chi(G) = n - 4$, G has MCS with $|V(MCS)| = n - 4$. Let $C = V - V(MCS) = \{v_1, v_2, v_3, v_4\}$. Then the subgraph induced by $< C >$ has many different structures. By considering all possible various subgraph induced by $< C >$, the entire proof follows from Lemmas 3.1 to 3.6. All the graphs stated in the family $\mathcal{G}_3$ satisfies the hypothesis and hence the converse.

Consider the families of graphs $\mathcal{G}_4 = \{K_4(P_3), K_5(P_3)\ K_6(u(2P_2)), K_5(u(2P_2)), K_6(u(2P_2)), K_6(P_2, P_2, 0,0,0,0,)$. $K_6(P_2, P_2, 0,0,0,0,)\ P_4(K_{1,2}), P_6, C_5(P_2), C_3(P_3, P_2), C_4(P_3), P_3P_2$ and the graphs H_i for $i = 1$ to 4 as stated in Proposition 3.8$\}$.

Proposition 3.8 Let G be a connected graph with $\chi = n - 2$. Then $\gamma_{atop} = n - 4$ if and only if G is any one of the graph in $\mathcal{G}_4$.

Proof Let G be a connected graph with $\chi = n - 2$. Assume $\gamma_{atop} = n - 4$. Since $\chi(G) = n - 2$, G has MCS with $|V(MCS)| = n - 2$. Let $V(MCS) = \{u_1, u_2, \ldots\ldots u_{p-3}\}$ and let $C = V - V(MCS) = \{v_1, v_2\}$. Then $< C > = K_2$ or $\overline{K_2}$.

Subcase 1 Let $< C > = K_2 = (v_1, v_2)$. Since G is connected, let v_1 be adjoining to u_1 in MCS. Suppose |V(MCS)| is even. Then $\{v_2, u_2\}$ is an γ_{atop} set of G. Hence $\gamma_{atop} = 2$ which gives $p = 6$. Thus MCS must be K_4. If $i(v_1) = 2$ and $i(v_2) = 1$, then $G \cong K_4(P_3)$. By raising the degrees of $v_1$and v_2, we get all different types of graphs $G \cong H_1$, where H_1 is obtained by increasing the degrees of v_1 and v_2 from $K_4(P_3)$, without changing $\chi(G) = n - 2$ and $\gamma_{atop} = n - 4$.The general Structure of H_1 is.

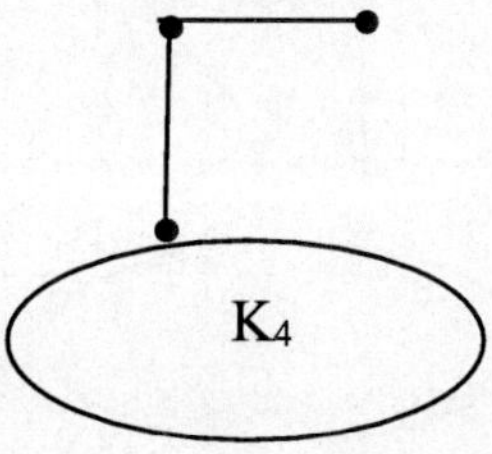

H_1

Suppose |V(MCS)| is odd. Then $\{v_2, u_2, u_3\}$ is an γ_{atop} set of G. Hence $\gamma_{atop} = 3$ which gives p = 7. Thus MCS = K_5. If $i(v_1) = 2$ and $i(v_2) = 1$ then $G \cong K_5(P_3)$. By raising the degrees of v_1 and v_2, we get all different types of graphs $G \cong H_2$, where H_2 is obtained by increasing the degrees of v_1 and v_2 from $K_5(P_3)$, without changing $\chi(G) = n - 2$ and $\gamma_{atop} = n - 4$.The general Structure of H_2 is.

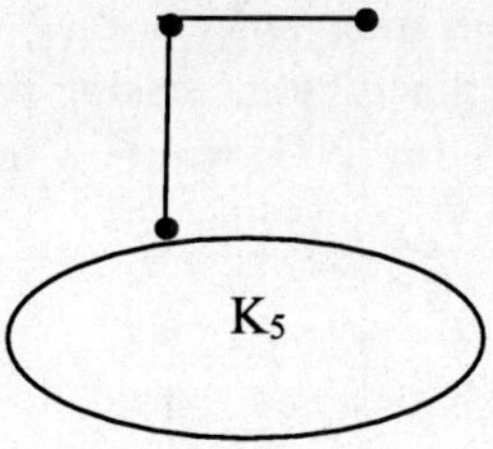

H_2

Subcase 2 Let $< C > = \overline{K_2} = (v_1,v_2)$. Since G is connected, both the nodes v_1 v_2 are adjoining to the same node u_1 in MCS or v_1 is adjoining to $u_1$in MCS and v_2 is adjoining to u_2 in MCS.

Let v_1 and v_2 be adjoining to the same node u_1 in MCS.

Suppose |V(MCS)| is even. Then $\{v_1, v_2, u_1\ u_2\}$ is an atopd-set of G. Hence γ_{atop} = 4 which gives n = 8. Thus MCS must be K_6. If $i(v_1) = 2$ and $i(v_2) = 1$, then $K_6(u(2P_2))$. If we increase the intensity, atop is less than 4, which is a contradiction.

Suppose |V(MCS)| is odd. Then $\{v_1, v_2, u_1\}$ is an atopd-set of G. Hence $\gamma_{atop} =$ 3 which gives p = 7. Thus MCS = K_5.

If $i(v_1) = 2$ and $i(v_2) = 1$ then $G \cong K_5(u(2P_2))$. By raising the degrees of v_1 and v_2, we get all different types of graphs $G = H_3$, where H_3 is obtained by increasing the degrees of v_1 and v_2 from $K_6(u(2P_2))$, without changing $\chi(G) = n - 2$ and γ_{atop} = n − 4.The general Structure of H_3 is.

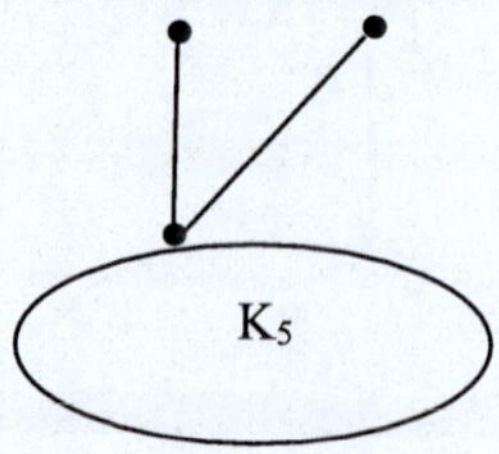

H_3

Let v_1 be adjoining to u_1 in MCS and v_2 be adjoining to u_2 in MCS.

Suppose $|V(MCS)|$ is even. Then $\{v_1, v_2, u_1\ u_2\}$ is an γ_{atop} set of G. Hence $\gamma_{atop} = 4$ which gives n = 8. Thus MCS must be K_6. If $i(v_1) = 2$ and $i(v_2) = 1$, then $K_6(P_2, P_2,0,0,0,0,)$. If we increase the intensity, atop is less than 4, which is a contradiction.

Suppose $|V(MCS)|$ is odd. Then $\{v_1, v_2, u_1\}$ is an γ_{atop} set of G. Hence $\gamma_{atop} = 3$ which gives p = 7. Thus MCS = K_5. If $i(v_1) = 2$ and $i(v_2) = 1$ then $G \cong K_5(P_2,P_2,0,0,0)$. By raising the degrees of v_1 and v_2, we get all different types of graphs $G = H_4$, where H_4 is obtained by increasing the degrees of v_1 and v_2 from $K_6(P_2, P_2,0,0,0,0,)$, without changing $\chi(G) = n - 2$ and $\gamma_{atop} = n - 4$.The general Structure of H_4 is.

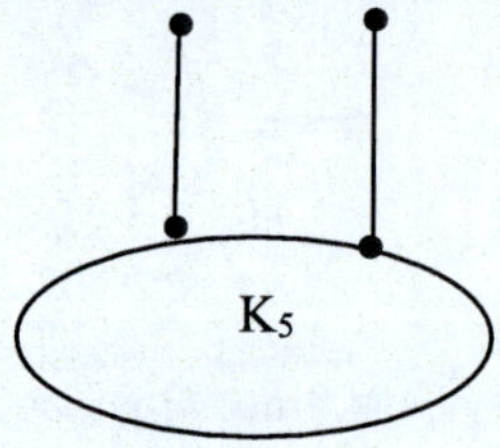

H_4

Converse is trivial.

Proposition 3.9 Let G be a connected graph with $\chi = n$. Then $\gamma_{atop} = n - 6$ if and only if G is any one of the graphs in the family $\mathcal{G}_6 = \{K_7, K_8\}$.

Proof Since $\chi = n$, G is complete graph K_n. If n is even, $\gamma_{atop} = 2$ so that $G = K_8$ and if n is odd, $\gamma_{atop} = 1$ so that $G = K_7$.

Theorem 3.10 For any connected graph G with $p \geq 3$ nodes, γ_{atop} (G) $+\chi(G) = = 2n - 6$ if and only if, G is any one of the graph in the families $\mathcal{G}_3, \mathcal{G}_4, \mathcal{G}_5$.

Proof If G is any one of the graph stated in the theorem, it can be verified with the equality. Conversely, assume that γ_{atop} (G) $+\chi(G) = = 2n - 6$. Then, all the possible structures are (i) $\gamma_{atop} = n - 2$ and $\chi = n - 4$ (ii) $\gamma_{atop} = n - 3$ and $\chi = n - 3$ (iii) $\gamma_{atop} = -4$ and $\chi = n - 2$ (iv)$\gamma_{atop} = n - 5$ and $\chi = n - 1$ (vi) γ_{atop}

n-6 and χ = n. As per the previous argument, the cases, γ_{atop} = n − 3 and χ n − 3 and γ_{atop} = n − 5 and χ= n − 1 are excluded. For all the remaining cases, the proof follows from Propositions 3.7, 3.8 and 3.9. Conversely, if G is any one of the graph in the family $\mathcal{G}_3$, $\mathcal{G}_4$, $\mathcal{G}_5$, it is clear that their sum of atop and chromatic number is 2n-6 for any n > 3.

2 Conclusion

In this paper, the author has studied all possible graphs, for colouring, with Atmost Twin outer perfect domination number 2n-6 for $n \geq 4$. The author is in the process of investigation for next higher and higher order of the above said enactment of colouring which will be reported in subsequent papers.

References

1. Harary, F.: Graph Theory. Addison Wesley Reading Mass (1972)
2. Mahadevan, G., Basira, A.I., Sivagnanam, C.: At most twin outer perfect domination and colouring of a graph. Int. J. Pure Appl. Math. **117**(13), 109–116 (2017)
3. Mahadevan, G., Basira, A.I., Sivagnanam, C.: At most twin outer perfect domination and colouring of a graph. Accepted for publication in the Int. J. Pure Appl. Math. (2018)
4. Mahadevan, G., Basira, A.I., Sivagnanam, C.: Characterization of chromatic number with complementary connected perfect domination number of a graph. Int. J. Appl. Eng. Res. **10**(92), 404–409 (2015)
5. Mahadevan, G., Basira, A.I., Sivagnanam, C.: Further results on chromatic number with complementary connected perfect domination number of a graph. Int. J. Eng. Sci. Adv. Comput. Bio Technol. **8**(1), 17–23 (2017)
6. Mahadevan, G., Avadayappan, S., Joseph, J.P., Ayisha, B., Subramanian, T.: Complementary triple connected domination number of a graph. Int. J. Adv. Appl. Discret. Math. **12**, 39–54 (2013)
7. Mahadevan, G.S., Bibi, A.M. (2013) A Complementary perfect triple connected domination number of a graph. Int. J. Eng. Res. Appl. **2**, 260–265 (2013)
8. Mahadevan, G., Basira, A.I., Sivagnanam, C.: Complementary connected perfect domination number of a graph, International Journal of pure and Applied Mathematics, Vol. 106 , pp 17–24, (2016).
9. Mahadevan, G., Basira, A.I., Sivagnanam, C.: Characterization of chromatic number with complementary connected perfect domination number of a graph. Int. Organ. Sci. Res. Dev. **10**, 404 – 410 (2016)
10. Mustapha, C. et al.: [1,2] sets in graphs. Discret. Appl. Math. **161** 2885–2893 (2013)
11. Joseph J.P., Mahadevan, G., Selvam, A.: On complementary perfect domination number of a graph. Acta Ciencia Indica **XXXI M**(2), 846–854 (2006)
12. Joseph J.P., Arumugam, S.: Domination and colouring in graphs. Int. J. Manag. Syst. **8**(1), 37–44 (1997)
13. Haynes, W., Hedetniemi, S.T., Slater, P.J.: Fundamentals of Domination in Graphs. Marcel Dekker Inc., New York (1998)
14. Yang, X., Baoyindureng, Wu.: [1, 2]-domination in graphs. Discret. Appl. Math. **175**, 79–86 (2014)

Computing Modelling of Quotient Supra Spaces

M. Sathyabama and A. Gracy

1 Introduction

Quotient topology is one of the foremost present constructions in algebraic, combinatorial and differential topology. The concept of quotient topology permits us to construct a topology on a set from the one that we are familiar with by the aid of quotient maps. Precisely, quotient topology is the collection of all sets with an open preimage under the quotient map. Quotient space is a quotient set together with quotient topology induced by the canonical projection map that maps each element to its equivalence class. The quotient mapping and a product of quotient spaces was discussed by [1] and [2]. The quotient space theory plays a role in problem solving [3]. In 2011, De Arruda Saraiva [4] defined generalized quotient topology and studied its characterization. In 1983, Mashhour et al. [5] introduced the concept of supra topological spaces by dropping a finite intersection property of topological spaces. They also introduced the notion of S^*—continuous and studied some of its properties. Binary structure and some results based on supra topological space were studied by [6] and [7]. Hierarchical network models utilize iterative algorithms for creating networks. They can reduce the complexity of the unique properties of the scale-free topology which is solved through high clustering of the nodes simultaneously. The above characteristics are widely captured in nature, from biology to language through to some social networks. Hierarchical network models are based in an iterative type of replication of the initial cluster of the network conditionally followed by certain rules.

M. Sathyabama (✉)
Department of Mathematics, Periyar University Constituent College of Arts and Science, Idappadi, Salem, India

A. Gracy
Research Scholar, Department of Mathematics, School of Mathematics, Madurai 625021, India

P. Srivastava et al. (eds.), *Mathematical, Computational Intelligence and Engineering Approaches for Tourism, Agriculture and Healthcare*, Lecture Notes in Networks and Systems 214, https://doi.org/10.1007/978-981-16-3807-7_14

The motivation behind this paper is to present the new idea of quotient supra space and using this space to represent the concept of hierarchy for the hierarchy network model. In literature there are some efficient measurement and network intrusion based on hierarchical network were done by Zhang [8] and Kao [9].

In Sect. 2 we have some basis notes regarding this paper. In Sect. 3 we provide a brief idea for constructing a supra topology on a non-empty set using supra quotient mapping and calling it quotient supra topology. Along with this, some of its basic results are discussed. Section 4, illustrates the quotient supra space by defining a quotient supra topology on a quotient set, as well as providing some theorems and suitable examples. In Sect. 5 we originate a vivid supra topology on a non-empty set X generated by arbitrary surjective functions. We also prove the existence and uniqueness theorem of vivid supra topology. Moreover, we tend to establish its universal properties. In the final section, we present a new approach to the representation of concept hierarchy based on quotient supra space.

2 Preliminaries

In this section we set out some definitions, theorems and results which are very helpful for the rest of this paper. Within this paper we have a tendency to simply let X be any non-empty set and R be an equivalence relation defined on X, and for $|X|^R$ to be the family of all equivalence classes of X. Then $|X|^R$ is termed the quotient set of X with the relation R. Now consider the map $p : X \to |X|^R$ given by p(x) $= R[X]$ for all x in X, where $R[X]$ is an equivalence class of x. Since each x in X belongs to exactly one equivalence class, p is well defined and surjective. In this case p is called the projection of X onto $|X|^R$.

Definition 2.1 ([10]) Let (X, τ) and (Y, τ ‘) be a topological space. Let f : $X \to Y$ be a onto. Then the map is said to be a quotient map provided a subset U of Y is open in Y if and only if $p^{-1}(U)$ is open in X.

Definition 2.2 ([10, 11]) Let $A \subseteq X$. Define the following.

$R[X]$={$y \in X : (x, y) \in R$}.
$R[X]$= {$y \in X : (x, y) \in R for some x \in A$}

Remark 2.3 Let X be any set and R be a equivalence relation defined on X. Suppose p is a projection map. Then for any $A \subseteq X$, we have the following.

(i) $R[X] = p^{-1}(p(A))$.
(ii) $R[X] = \bigcup\{D : D \in |X|^R and D \bigcap A \neq \phi\}$.

Definition 2.4 ([5]) Let X be a non-empty set. A subclass $\mu \subseteq \mathcal{P}(X)$ where $\mathcal{P}$(X) is the power set of X is called supra topology on X if $X \in \mu$ and μ are closed under arbitrary union. The space (X, μ) is called supra topological space. The members of μ are called supra open sets. The complement of a supra open set is called the supra closed set.

Definition 2.5 ([5]) Let (X, τ) be a topological space and μ be a supra topology on X. We call μ as a supra topology associated with τ if $\tau \subseteq \mu$.

Many results and notions of topological spaces can be defined in supra topological spaces in analogy with topological spaces.

Definition 2.6 ([5]) Let (X, τ) and $(Y, \tau^{'})$ be two topological spaces and μ is supra topology associated with τ. If $\mathfrak{f} : X \to Y$ is S -continuous if the inverse image of each open set in Y is μ supra open in X.

Definition 2.7 ([5]) Let (X, τ) and $(Y, \tau^{'})$ be two topological spaces and μ and $\mu^{'}$ are their corresponding supra topologies. If $\mathfrak{f} : X \to Y$ is S^* -continuous if the inverse image of each supra open set in Y is supra open in X.

Remark 2.8 ([5]) Let (X, τ) and $(Y, \tau^{'})$ be two topological spaces and μ and μ' are their corresponding supra topologies and $\mathfrak{f} : X \to Y$,

(i) If $\mathfrak{f}$ is continuous then f is S-continuous.
(ii) If $\mathfrak{f}$ is S^*-continuous then f is S-continuous.

We note that the concepts of continuous and S^*-continuous are independent.

Definition 2.9 ([12]) Let (X, τ) and $(Y, \tau^{'})$ be two topological spaces and μ and $\mu^{'}$ are their corresponding supra topologies. If $\mathfrak{f} : X \to Y$ is S^*-open if the image of each S^*-open set of X is supra open in Y.

Theorem 2.10 Composition of two S^*-continuous functions is S^*-continuous.

Definition 2.11 ([12]) Let (X, τ) and $(Y, \tau^{'})$ be two topological spaces and μ and $\mu^{'}$ be supra topologies associated with τ and $\tau^{'}$ respectively. Let $\mathfrak{f} : X \to Y$ be a surjection map. Then f is said to be a supra quotient provided that a subset S is supra open in Y if and only if $\mathfrak{f}^{-1}(S)$ is supra open in$X$.

3 Quotient Supra Topology

Our principle point of this area is to present the new idea of quotient supra topology. We define supra topology on a non-empty set with the help of supra quotient mappings. Then some of its basic results are discussed.

Definition 3.1 Let X be a topological space with topology τ and μ be a supra topology associated with τ and Y be any non-empty set. Suppose $\mathfrak{f} : X \to Y$ be any surjective map. Then $\mu_{\mathfrak{f}} = \{A \subseteq Y : \mathfrak{f}^{-1}(A) is supra open in X\}$ is a supra topology on Y relative to which $\mathfrak{f}$ is a supra quotient map and called $\mu_{\mathfrak{f}}$ as Quotient Supra topology.

Remark 3.2 Let μ be a supra topology associated with the topological space (X, τ) and consider Y as any non-empty set. Suppose$\mathfrak{f}$: $X \to Y$ be any surjective map and $\mu_{\mathfrak{f}}$ be a collection of all subsets A of Y such that $\mathfrak{f}^{-1}(A)$ is supra open in X. Thus we have $\mu_{\mathfrak{f}} = \{A \subseteq Y : \mathfrak{f}^{-1}(A) is supra open in X\}$. We have to show that a collection $\mu_{\mathfrak{f}}$ forms a supra topology on Y. For,

($\mathcal{P}1$) Since $\mathfrak{f}^{-1}(\phi) = \phi$ and $\mathfrak{f}^{-1}(Y) = X$, ϕ and Y are the elements of $\mu_{\mathfrak{f}}$.

($\mathcal{P}$ 2) Let $\{S_i\}_{i \in I}$ be a arbitrary collection in $\mu_{\mathfrak{f}}$ for some indexed family I. Since $S_i \in \mu_{\mathfrak{f}}$ for each i, $\mathfrak{f}^{-1}(S_i) \in \mu$. Then $\cup_{i \in I} \mathfrak{f}^{-1}(S_i) \in \mu$. Which implies that $\mathfrak{f}^{-1}(\bigcup_{i \in I} S_i) \in \mu$. That is $\mathfrak{f}^{-1}(\bigcup_{i \in I} S_i)$ is supra open in X. Hence $\bigcup_{i \in I} S_i$ is in $\mu_{\mathfrak{f}}$. Thus the collection $\mu_{\mathfrak{f}}$ forms a supra topology on Y.

Example 3.3 Let $X = \{\alpha, \beta, \gamma, \delta\}$ be topological space and the topology is given by $\tau = \{\phi, X, \{\gamma\}, \{\alpha, \gamma\}\}$ and $Y = \{a, b, c\}$ and $\mu = \{\phi, X, \{\gamma\}, \{\alpha, \gamma\}, \{\alpha, \beta\}, \{\alpha, \beta, \gamma\}\}$. Define $\mathfrak{f}$: X $\to$ Y by $\mathfrak{f}(\alpha) = \mathfrak{f}(\beta) = a$, $\mathfrak{f}(\gamma) = b$, $\mathfrak{f}(\delta) = c$. Thus we have $\mu_{\mathfrak{f}} = \{\phi, Y, \{a\}, \{b\}, \{a, b\}\}$ is a quotient supra topology on Y.

Example 3.4 Let us consider the topological space X = {a,b,c,d,e} with topology τ = {ϕ,X,{a,e}} and Y = {p,q,r}, μ = {ϕ,X,{d,e},{a,e},{c,d},{a,d,e},{c,d,e},{a,c,d,e}}. Define $\mathfrak{f} : X \to Y$ by $\mathfrak{f}(a) = \mathfrak{f}(d) = r$, $\mathfrak{f}(b) = \mathfrak{f}(e) = p$, $\mathfrak{f}(c) = q$. Then $\mu_{\mathfrak{f}} = \{\phi, Y, \{p\}\}$ is a quotient supra topology on Y.

Lemma 3.5 The quotient supra topology $\mu_{\mathfrak{f}}$ is the finest supra topology on Y for which the surjective map $\mathfrak{f} : X \to Y$ is S^*-continuous.

Proof Let μ' be any other topology on Y such that the surjective map $\mathfrak{f} : X \to Y$ is S^*-continuous. Let $S \in \mu'$. Since μ' is S^*-continuous, $\mathfrak{f}^{-1}(S)$ is supra open in X. By Definition 3.1, $S \in \mu_{\mathfrak{f}}$. Hence $\mu' \subseteq \mu_{\mathfrak{f}}$. Thus $\mu_{\mathfrak{f}}$ is the finest supra topology on Y.

Theorem 3.6 Let (X,τ) and (Y, τ') be two topological spaces and μ, μ' are their corresponding supra topologies. If $\mathfrak{f} : X \to Y$ is S^*-continuous S^*-open surjection, then $\mu' = \mu_{\mathfrak{f}}$, the quotient supra topology on Y.

Proof By lemma 3.5, $\mu' \subseteq \mu_{\mathfrak{f}}$.We have to prove that $\mu_{\mathfrak{f}} \subseteq \mu'$. For this let $A \in \mu_{\mathfrak{f}}$. Then $\mathfrak{f}^{-1}(A)$ is supra open in X. Since $\mathfrak{f}$ is S^*-open, $\mathfrak{f}(\mathfrak{f}^{-1}(A))$ is μ' supra open in Y. Also $\mathfrak{f}$ is a surjective function $\mathfrak{f}(\mathfrak{f}^{-1}(A)) = A$ is μ' supra open in Y. Hence we have $\mu_{\mathfrak{f}} \subseteq \mu'$. Thus $\mu' = \mu_{\mathfrak{f}}$.

Corollary 3.7 Let (X, τ) and (Y, τ') be two topological spaces and μ and μ' are their corresponding supra topologies. If $\mathfrak{f} : X \to Y$ is S^*-continuous S^*-closed surjective map, then $\mathfrak{f}$ is an supra quotient map.

Theorem 3.8 Let X, Y and Z be supra topological spaces. Let $\mathfrak{f} : X \to Y$ be a S^*-continuous surjective map and Y have a quotient supra topology. Then the function g: Y $\to$ Z is S^*-continuous if and only if $go\mathfrak{f} : X \to Z$ is S^*-continuous.

Proof The S^*-continuity of f ang g implies that $gof : X \to Z$ is S^*-continuous. Conversely assuming that $gof : X \to Z$ and f are S^*-continuous functions. we are going to show that g is S^*-continuous. For this let A be a supra open set in Z. Since gof is S^*-continuous, $(gof)^{-1}(A)$ is supra open in X. which implies $f^{-1}(g^{-1}(A))$ is supra open in X. Since f is a supra quotient map, $g^{-1}(A)$ is supra open in Y. Thus we have g as an S^*-continuous map. Hence the theorem.

4 Quotient Supra Spaces

Now we will build up a supra topology on a supra topology on a quotient set $|X|^R$ with the assistance of quotient supra topology. Additionally, we will prove a quotient supra theorem.

Definition 4.1 Let X be a supra topological space and R be a equivalence relation on X. Now consider a map p: X→ $|X|^R$ given by $R[X]$, that is p is a projection map of X into $|X|^R$. Then there exists a quotient supra topology on $|X|^R$ induced by p, and the set $|X|^R$ under this supra topology is called the quotient supra space of X by the equivalence relation R.

Example 4.2 Let $X=\{\alpha, \beta, \gamma, \delta, \eta\}$ be topological space and the topology of X is $\tau = \{\phi, X, \{\alpha\}, \{\alpha, \eta\}\}$ $\mu = \{\phi, X, \{\alpha\}, \{\beta, \gamma\}, \{\alpha, \eta\}, \{\alpha, \beta, \gamma\}, \{\beta, \gamma, \eta\}, \{\alpha, \beta, \gamma, \eta\}\}$ and . Suppose $|X|^R=\{\{\beta\}, \{\gamma, \delta\}, \{\alpha, \eta\}\}$. Define $f{:}X \to |X|^R$ by $f(\alpha) = f(\eta) = \{\alpha, \eta\}$, $f(\beta) = \{\beta\}$, $f(\gamma) = f(\delta) = \{\gamma, \delta\}$, as a canonical projection of X onto $|X|^R$. Then the quotient supra topology induced on $|X|^R$ by f is $\{\varnothing, \{\alpha, \eta\}, |X|^R\}$

Theorem 4.3 Let $|X|^R$ be a quotient supra space of a supra topological space X with the projection p: X→ $|X|^R$. For A ⊆ X, let $R[X]$ and $S[A] = \bigcup\{D \in |X|^R : D \subseteq A\}$. Then the following statements are equivalent:

($\mathcal{A}$1) p is an S*-open map.
($\mathcal{A}$2) $R[A]$ is supra open for every supra open subset A of X.
($\mathcal{A}$3) S[A] is supra closed for every supra closed subset A of X.

Proof ***mathcal*A1)implies** *boldsymbol*($\mathcal{A}$2):

Let p: X→ $|X|^R$ be an S*-open map. Suppose A be supra open in X. By hypothesis, p(A) is supra open in$|X|^R$. Since p is S*-continuous, p^{-1} (p(A)) is supra open in X. By remark 2.3, R[A] is supra open.

(**$\mathcal{A}$2**) implies (**$\mathcal{A}$1**):

Suppose that $R[X]$ is supra open for every supra open subset A of X. We have to prove p is S*-open. Let S be a supra open subset of X. Then R[S] is supra open. By remark 2.3, $p^{-1}(p(S))$ is supra open in X. Then p(S) is supra open in $|X|^R$ since $|X|^R$ is a supra quotient space of X. Thus p is S*-open.

($\mathcal{A}2$) implies ($\mathcal{A}$ 3):

Let $R[X]$ be supra open for every supra open subset A of X. We have to show that S[A] is a supra closed subset for every supra closed subset A of X. For each A $\subseteq$ X, $R[X]$= $\bigcup$\{D:D$\in|X|^R$ and D$\bigcap$(X-A) $\neq \phi$\} and so$S[A] = \bigcup\{D \in |X|^R : D \subseteq A\}$ which equals to$X - R[X - A]$. Now let A be a supra closed subset of X. Then X-A is supra open in X. By hypothesis $R[X]$ is supra open. Which implies X-$R[X - A]$ is supra closed. Therefore we get S[A] is supra closed for every supra closed subset A of X.

($\mathcal{A}3$) implies ($\mathcal{A}2$):

Let S[A] be supra closed for every supra closed subset A of X. Let B be a supra open subset of X. Then X-B is a supra closed subset of X. Which implies S[X-B] is supra closed using hypothesis. which implies that X-$R[X - (X - B)]$ is supra closed. That is, $R[B]$ is supra open.

Hence we have proved the theorem.

Corollary 4.4 Let $|X|^R$ be a quotient supra space of a supra topological space X with the projection p: X $\rightarrow$ $|X|^R$. Then the following statements are equivalent:

($\mathcal{C}1$) p is an S^*-closed map.

($\mathcal{C}2$) $R[A]$ is supra closed for every supra closed subset A of X.

($\mathcal{C}$ 3) S[A] is supra open for every supra open subset A of X.

Theorem 4.5. (Quotient supra theorem) Let (X, τ) and (Y,τ') be two topological spaces and μ and μ' be supra topologies associated with τ and τ ' respectively. Let R and R' be equivalence relations defined on X and Y respectively. If $\mathfrak{f} : X \rightarrow Y$ is S^* -continuous and a relation preserving map, then the function $\mathfrak{f}' : |X|^R \rightarrow |Y|^{R'}$ given by $\mathfrak{f}'(R[X]) = R'[\mathfrak{f}(x)]$ for every x in X is S^*-continuous.

Proof Let $\mathfrak{f} : X \rightarrow Y$ is S^*-continuous and $|X|^R$ and $|Y|^{R'}$ be quotient supra spaces of X and Y respectively. Also $p : X \rightarrow |X|^R$ and $p' : Y \rightarrow |Y|^{R'}$ are their projection maps. We have $\mathfrak{f}'(p(x)) = \mathfrak{f}'(R[X]) = R'[\mathfrak{f}(x)]$ for every x in X. Also we have $p'(\mathfrak{f}(x)) = p'(\mathfrak{f}(x)) = R'[\mathfrak{f}(x)]$ for every x in X. Hence we have $\mathfrak{f}' op = p' o\mathfrak{f}$. Since $\mathfrak{f}$ and p' are S^*-continuous by Theorem 2.10, $p' o\mathfrak{f}$ is S^* -continuous. We have p is S^*-continuous surjective function and $|X|^R$ is a quotient supra space of X. By Theorem 3.8, $\mathfrak{f}'$ is S^*-continuous. Hence the theorem.

5 Vivid Supra Topology

In this section we define the vivid supra topology on a non-empty set X generated by q_i 's where q_i maps X_i onto X and $(X_i, \tau_i)_{i \in I}$ be a family of topological spaces indexed by some I and μ_i 's are supra topologies associated with τ_i 's. We also prove its existence and uniqueness theorem. Moreover, we constitute its universal property.

Definition 5.1 let X be any non-empty set and $(X_i, \tau_i)_{i \in I}$ be a family of topological spaces indexed by some I and μ_i 's are supra topologies associated with τ_i 's. Suppose $q_i : X_i \to X$ be a surjective map. Then the supra topology μ on X is called vivid supra topology generated by q_i 's if and only if.

($C1$) μ makes all q_i 's S^*-continuous.

($C2$) If $\mu^{'}$ is a supra topology on X that makes all q_i 's are S^*-continuous then $\mu^{'} \subseteq$ μ.

Example 5.2 Let $X_1 = \{p, q, r, s\}$ and $X_2 = \{a, b, c, d\}$ be any two topological spaces. Let $\tau_1 = \{\varnothing, X_1, \{p\}, \{p, q\}\}$, $\tau_2 = \{\varnothing, X_2, \{a, b, c\}\}$ be a topologies of X_1 and X_2 respectively $\mu_1 = \{\varnothing, X_1, \{p\}, \{p, q\}, \{r\}, \{p, r\}, \{p, q, r\}\}$ be a supra topology associated with $X_1 and \mu_2 = \{\varnothing, X\_\{2\}, \{a\}, \{b, c\}, \{a, b\}, \{a, b, c\}\}$ be supra topologies associated with τ_1 and τ_2 respectively. Take $X = \{\alpha, \beta, \gamma, \delta\}$. Consider the function $q_1 : X_1 \to X$ is defined by $q_1(p) = \alpha; q_1(p) = \beta; q_1(r) = \delta; q_1(s) = \gamma$. Also the map $q_2 : X_2 \to X$ is defined by $q_2(a) = \beta; q_2(b) = \alpha; q_2(c) = \delta; q_2(d) = \gamma$. Then the inverse image of $\{\alpha, \beta\}$ under q_1 is $\{p, q\} \in \mu_1$ and the inverse image of {α,β} under q_2 is $\{a, b\} \in \mu_2$. Therefore {α,β} is in the vivid supra topology μ on X. Likewise {α,δ} and {α,β,δ} are in μ. On the other hand, the inverse image of {γ,δ} under q_1 is {r,s} which is not supra open in X_1. Thus {γ,δ} is not a supra open set in X. Hence μ = { Ø, X, {α,β}, {α,δ}, {α,β,δ}} is the vivid supra topology on X.

Theorem 5.3 (Existence and Uniqueness theorem) let X be any non-empty set and $(X_i, \tau_i)_{i \in I}$ be a family of topological spaces indexed by some I and μ_i 's are supra topologies associated with τ_i 's. Suppose $q_i : X_i \to X$ be a surjective map. Then there is a supra topology μ on X which is finest with respect to q_i 's. Moreover, this topology is unique and is given by $\mu = \{A \subseteq X : q_i^{-1}(A) \in \mu_i forall i \in I\}$.

Proof It is clear that the empty set and X are in μ also pre images are compatible with arbitrary union. Hence μ is indeed a supra topology on X. Now let A ∈ μ, by d efinition $q_i(A)$ is supra open in X_i for all i, so that indeed μ makes all q_i's continuous. If $\mu^{'}$ is another such supra topology. Let A be in $\mu^{'}$. Then for all i, we have as q_i continuous that $q_i^{-1}(A)$ is supra open in X_i. But this means exactly that A ∈ μ. Which implies $\mu^{'} \subseteq \mu$. Let μ and λ be two supra topologies which satisfies ($C1$) and ($C2$). Thus we have $\mu \subseteq$ λ and λ $\subseteq \mu$. Hence λ $= \mu$.

Remark 5.4 Suppose μ is a vivid supra topology generated by a single map q. Then μ coincides with quotient supra topology on X.

Theorem 5.5 (Universal Property). The vivid supra topology μ on X satisfies the universal property that a map $\mathfrak{f} : X \rightarrow Y$ into a supra topological space (Y, λ) isS^*- continuous if and only if $\mathfrak{f}oq_i : X_i \rightarrow Y$ is S^*-continuous for all i ∈ I.

Proof let μ be supra topology on X such that all the q_i 's becomes S^*-continuous and $\mathfrak{f} : X \rightarrow Y$ be S^*-continuous. Suppose A ⊆ Y is supra open in Y. Since $\mathfrak{f}$ is a S^*-continuous function, $\mathfrak{f}^{-1}(A)$ is supra open in X. By hypothesis, $q_i^{-1}(\mathfrak{f}^{-1}(A))$ is supra open in X_i for all i ∈ I. Which implies that $(\mathfrak{f}oq_i)^{-1}(A)$ is supra open in X_i for all i ∈ I. Hence we have $\mathfrak{f}oq_i : X_i \rightarrow Y$ is S^*-continuous for all i ∈ I. Conversely, assume that $\mathfrak{f}oq_i : X_i \rightarrow Y$ is S^*-continuous for all i ∈ I. Let A be a supra open set in Y. Then $(\mathfrak{f}oq_i)^{-1}(A)$ is supra open in X_i for all i ∈ I. We have $q_i^{-1}(\mathfrak{f}^{-1}(A))$ is supra open in X_i for all i ∈ I. which implies $\mathfrak{f}^{-1}(A)$ is a supra open subset of X. Therefore $\mathfrak{f}$ is S^*-continuous.

6 Computing Approach of Representation of Concept Hierarchy Based on Quotient Supra Space

Consider the triplet $(\mathcal{N}, \mathbb{K}, S)$ which describes the problem; here $\mathcal{N}$ is the domain and $\mathbb{K}$ is the attribute of the elements in the domain, $\mathbb{K} : \mathcal{N} \rightarrow I$ where I be any set and S is its structure. Now find a supra topology on $(\mathcal{N}, \mathbb{K}, S)$ with the help of Definition 3.1 and $(\mathcal{N}, \mathbb{K}, S)$ together with this supra topology is denoted by $\langle\mathcal{N}\rangle$. In various problems we get a distinctive sort of equivalence relation and this relation is accustomed to partitioning the domain $(\mathcal{N}, \mathbb{K}, S)$. In the literature there are some computing techniques used for partitioning the domain. After utilizing any of the dividing techniques on domain $\mathcal{N}$ we get the quotient set $|\mathcal{N}|$. Now using Definition 4.1, find quotient supra space $\langle|\mathcal{N}|\rangle$. Rehash this procedure until the broadest level is gained.

Different quotient set $|\mathcal{N}|$ and quotient supra topological space $\langle|\mathcal{N}|\rangle$ which correspond to the problem $(\mathcal{N}, \mathbb{K}, S)$ can be constructed to represent concept hierarchies. Thus we can represent concept hierarchies by quotient supra space.

7 Conclusion

In this paper, we originated the new ideas of quotient supra topology and vivid supra topology on a non-empty set, together with investigating their behaviors and applied quotient supra concept to represent hierarchy. Using quotient supra topology, we developed the quotient supra space. We illustrated that the vivid supra topology generated by a single map becomes quotient supra topology. In the final section we proposed the new approach to representing the hierarchy; however, we just propose the representation model to depict the idea's pecking order. In the future we will set up its pragmatic application. Also, our work can be further extended to the generalization of quotient supra topology.

Acknowledgements We are grateful to Professor M. Lellis Thivagar for inspiring discussions and valuable advice regarding this paper.

References

1. Himmelberg, C.J.: On the product of quotient space. Am. Math. Mon. **72**(10), 1103–1106 (1965)
2. Lellis Thivagar, M.: A note on quotient mappings. Bull. Malays. Math. Soc. (2) **14**(1), 21–30 (1991)
3. Zhang, L., Zhang, B.: The quotient space theory of problem solving. Fundam. Informatcae **59**, 287–298 (2004)
4. De Arruda Saraiva, L E.: Generalized quotient topologies. Acta Math. Hungar. **132** (1–2), 168–173 (2011)
5. Mashhour, A.S., Allam, A.A., Mahmoud, F.S., Khedr, F. H.: On supratopological spaces. Indian, J. Pure Appl. Math. **14**(4), 502–510 (1983)
6. Lellis Thivagar, M., Kavitha, J.: On binary structure of supra topological spaces. Bol. Soc. Paran. Mat. **35**(3), 25–37 (2017)
7. Al-shami, T.M.: Some results related to supra topological spaces. J. Adv. Stud. Topol. **7**(4), 283–294 (2016)
8. Zhang, Y., Chen, X., Jin, L., Wang, X., Guo, D.: Network intrusion detection: based on deep hierarchical network and original flow data. IEEE Access **7**, 37004–37016 (2019)
9. Kao, C.: Efficiency measurement for hierarchical network systems. Omega **51**, 121–127 (2015)
10. Munkres, J.R.: Topology, 2nd edn. Pearson India Education (2017)
11. Paulson, L.: Defining functions on equivalence classes. ACM Trans. Comput. Log. **7**(4), 658–675 (2006)
12. Kamaraj, M., Ramkumar, G., Ravi, O.: On supra quotient mappings. Int. J. Math. Arch. **3**(1), 245–252 (2012)

Design and Development of Intelligent Information System Using Hesitant Fuzzy Weighting Linguistic Term Sets for Computing with Words

Pankaj Srivastava and Rajkrishna Mondal

1 Introduction

The emergence of artificial intelligence-based approach in a day-to-day life is playing key role, although its application under imprecise [15] scenario concerned with decision-making is a challenging one.

There are a number of situations occurs in nature where imprecise, ambiguous and complex situations appear simultaneously anytime and due to that decision-making phenomena [7] becomes more complex. In order to handle such situations with appropriate decisions, many experts have different opinions and that's why a kind of dilemma face appears due to that a concrete decision does not come in picture. As it is visible in a project where many experts involved with different opinions in linguistic word [16] selections to evaluate an acceptable decision with ordinary fuzzy tools. In such situation, better results are developed with the help of hesitant fuzzy sets [10], since in nature, qualitative is everywhere rather than quantitative like uncertainty, vague, imprecise information. So, fuzzy tools is the most appropriate mathematical tools to decision on qualitative setting. The qualitative decisions in soft computing criteria sometimes gives the more appropriate result with less computational complexity using "Computing with Words" [4] technique.

The fuzzy linguistic approach [3, 6, 11, 12, 17]-based modelling have some limitations when the group of decision-makers' conflicting views appears with reference to linguistic terms either single or simple variable to provide a single linguistic term with knowledge base. Therefore, in order to overcome such limitations Hesitant Fuzzy Linguistic Term Set [8] (HFLTS) introduced. To provide a scientific platform about the rating of decisions and to express the knowledge base in a hesitant situation

P. Srivastava · R. Mondal (✉)
Department of Mathematics, Motilal Nehru National Institute of Technology Allahabad, Prayagraj 211004, India
e-mail: rajkrishna@mnnit.ac.in

P. Srivastava et al. (eds.), *Mathematical, Computational Intelligence and Engineering Approaches for Tourism, Agriculture and Healthcare*, Lecture Notes in Networks and Systems 214, https://doi.org/10.1007/978-981-16-3807-7_15

for decision-makers, we are introducing Hesitant Fuzzy Weighting Linguistic Term (HFWLTS) Set, i.e., more powerful than HFLTS in decision-making.

In this article, we propose the properties and operations of HFWLTS with design and development of an Intelligent Information System Computing with words.

We have designed this study as follows:

- **Section** 2: Explain briefly some preliminary definitions, formula operations. Here, we have first introduced the concept of HFWLTS.
- **Section** 3: Formulate the problem and model it using the HFWLTS.
- **Section** 4: It is shown that our result is more sharper than HFLTS by using HFWLTS.
- **Section** 5: Discussed the advantage of HFWLTS
- **Section** 6: Draw a proper conclusion of this study.

2 Preliminary

2.1 *Hesitant Fuzzy Set (HFS) [10]*

Let X be a subset of a universal set U. Then X is called a HFS over U which is denoted by $Hes_U(X)$ and defined by

$$Hes_U(X) = \{< u, h_X(u) >\mid u \in U\} \tag{1}$$

where $h_X(u)$ is set of values in [0, 1] to the set X.

2.2 *Hesitant Fuzzy Linguistic Term Set (HFLTS) [8]*

Let, $S = \{s_0, s_1, \ldots, s_n\}$(an order set) be a universal set for a state of system. Then a set of finite-order consecutive terms of S is called an HFLTS over S and is denoted by H_S.

It is briefly discussed through an example.

Example 1 Let us consider a linguistic set "**Age**", whose linguistic terms are s_1: $VeryYoung$, s_2: $Young$, s_3: $MiddleAged$, s_4: $Aged$, s_5: $VeryAged$, s_6: Old, s_7: $VeryOld$.

χ_1 = "*Between Middle Aged and Very Aged*" and χ_2 = "*greater than very aged*" be different linguistic expressions over S. For these two expressions, HFLTS will be $H_S^1 = H_S(\chi_1) = \{s_3$: $MiddleAged$, s_4: $Aged$, s_5: $VeryAged\}$ $H_S^2 = H_S(\chi_2) =$ $\{s_5$: $VeryAged$, s_6: Old, s_7: $VeryOld\}$

Let H_S, H_S^m, H_S^n, where $m \neq n (\in \mathbb{N})$, three different HFLTS,

1. Complement: $(H_S)^c = \{s_i \mid s_i \in S \text{ and } s_i \notin H_S\}$
2. Union: $H_S^m \cup H_S^n = \{s_i \mid s_i \in H_S^m \text{ or } s_i \in H_S^n\}$
3. Intersection: $H_S^m \cap H_S^n = \{s_i \mid s_i \in H_S^m \text{ and } s_i \in H_S^n\}$
4. Envelope: $\text{env}(H_S) = [H_{S^-}, H_{S^+}]$,
   where $H_{S^-} = min(s_i)$, $H_{S^+} = max(s_i)$, $s_i \in H_S$

2.3 *Hesitant Fuzzy Weighting Linguistic Term Set (HFWLTS)*

Let two linguistic expressions χ_1 and χ_2 over S which are different from each other. Then HFWLTS is denoted by HW_S and defined by

$$HW_S(\chi_1) = \left\{ \frac{s_m}{w_1^1}, \frac{s_{m+1}}{w_2^1}, \ldots, \frac{s_{m+l_1}}{w_{l_1-1}^1} \right\}, \; m + l_1 \leq n$$

$$HW_S(\chi_2) = \left\{ \frac{s_p}{w_1^2}, \frac{s_{p+1}}{w_2^2}, \ldots, \frac{s_{p+l_2}}{w_{l_2-1}^2} \right\}, \; p + l_2 \leq n$$

where $l_i - 1 = length(HW_S(\chi_i)), i = 1, 2$
and $m, p \in \{0, 1, \ldots, n\}$ and
$w_j^i \in [0, 1]; i = 1, 2; j = length(HW_s(\chi_i))$

Example 2 If we consider the example 1, then HEWLTS will be

$$HW_S(\chi_1) = \left\{ \frac{s_3}{0.4}, \frac{s_4}{0.2}, \frac{s_5}{0.7} \right\}$$

$$HW_S(\chi_2) = \left\{ \frac{s_5}{0.5}, \frac{s_6}{0.7}, \frac{s_7}{0.6} \right\}$$

2.4 *Context-Free Grammar [1, 2]*

Context-Free Grammar (G) is generated through a linguistic expression. It is denoted by 4-tuples (V_N, V_T, I, P)

- V_N: Non-Terminal Symbol
- V_T: Terminal Symbol
- I: Starting Symbol
- P: Production Rules

P: Production Rules	**Examples**
• Primary term ($s_i \in S$)	Aged, Old, Very Old *etc.*
• Unary Relation *(Lower than* or *Greater Than)* • Primary Term ($s_i \in S$)	Lower than Aged Greater than Very Aged *etc.*
• Binary Relation *(Between)* • Primary Term ($s_i \in S$) • Conjunction *(And)* • Primary Term ($s_i \in S$)	Between Middle Aged and Very Aged *etc.*

2.5 Linguistic Expressions into HFWLTS Transformation

Rodriguez et al. [8] introduced the concept of how to transform the linguistic expression into a HFLTS, using the same concept, here we have transformed the linguistic expression into a HFWLTS.

Θ_i, a linguistic expressions which is generated by G is transformed into HFWLTS by the function E_{GW}.

Let $\Theta = \{\theta_i \mid i \in \mathbb{N}\}$ be the set of linguistic expressions and H_{GW} be the set of HFWLTS on S. Then the function $E_G : \Theta \mapsto H_{SW}$ is defined by

$$E_{GW}(\theta_i) = \left\{ \frac{s_k}{w_1^i}, \frac{s_{k+1}}{w_2^i}, \ldots, \frac{s_{k+l}}{w_{l-1}^i} \mid s_k, \ldots, s_{k+l} \in S \right\}$$

Where E_{GW} follows these rules:

1. $E_{GW}(s_m) = \left\{ \frac{s_m}{w_m} \mid s_m \in S \right\}$
2. $E_{GW}(LessThan\ s_m) = \left\{ \frac{s_p}{w_p} \mid \forall s_p \in S \text{ and } s_p \preceq s_m \right\}$
3. $E_{GW}(GreaterThan\ s_n) = \left\{ \frac{s_p}{w_p} \mid \forall s_p \in S \text{ and } s_p \succeq s_n \right\}$
4. $E_{GW}(Between\ s_m\ And\ s_n) = \left\{ \frac{s_k}{w_k} \mid s_m \preceq s_k \preceq s_n \right\}$

Example 3 As on example 1, let "Age" is a linguistic variable and it is classified into various linguistic terms$S = \{s_i \mid i = 1 : 7\}$ which is defined in example 1.

Three different linguistic expressions $\Theta_i, i = 1, 2, 3$ have been taken here.

- $\Theta_1 = Aged$
- $\Theta_2 = LowerThan\ Aged$
- $\Theta_3 = Between\ Aged\ and\ Old$

After applying the E_{GW} on $\Theta_i, i = 1, 2, 3$, the output will be

- $E_{GW}(\Theta_1) = E_G(Aged) = \{\frac{s_4}{1}\}$
- $E_{GW}(\Theta_2) = E_G(LowerThan\ Aged) = \{\frac{s_1}{0.2}, \frac{s_2}{0.5}, \frac{s_3}{0.4}, \frac{s_4}{0.8}\}$
- $E_{GW}(\Theta_3) = E_G(Between\ Aged\ and\ Old) = \{\frac{s_4}{0.5}, \frac{s_5}{0.8}, \frac{s_6}{0.7}\}$

2.6 HFWLTS Aggregation Operators [13]

In general, type of human nature is linguistic. Naturally, these kind of expressions are two types, one is **Optimistic** and another one is **Pessimistic**. To balance these kind of linguistic expressions, we used two aggregation operators on **HFWLTS**, one is **Min-Upper** and another one is **Max-Lower**.

Let Linguistic term set $S = \{s_i \mid i \in \mathbb{N}\}$ and $HW_S = \{HW_S^i \mid i = 1, 2, \dots, n\}$ be a set of HFWLTS over S, where $HW_S^i = \{\frac{s_{k_i}}{w_1^i}, \frac{s_{k_{i+1}}}{w_2^i}, \dots, \frac{s_{k_{l(HW_S^i)}}}{w^i_{l(HW_S^i)}} \mid \forall s_{k_i} \in S; k_i, k_{i+1}, \dots, k_{l(HW_S^i)} \in \mathbb{N}\}$ where $l(HW_S^i)$ is the length of HW_S^i, $i = 1 : n$.

Min-Upper:

1. $HW_{S+}^i = \frac{s_j}{w_j}$; $s_j = Max\{s_{k_i}, s_{k_{i+1}}, \dots, s_{k_{l(H_S^i)}}\}, i = 1, 2, \dots, n$
2. $HW_{S^+_{Min}} = Min\{H_{S+}^1, H_{S+}^2, \dots, H_{S+}^n\}$ with respect to s_i

Max-Lower:

1. $HW_{S-}^i = \frac{s_j}{w_j}$; $s_j = Min\{s_{k_i}, s_{k_{i+1}}, \dots, s_{k_{l(H_S^i)}}\}, i = 1, 2, \dots, n$
2. $HW_{S^-_{Max}} = Max\{H_{S-}^1, H_{S-}^2, \dots, H_{S-}^n\}$ with respect to s_i

Example 4 Let $HW_S^1 = \{\frac{s_1}{0.2}, \frac{s_2}{0.5}, \frac{s_3}{0.6}, \frac{s_4}{0.4}\}$ and $HW_S^2 = \{\frac{s_4}{0.3}, \frac{s_5}{0.9}, \frac{s_6}{0.5}\}$ be two different HFWLTSs.

Then, $HW_{S+}^1 = \{\frac{s_4}{0.4}\}$, $HW_{S+}^2 = \{\frac{s_6}{0.5}\}$ and $HW_{S-}^1 = \{\frac{s_1}{0.2}\}$, $HW_{S-}^2 = \{\frac{s_4}{0.3}\}$
So, $HW_{S^+_{Min}} = \{\frac{s_4}{0.4}\}$ and $HW_{S^-_{Max}} = \{\frac{s_4}{0.3}\}$.

3 Intelligent Information System

3.1 Problem Formulation

Let there have n alternatives $X = \{x_1, x_2, \dots, x_n\}$ and, for each alternative, there have m criteria $C = \{c_1, c_2, \dots, c_m\}$. According to the condition, let there have set of linguistic terms $S = \{s_1, s_2, \dots, s_g\}$.

G produced through the linguistic expressions $\Theta_{ij}(x_i, c_j)$. The graphical picture has been shown in Fig. 1.

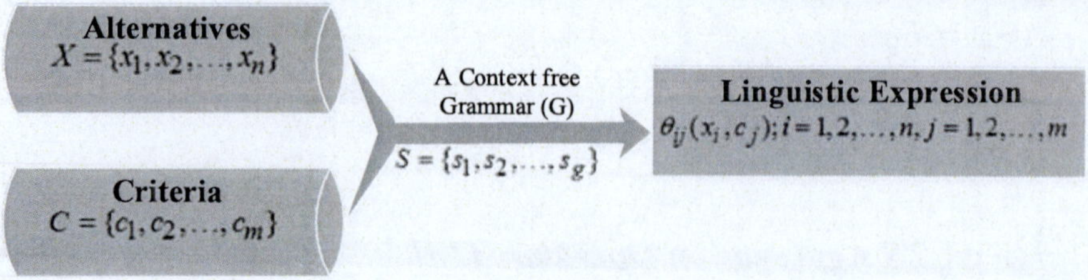

Fig. 1 Generating the linguistic expressions in diagnosis model

3.2 Model Designing

This model is phased into three parts: (1)Transformation, (2)Aggregation, (3)Exploitation.

Transformation Phase: This phase is computed into two steps:

1. According to the experts opinion G *i.e.* G produces the linguistic expressions $\Theta_{ij}(x_i, c_j)$.
2. Using the transformation function E_{GH}, each $\Theta_{ij}(x_i, c_j)$ is converted into HFWLTS.

$$E_{GW}(\Theta_{ij}(x_i, c_j)) = \{\frac{s_k}{w_k} \mid s_k \in S\}$$

Aggregation Phase: Since there have two types of experts opinion, one is *Optimistic* and another one is *Pessimistic* [9]. So, to balance it we have used aggregation operators **Min-Upper** and **Max-Lower**.

Let $E_{GW}(\Theta_{ij}(x_i, c_j)) = H^j_{SW}(x_i); i = 1 : n; j = 1 : m$.

- **Min-Upper:**

(a) $H_{SW^+}(x_i) = \{H^j_{SW^+} \mid j = 1 : m\}, i = 1 : n$
(b) $H_{SW^+_{Min}}(x_i) = Min\{H^j_{SW^+} \mid j = 1 : m\}, i = 1 : n$

- **Max-Lower:**

(a) $H_{SW^-}(x_i) = \{H^j_{SW^-} \mid j = 1 : m\}, i = 1 : n$
(b) $H_{SW^-_{Max}}(x_i) = Max\{H^j_{SW^-} \mid j = 1 : m\}, i = 1 : n$

Using the above aggregation operators, linguistic interval is created for each alternatives x_i, $H^*(x_i); i = 1 : n$

$$H^*(x_i) = [H^*_{Min}(x_i), H^*_{Max}(x_i)], i = 1 : n$$

where $H^*_{Max} = Max\{H_{SW^+_{Min}}, H_{SW^-_{Max}}\}$, $H^*_{Min} = Min\{H_{SW^+_{Min}}, H_{SW^-_{Max}}\}$

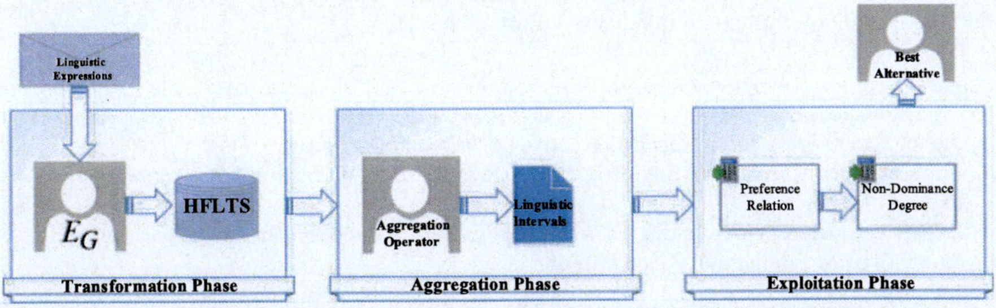

Fig. 2 Decision-making model

Exploitation Phase: Finally, the best alternative is decided in this phase. It is performed through three steps:

Step 1: Preference Relation Matrix P is built for each pair of alternatives (x_i, x_j) [7, 13].

Let $x_i = [\frac{s_{iL}}{w_{iL}}, \frac{s_{iR}}{w_{iR}}], x_j = [\frac{s_{jL}}{w_{jL}}, \frac{s_{jR}}{w_{jR}}]; i \neq j$

Where $s_{iL}, s_{iR}, s_{jL}, s_{jR} \in S$ and $w_{iL}, w_{iR}, w_{jL}, w_{jR} \in [0, 1]$, be two weighted linguistic intervals for alternatives x_i and x_j and $ind(x_i) = i$ be the indexing function.

$$P = \begin{pmatrix} - & p_{12} & \cdots & p_{1n} \\ p_{21} & - & \cdots & p_{2n} \\ \vdots & \vdots & \ddots & \vdots \\ p_{n1} & p_{n2} & \cdots & - \end{pmatrix}_{n \times n}$$

where $p_{ij} = P(x_i \succeq x_j) = \frac{Max\{0, ind(S_{iR}) - ind(S_{jL})\} - Max\{0, ind(S_{iL}) - ind(S_{jR})\}}{[ind(S_{iR}) - ind(S_{jL})] - [ind(S_{iL}) - ind(S_{jR})]}$

Step 2: Now, it is calculated the non-dominance degree of each alternative x_i such that

$$NDD_i = Min\{1 - p_{ji}^S \mid i \neq j, j = 1 : n\}, i = 1 : n,$$

where $p_{ji}^S = max\{p_{ji} - p_{ij}, 0\}$

Step 3: Finally, the best alternative is selected through the formula

$$x^{Best} = \{x_j \mid NDD_j = Max(NDD_i \mid i = 1 : n)\}$$

Step 4: If $x^{Best} = \{x_m, x_n\}$, then we have chosen best alternative using the likelihood-based comparison relation $L(w_m \succeq w_n) = Max\left\{1 - Max\left(\frac{w_{rn} - w_{lm}}{l(w_m) + l(w_n)}, 0\right), 0\right\}$, where $w_m = [w_{lm}, w_{rm}]$, $w_n = [w_{ln}, w_{rn}]$ are the corresponding interval weights of the linguistic interval of the alternatives x_m, x_n and $l(w_i) = w_{ri} - w_{li}$.

A graphical overview of each phase has been shown in Fig. 2.

Algorithm and Flow chart of the above decision-making model has been shown in Table 1 and Fig. 3, respectively.

Table 1 Algorithm for proposed intelligent information system

INPUT: 1. No. of Alternatives and for each alternative corresponding criteria. 2. Form expert system taking linguistic expressions for each alternatives corresponding to each criteria. 3. Expert confusion degree taking as weighting values of the linguistic terms for each alternative corresponding to each criteria. **OUTPUT:** Ranking of the alternatives according to realistic decision
METHODOLOGY **1) Transformation Phase:** $E_{GW}\left(\theta_{ij}(x_i, c_j)\right) \mapsto HFWLTS$ where i=1:no. of alternatives(n) and j=1:no. of criteria(m) **2) Aggregation Phase:** a. Calculate $H_{SW^+_{Min}}$ and $H_{SW^-_{Max}}$ for each alternative. b. $H^*_{Max} = Max\{H_{SW^+_{Min}}, H_{SW^-_{Max}}\}$ $H^*_{Min} = Min\{H_{SW^+_{Min}}, H_{SW^-_{Max}}\}$ c. Calculate Linguistic Interval $H^*(x_i) = [H^*_{Min}(x_i), H^*_{Max}(x_i)], i = 1, \ldots, n$ for each alternative. **3) Exploitation Phase:** a. Calculate P where $p_{ij} = P(x_i \succeq x_j) = \frac{Max\{0, ind(S_{iR}) - ind(S_{jL})\} - Max\{0, ind(S_{iL}) - ind(S_{jR})\}}{[ind(S_{iR}) - ind(S_{jL})] - [ind(S_{iL}) - ind(S_{jR})]}$ b. calculated non-dominance degree for each alternative x_i such that $NDD_i = Min\{1 - p^S_{ji} \mid i \neq j,$ $j = 1:m\}, i = 1:n$ where $p^S_{ji} = max\{p_{ji} - p_{ij}, 0\}$ c. Best Alternative=$\{x_j \mid NDD_j = Max(NDD_i \mid i = 1:n)\}$ **4)** If Best Alternative= $\{x_m, x_n\}$ then calculate $L(w_m \succeq w_n)$ and $L(w_n \succeq w_m)$ according to Max value of L decide best alternative between x_m and x_n.

4 Sharpness of the Intelligent Information System

Here, we have taken an example that has been taken in Lee and Chen [5].

Assume that we have three alternatives x_1, x_2, x_3 and for each alternative there have three criteria c_1, c_2, c_3

Let the linguistic terms set is

$S = \{s_0$: $Nothing(n)$, s_1: $VeryLow(vl)$, s_2: $Low(l)$, s_3: $Medium(m)$, s_4: $High(h)$, s_5: $VeryHigh(vh)$, s_6: $perfect(p)\}$.

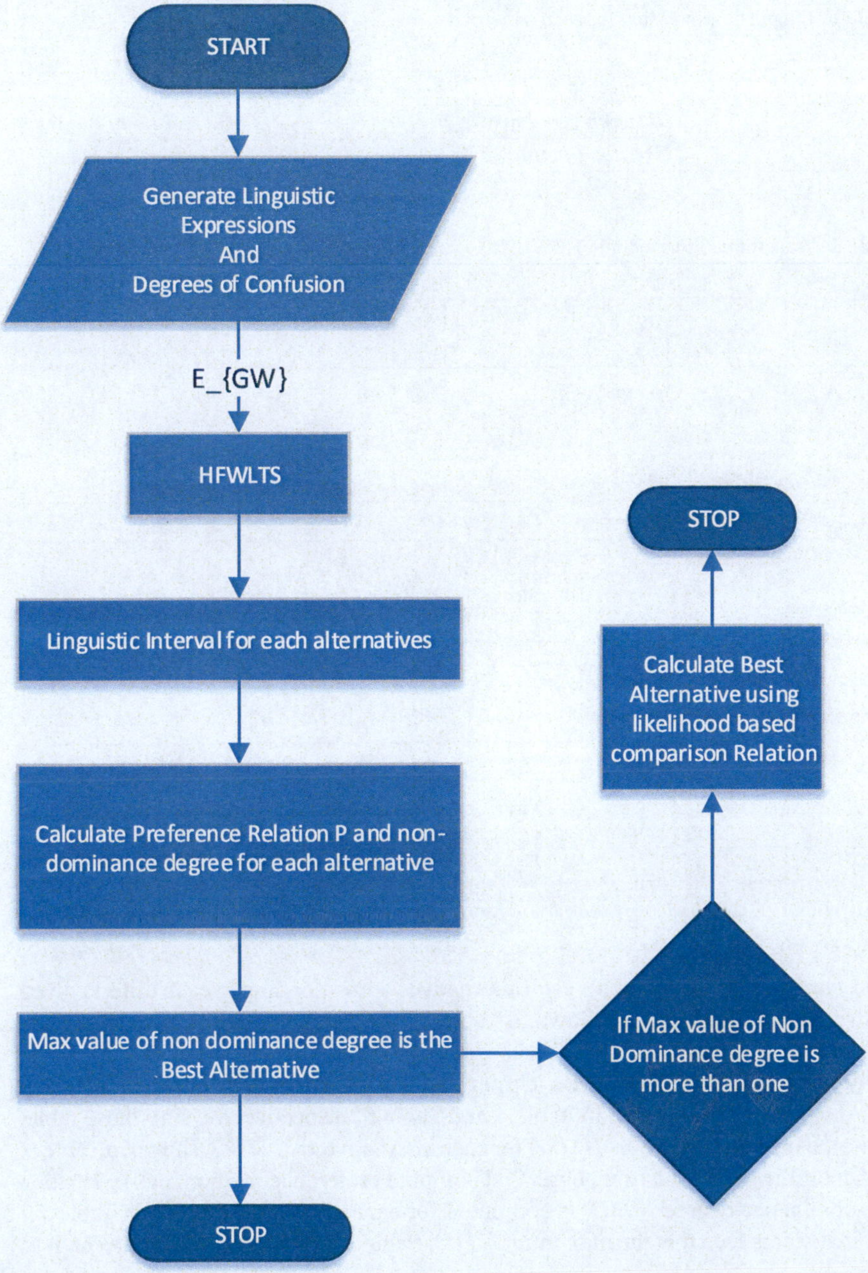

Fig. 3 Flowchart of the proposed system

Table 2 Linguistic expression for each patients

Θ	c_1	c_2	c_3
x_1	$Between\ n\ and\ l$	$Between\ h\ and\ vh$	h
x_2	$Between\ l\ and\ m$	m	$LowerThan\ l$
x_3	$Greater\ Than\ h$	$Between\ vl\ and\ l$	$Greater\ Than\ h$

Table 3 Transformed formed into HFWLTS

$H^j_{SW}(p_i)$	c_1	c_2	c_3
x_1	$\left\{\frac{n}{0.8}, \frac{vl}{0.6}, \frac{l}{0.7}\right\}$	$\left\{\frac{h}{0.6}, \frac{vh}{0.5}\right\}$	$\left\{\frac{h}{1}\right\}$
x_2	$\left\{\frac{l}{0.4}, \frac{m}{0.7}\right\}$	$\left\{\frac{m}{0.8}\right\}$	$\left\{\frac{n}{0.4}, \frac{vl}{0.5}, \frac{l}{0.2}\right\}$
x_3	$\left\{\frac{h}{0.7}, \frac{vh}{0.8}, \frac{p}{0.8}\right\}$	$\left\{\frac{vl}{0.7}, \frac{l}{0.8}\right\}$	$\left\{\frac{h}{0.6}, \frac{vh}{0.7}, \frac{p}{0.8}\right\}$

Table 4 $H_{SW^+_{Min}}$ calculation table

$H^j_{SW^+}(x_i)$	c_1	c_2	c_3	$H_{SW^+_{Min}}(x_i)$
x_1	$\left\{\frac{l}{0.7}\right\}$	$\left\{\frac{vh}{0.5}\right\}$	$\left\{\frac{h}{1}\right\}$	$\left\{\frac{l}{0.7}\right\}$
x_2	$\left\{\frac{m}{0.7}\right\}$	$\left\{\frac{m}{0.8}\right\}$	$\left\{\frac{l}{0.2}\right\}$	$\left\{\frac{l}{0.2}\right\}$
x_3	$\left\{\frac{p}{0.8}\right\}$	$\left\{\frac{l}{0.8}\right\}$	$\left\{\frac{p}{0.8}\right\}$	$\left\{\frac{l}{0.8}\right\}$

Table 5 $H_{SW^-_{Max}}$ calculation table

$H^j_{SW^-}(x_i)$	c_1	c_2	c_3	$H_{SW^-_{Max}}(x_i)$
x_1	$\left\{\frac{n}{0.8}\right\}$	$\left\{\frac{h}{0.6}\right\}$	$\left\{\frac{h}{1}\right\}$	$\left\{\frac{h}{0.6}\right\}$
x_2	$\left\{\frac{l}{0.4}\right\}$	$\left\{\frac{m}{0.8}\right\}$	$\left\{\frac{n}{0.4}\right\}$	$\left\{\frac{m}{0.8}\right\}$
x_3	$\left\{\frac{h}{0.7}\right\}$	$\left\{\frac{vl}{0.7}\right\}$	$\left\{\frac{h}{0.6}\right\}$	$\left\{\frac{h}{0.6}\right\}$

Let linguistic expression for each alternatives corresponding to each criteria given by the experts that has been shown in Table 2.

By using E_{GW} on Table 2, it is transformed into HFWLTS (shown in Table 3).Then apply aggregation operators (**Min-Upper** and **Max-Lower**) to calculate $H_{SW^+_{Min}}$ and $H_{SW^-_{Max}}$, that has been given in Table 4 and Table 5, respectively. Using these table, create the linguistic intervals $H^*(x_i)$ for each alternative x_i, $i = 1{:}3$, shown in Table 6.

According to exploitation phase, first compute preference relation matrix P^S then non-dominance degree NDD_i is computed for each alternative (shown in Table 7).

Finally, it is seen that the maximum NDD_i value is 1, which arises for alternatives x_1, x_3.

Now according to step 4, here $w_1 = [0.6, 0.7]$ and $w_3 = [0.6, 0.8]$ for x_1 and x_3, respectively. Now $L(w_1 \succeq w_3) = 0.3333$ and $L(w_3 \succeq w_1) = 0.6667$. So, x_3 is the best from x_1.

Therefore, finally, we can conclude that $x_3 > x_1 > x_2$.

Table 6 Linguistic intervals for each alternatives

	x_1	x_2	x_3
$H^*(x_i)$	$\frac{[l,h]}{[0.6,0.7]}$	$\frac{[l,m]}{[0.2,0.8]}$	$\frac{[l,h]}{[0.6,0.8]}$
In $s_i \in S$ format	$[s_2, s_4]$	$[s_2, s_3]$	$[s_2, s_4]$

Table 7 Preference relation matrix P and NDD value

P^S	x_1	x_2	x_3
x_1	NaN	0.3333	0
x_2	0	NaN	0
x_3	0	0.3333	NaN
NDD_i	1	0.6667	1

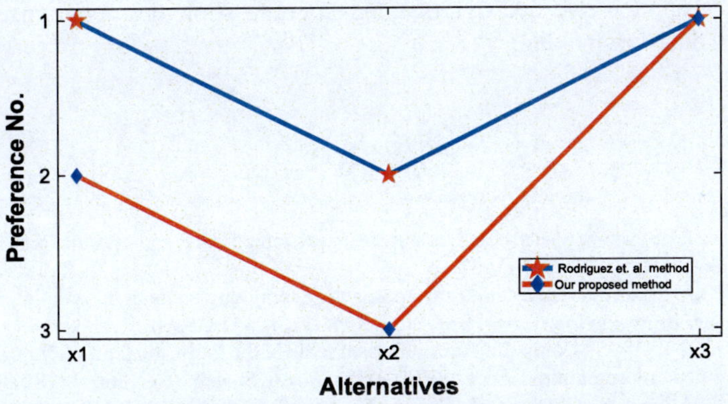

Fig. 4 Comparison between Rodriguez et al.'s method versus our method

If taking the same example in Rodriguez et al.'s method [8] and Wei et al.'s pessimistic method [14], then Rodriguez et al.'s method can't distinguish between x_1 and x_3 and Wei et al.'s pessimistic method can't distinguish between x_1 and x_2. But using the same example Lee and Chen [5] got the same result, i.e. $x_3 > x_1 > x_2$. It is shown in Fig. 4.

5 Advantage

The main advantage of this proposed method is the concept of HFWLTS. In real situation of human decision-making, hesitation and confusion always appeared. The intelligent information system designed on the concept of HFWLTS may be to overcome such hesitant situation with confusion through the weighting vector of the elements of HFLTS. Using this HFWLTS, we can get the more realistic result where Rodriguez and Wei can't solve such situations.

6 Conclusion

In this article, the concept of HFWLTS is introduced by us, makes an intelligent information system with flexibility and more realistic as per human nature. Sometimes, there have such situations where HFLTS are not able to take a proper decision, but through HFWLTS, it is more realistic to take a proper and better decisions which we will discussed through an example. In multicriteria linguistic decision-making model where experts are hesitated and confused to provide their knowledge by using linguistic expressions and for their confusion they can easily use weighting vector that has been applied to a decision-making model to show the importance of the HFWLTS in decision-making.

References

1. Bonissone, P.P.: A fuzzy sets based linguistic approach: theory and applications. Approx. Reason. Decis. Anal. 329–339 (1982)
2. Bordogna, G., Pasi, G.: A fuzzy linguistic approach generalizing boolean information retrieval: A model and its evaluation. J. Am. Soc. Inf. Sci. **44**(2), 70–82 (1993)
3. Dong, Y., Xu, Y., Yu, S.: Computing the numerical scale of the linguistic term set for the 2-tuple fuzzy linguistic representation model. IEEE Trans. Fuzzy Syst. **17**(6), 1366–1378 (2009)
4. Herrera, F., Alonso, S., Chiclana, F., Herrera-Viedma, E.: Computing with words in decision making: foundations, trends and prospects. Fuzzy Optim. Decis. Mak. **8**(4), 337–364 (2009)
5. Lee, L.w., Chen, S.M.: Fuzzy decision making based on likelihood-based comparison relations of hesitant fuzzy linguistic term sets and hesitant fuzzy linguistic operators. Inf. Sci. **294**, 513–529 (2015). https://doi.org/10.1016/j.ins.2014.09.061
6. Mendel, J.M.: An architecture for making judgments using computing with words. Int. J. Appl. Math. Comput. Sci. **12**, 325–335 (2002)
7. Orlovsky, S.: Decision-making with a fuzzy preference relation. Fuzzy Sets Syst. **1**(3), 155–167 (1978)
8. Rodriguez, R.M., Martinez, L., Herrera, F.: Hesitant fuzzy linguistic term sets for decision making. IEEE Trans. Fuzzy Syst. **20**(1), 109–119 (2012)
9. Sengupta, A., Pal, T.K.: On comparing interval numbers. Eur. J. Oper. Res. **127**(1), 28–43 (2000)
10. Torra, V.: Hesitant fuzzy sets. Int. J. Intell. Syst. **25**(6), 529–539 (2010)
11. Türkşen, I.B.: Type 2 representation and reasoning for cww. Fuzzy Sets Syst. **127**(1), 17–36 (2002)
12. Wang, J.H., Hao, J.: A new version of 2-tuple fuzzy linguistic representation model for computing with words. IEEE Trans. Fuzzy Syst. **14**(3), 435–445 (2006)
13. Wang, Y.M., Yang, J.B., Xu, D.L.: A preference aggregation method through the estimation of utility intervals. Comput. Oper. Res. **32**(8), 2027–2049 (2005)
14. Wei, C., Zhao, N., Tang, X.: Operators and comparisons of hesitant fuzzy linguistic term sets. IEEE Trans. Fuzzy Syst. **22**(3), 575–585 (2014)

15. Zadeh, L.: Fuzzy sets. Inf. Control. **8**(3), 338–353 (1965). https://doi.org/10.1016/S0019-9958(65)90241-X
16. Zadeh, L.A.: The concept of a linguistic variable and its application to approximate reasoning. Part 1 Inform. Sci. **8**, 199–249 (1975)
17. Zhou, S.M., John, R.I., Chiclana, F., Garibaldi, J.M.: On aggregating uncertain information by type-2 owa operators for soft decision making. Int. J. Intell. Syst. **25**(6), 540–558 (2010)

Totally $\xi_{\Im}$-Continuous Maps in $I\xi_T S$

Nazir Ahmad Ahengar, Jitendra Kumar Maitra, Roshani Sharma, Sujeet Chaturvedi, and Mudasir Ahmad

1 Introduction

The importance of *general topology* is quickly increasing in both the pure and applied domains. In data mining, it plays a significant role [16, 26]. Information systems are fundamental instruments for generating information understanding in any real-life sector, and topological information collection structures are useful and appropriate mathematical models for both quantitative and qualitative information mathematics. In computer science and digital topology [14–14], computer topology for geometric and molecular design [23], quantum physics, high-energy physics, particle physics and superstring theory [17, 29], one can observe the impact of particular *topological spaces* as well.

Initially, the concept of ideal topological space was initiated by Kuratowski [15]. The concepts of *semi-open* and *semi-continuities* were introduced and studied by Levine [19]. Further, the same author [20] studied generalized closed sets in topology and investigated various properties. Jankovic and Hamlett [11] studied the concept of a local function and obtained the significant properties of these functions. Meanwhile, Dontchev [5] verified the certain properties of *pre-I-open sets* and decomposition of *I-continuity.* The notion of generalized topology was introduced by Csaszar [3]. Hatir and Noiri [8] verified several properties of $\beta-$*I-open sets* and studied the several results of *almost-I-continuities*. Jafari1 and Rajesh [10] studied the concept

N. A. Ahengar (✉) · J. K. Maitra · S. Chaturvedi
Department of Mathematics and Computer Sciences, Rani Durgavati University Jabalpur, Jabalpur, Madhya Pradesh, India

R. Sharma
Department of Mathematics, LNCT, Bhopal, Madhya Pradesh, India

M. Ahmad
Department of Physical Sciences, S.B.B.S University Jalandhar, Jalandhar, Punjab, India

P. Srivastava et al. (eds.), *Mathematical, Computational Intelligence and Engineering Approaches for Tourism, Agriculture and Healthcare*, Lecture Notes in Networks and Systems 214, https://doi.org/10.1007/978-981-16-3807-7_16

of *generalized closed sets with respect to an ideal* and studied the various characterizations. Njastad [24] has introduced several classes of *nearby open sets*. Rodyna and Deena [27] studied several generalized forms of *open sets with respect to ideal* categories them suitably. Michael [23] has studied the concept of *semi-open sets* in ideal topological spaces. An ideal in generalized topological space is further defined by Maitra and Tripathi [21]. Maitra and Tripathi [21] obtained important properties of local function in generalized ideal topological spaces.

Hatir and Noiri [9] studied *d–b-continuous functions* and obtained several results related to continuity. Continuity is the most important concept in Mathematics. Levine [18] introduced weakly continuous functions and established some new results. Further, Son et al. [28] introduced *weakly clopen* and *almost clopen functions*. These authors [28] investigate that *almost clopen functions* are the generalized forms of *perfectly continuous, regular set-connected* and *clopen functions. Devi, Balachandran and Maki* [4] *introduced the concept of semi-generalized closed maps and generalized semi-closed maps.*

Egenhofer [6] discussed the very useful concept for *binary topological relations.* Gevorgyan [7] studied the group of *continuous binary operations* on a topological space and established its relationship with the group of homeomorphisms. Chen et al. [2] demonstrated the dynamics of *binary relations* over *topological spaces.* Nithyanantha and Thangavelu [24] studied the concept of *binary topology* and investigated various properties. Nazir Ahmad Ahengar and J.K. Maitra [1] studies the various types of *continuities in generalized binary topological spaces* and establish various relationships with the help of some counter examples.

1.1 Our Contribution

As outline we developed the concept of totally $\xi_{\mathfrak{I}}$-continuous maps, totally $\xi^*_{\mathfrak{I}}$-continuous, totally $\xi^*_{\mathfrak{I}}$-semi-continuous and totally $\xi^*_{\mathfrak{I}}$-pre-continuous. Several relationships are established and verified by making use of examples.

1.2 Organization

The rest of the paper is structured as follows: Some required basic definitions, concepts of ξ-topological, ideal ξ-topological spaces and notations are discussed in Sect. 2. The concept of totally $\xi_{\mathfrak{I}}$-continuous maps is demonstrated in Sect. 3. The concept of totally $\xi^*_{\mathfrak{I}}$-continuous maps is discussed in Sect. 4. The concept of totally $\xi^*_{\mathfrak{I}}$-semi-continuous maps is discussed in Sect. 5. Finally, the conclusion of this paper is given in Sect. 6.

2 Preliminaries

In this portion, we discussed a few required and important definitions, concepts of ξ-*topological*, *Ideal* ξ*-topological Space* ($I\xi_T S$) and notations. Throughout the paper, $\wp(Y)$ denotes the power set of Y.

Definition 2.1: Suppose Υ_1 and Υ_2 are any two non-void sets. Then ξ-topology (ξ_T) from Υ_1 to Υ_2 is a binary structure $\xi \subseteq \wp(\Upsilon_1) \times \wp(\Upsilon_2)$ satisfying the conditions, i.e.,$(\emptyset, \emptyset)$, $(\Upsilon_1, \Upsilon_2) \in \xi$ and if $\{(L_\alpha, M_\alpha); \alpha \in \Gamma\}$ is a family of elements of ξ, then $\left(\bigcup_{\alpha\in\Gamma} L_\alpha, \bigcup_{\alpha\in\Gamma} M_\alpha\right) \in \xi$. If ξ is ξ_T from Υ_1 to Υ_2, then $(\Upsilon_1, \Upsilon_2, \xi)$ is known as ***ξ-topological space*** $(\xi_T S)$ and the elements of ξ are known as ***ξ-open subsets*** of $(\Upsilon_1, \Upsilon_2, \xi)$. The elements of $\wp(\Upsilon_1) \times \wp(\Upsilon_2)$ are known as simply ***ξ-points***.

Definition 2.2: Suppose Υ_1 and Υ_2 are any two non-void sets and (L_1, M_1), (L_2, M_2) are the elements of $\wp(\Upsilon_1) \times \wp(\Upsilon_2)$. Then $(L_1, M_1) \subseteq (L_2, M_2)$ only if $L_1 \subseteq L_2$ and $M_1 \subseteq M_2$.

Remark 2.1: Suppose $\{T_\alpha; \alpha \in \Lambda\}$ is the family of ξ_T from Υ_1 to Υ_2. Then, $\bigcup_{\alpha\in\Lambda} T_\alpha$ is also ξ_T from Υ_1 to Υ_2. But $\cup_{\alpha\in\Lambda} T_\alpha$ need not be a ξ_T.

Definition 2.3: Suppose $(\Upsilon_1, \Upsilon_2, \xi)$ is $\xi_T S$ and $L \subseteq \Upsilon_1, M \subseteq \Upsilon_2$. Then (L, M) is called ***ξ-closed*** in $(\Upsilon_1, \Upsilon_2, \xi)$ if $(\Upsilon_1 \backslash L, \Upsilon_2 \backslash M) \in \xi$.

Proposition 2.1: Suppose $(\Upsilon_1, \Upsilon_2, \xi)$ is $\xi_T S$. Then (Υ_1, Υ_2) and $(\emptyset, \emptyset)$ are ξ-*closed sets*. Similarly if $\{(L_\alpha, M_\alpha) : \alpha \in \Gamma\}$ is a family of ξ-*closed sets*, then $\left(\bigcap_{\alpha\in\Gamma} L_\alpha, \bigcap_{\alpha\in\Gamma} M_\alpha\right)$ is ξ-*closed.*

Proposition 2.2: Suppose $(L, M) \subseteq (\Upsilon_1, \Upsilon_2)$. Then (L, M) is ξ-*open* in $(\Upsilon_1, \Upsilon_2, \xi)$ iff $(L, M) = I_\xi(L, M)$ and (L, M) is ξ-*closed* in $(\Upsilon_1, \Upsilon_2, \xi)$ iff $(L, M) = Cl_\xi(L, M)$.

Definition 2.4: The non-void collection $\Im$ of subsets of $\Upsilon_1 \times \Upsilon_2$ is known as an ***ideal*** only if it satisfies two important axioms, i.e., if $(L, M) \in \Im$ and $(P, Q) \subseteq (L, M)$ implies $(P, Q) \in \Im$ and if $(L, M) \in \Im$ and $(P, Q) \in \Im$ implies $(L \cup P, \ M \cup Q) \in \Im$. Suppose ξ is ξ_T and $\Im$ an ideal, then $(\Upsilon_1, \Upsilon_2, \xi, \Im)$ is said to be an ***ideal ξ-topological space*** $\left(I\xi_T S\right)$.

Example 2.2: Suppose $(\Upsilon_1, \Upsilon_2, \xi)$ is $\xi_T S$. The collection $\Im = \emptyset$ and $\Im = \wp(\Upsilon_1) \times \wp(\Upsilon_2)$ are also ideals on $\Upsilon_1 \times \Upsilon_2$.

Definition 2.5: Suppose $(\Upsilon_1, \Upsilon_2, \xi, \Im)$ is $I\xi_T S$ and $(L, M) \subseteq \Upsilon_1 \times \Upsilon_2$. Then the set $(L, M)^*(\Im) = \{(x, y) \in \Upsilon_1 \times \Upsilon_2/(U \cap L, \ V \cap M) \notin \Im$ for every nbd (U, V) of (x, y)} is known as the ***local function*** of (L, M) in the respect of $\Im$ and ξ. We normally denote $(L, M)^*$ instead of $(L, M)^*(\Im)$ to avoid any confusion.

Definition 2.6: Suppose $(\Upsilon_1, \Upsilon_2, \xi, \Im)$ is $I\xi_T S$ and $(L, M) \subseteq \Upsilon_1 \times \Upsilon_2$. Then (L, M) is known as ***$\xi_{\Im}$-semi-open set*** if there exists any ξ-*open set* (U, V) such that $(L, M)\backslash Cl_\xi(U, V) \in \Im$ whenever, $(U, V)\backslash(L, M) \in \Im$. Likewise (L, M) is known ***as $\xi_{\Im}$-α-open set*** if there exists any ξ-*open set* (U, V) such that $(L, M)\backslash I_\xi\left(Cl_\xi(U, V)\right) \in \Im$ whenever, $(U, V)\backslash(L, M) \in \Im$.

Definition 2.7: Suppose $(\Upsilon_1, \Upsilon_2, \xi, \Im)$ is $I\xi_TS$ and $(L, M) \subseteq \Upsilon_1 \times \Upsilon_2$. Then (L, M) is known as ***$\xi_\Im$-pre-open set*** if there exists any *ξ-open set* (U, V) such that $(U, V)\backslash Cl_\xi(L, M) \in \Im$ whenever, $(L, M)\backslash(U, V) \in \Im$. Likewise (L, M) is called ***$\xi_\Im$-β-open set*** if there exists any *ξ-open set* (U, V) such that $(U, V)\backslash I_\xi\left(Cl_\xi(L, M)\right) \in \Im$ whenever, $(L, M)\backslash(U, V) \in \Im$.

Definition 2.8: Suppose $(\Upsilon_1, \Upsilon_2, \xi, \Im)$ is $I\xi_TS$ and $(L, M) \subseteq \Upsilon_1 \times \Upsilon_2$. Then (L, M) is known as ***$\xi^*_\Im$-open set*** if $(U, V)\backslash(L, M) \in \Im$ whenever, $(L, M) \subseteq (U, V)$, where (U, V) is *ξ-open set*.

Definition 2.9: Suppose $(\Upsilon_1, \Upsilon_2, \xi, \Im)$ is $I\xi_TS$ and $(L, M) \subseteq \Upsilon_1 \times \Upsilon_2$. Then (L, M) is known as ***$\xi^*_\Im$-semi-open set*** if $(U, V)\backslash I_\xi(L, M) \in \Im$ whenever, $(L, M) \subseteq (U, V)$, where (U, V) is *ξ-open set*. Similarly (L, M) is known as ***$\xi^*_\Im$-α-open set*** if $(U, V)\backslash Cl_\xi\left(I_\xi(L, M)\right) \in \Im$ whenever, $(L, M) \subseteq (U, V)$, where (U, V) is *ξ-open set*.

Definition 2.10: Suppose $(\Upsilon_1, \Upsilon_2, \xi, \Im)$ is $I\xi_TS$ and $(L, M) \subseteq \Upsilon_1 \times \Upsilon_2$. Then (L, M) is known as ***$\xi^*_\Im$-pre-open set*** if $(U, V)\backslash Cl_\xi(L, M) \in \Im$ whenever, $(L, M) \subseteq (U, V)$, where (U, V) is *ξ-open set*. Similarly (L, M) is known as ***$\xi^*_\Im$-β-open set*** if $(U, V)\backslash I_\xi(Cl_\xi(L, M)) \in \Im$ whenever, $(L, M) \subseteq (U, V)$, where (U, V) is *ξ-open set*.

3 Totally $\xi_\Im$-Continuous Maps

In this section, we introduced the concept of totally $\xi_\Im$-continuous and in $I\xi_TS$ and established the relationships between these maps and some other maps.

Definition 3.1: Suppose $(Z, \mathcal{T})$ is G_T and $(\Upsilon_1, \Upsilon_2, \xi, \Im)$ is $I\xi_TS$. Then the mapping $\mathcal{F} : (Z, \mathcal{T}) \to \Upsilon_1 \times \Upsilon_2$ is known as ***$\xi_\Im$-semi-continuous ($\xi_\Im$-α-continuous) map*** if $\mathcal{F}^{-1}(L, M)$ is *$\mathcal{T}_\Im$-semi-open ($\mathcal{T}_\Im$-α-open)* in $(Z, \mathcal{T})$ for every *ξ-open set* (L, M) in $(\Upsilon_1, \Upsilon_2, \xi)$.

Definition 3.2: Suppose $(Z, \mathcal{T})$ is G_T and $(\Upsilon_1, \Upsilon_2, \xi, \Im)$ is $I\xi_TS$. Then the mapping $\mathcal{F} : (Z, \mathcal{T}) \to \Upsilon_1 \times \Upsilon_2$ is known as ***totally $\xi_\Im$-semi-continuous (totally $\xi_\Im$-α-continuous) map*** if $\mathcal{F}^{-1}(L, M)$ is *$\mathcal{T}_\Im$-semi-clopen ($\mathcal{T}_\Im$-α-clopen)* in $(Z, \mathcal{T})$ for every *ξ-open set* (L, M) in $(\Upsilon_1, \Upsilon_2, \xi)$.

Example 3.1: Suppose $Z = \{1, 2, 3\}$, $\Upsilon_1 = \{a_1, a_2\}$ and $\Upsilon_2 = \{b_1, b_2\}$. Then clearly $\mathcal{T} = \{\emptyset, \{1,3\}, \{2,3\}, Z\}$ is G_T, $\Im = \{\emptyset, \{3\}, \{2, 3\}\}$ is an ideal on Z and $\xi = \{(\emptyset, \emptyset), (\{a_1\}, \{b_1\}), (\{a_2\}, \{\Upsilon_2\}), (\Upsilon_1, \Upsilon_2)\}$ is ξ_T from Υ_1 to Υ_2. Now consider the mapping $\mathcal{F} : (Z, \mathcal{T}) \to \Upsilon_1 \times \Upsilon_2$ defined as $\mathcal{F}(1) = (a_1, b_1)$ and $\mathcal{F}(2) = (a_2, \Upsilon_2) = \mathcal{F}(3)$. Therefore $\mathcal{F}^{-1}(\emptyset, \emptyset) = \{\emptyset\}$, $\mathcal{F}^{-1}(a_1, b_1) = \{1\}$, $\mathcal{F}^{-1}(a_2, \Upsilon_2) = \{2, 3\}$, $\mathcal{F}^{-1}(\Upsilon_1, \Upsilon_2) = \{Z\}$. Thus $\mathcal{F}^{-1}(L, M)$ for every *ξ-open set* (L, M) in $(\Upsilon_1, \Upsilon_2, \xi)$ is *$\mathcal{T}_\Im$-semi-clopen ($\mathcal{T}_\Im$-α-clopen)* set in $(Z, \mathcal{T})$. Hence $\mathcal{F} : (Z, \mathcal{T}) \to \Upsilon_1 \times \Upsilon_2$ is *totally $\xi_\Im$-semi-continuous (totally $\xi_\Im$-α-continuous) map*.

Proposition 3.1: *ξ-continuous (ξ-semi-continuous, ξ-α-continuous, ξ-pre-continuous, ξ-β-continuous) map $\Rightarrow \not\Leftarrow$ totally $\xi_{\Im}$-semi-continuous (totally $\xi_{\Im}$-α-continuous) map.*

Proof The proof follows from the definitions, while the converse is explained in Example 3.2.

Example 3.2: Suppose Z = {1, 2, 3}, Υ_1 = {a_1, a_2} and Υ_2 = {b_1, b_2}. Then clearly $\mathcal{T}$ = {∅,{1}, {1,2}, {1,3}, {2,3}, Z} is G_T, $\Im$ = {∅, {1}, {2}, {1, 2}} is an ideal on Z and ξ = {(∅, ∅), ({a_2}, {b_2}), ({a_1}, {Υ_2}), (Υ_1, Υ_2)} is ξ_T from Υ_1 to Υ_2. Now consider the mapping $\mathcal{F}: (Z, \tau_g) \to \Upsilon_1 \times \Upsilon_2$ defined as $\mathcal{F}(1) = (a_2, b_2) = \mathcal{F}(3)$ and $\mathcal{F}(2) = (a_1, \Upsilon_2)$. Therefore $\mathcal{F}^{-1}(\emptyset, \emptyset) = \{\emptyset\}$, $\mathcal{F}^{-1}(a_2, b_2) = \{1, 3\}$, $\mathcal{F}^{-1}(a_1, \Upsilon_2) = \{2\}$, $\mathcal{F}^{-1}(\Upsilon_1, \Upsilon_2) = \{Z\}$. Thus $\mathcal{F}^{-1}$(L, M) for every *ξ-open set* (L, M) in (Υ_1, Υ_2, ξ) is *$\mathcal{T}_{\Im}$-semi-clopen set* in (Z, $\mathcal{T}$). Hence $\mathcal{F}: (Z, \mathcal{T}) \to \Upsilon_1 \times \Upsilon_2$ is *totally $\xi_{\Im}$-semi-continuous (totally $\xi_{\Im}$-α-continuous)* map but not *ξ-continuous (ξ-semi-continuous, ξ-α-continuous,ξ-pre-continuous, ξ-β-continuous) map* because $\mathcal{T}_{\Im}$-semi-clopen ($\mathcal{T}_{\Im}$-α-clopen) sets {2} and {1, 3} are not *ξ-open (ξ-semi-open, ξ-α-open,ξ-pre-open, ξ-β-open) sets in* (Z, $\mathcal{T}$).

Proposition 3.2: *Totally $\xi_{\Im}$-semi-continuous (totally $\xi_{\Im}$-α-continuous) map $\Rightarrow \not\Leftarrow$ $\xi_{\Im}$-semi-continuous ($\xi_{\Im}$-α-continuous) map.*

Proof Suppose (L, M) is *ξ-open set* in (Υ_1, Υ_2, ξ). Then as $\mathcal{F}: (Z, \mathcal{T}) \to \Upsilon_1 \times \Upsilon_2$ is *totally $\xi_{\Im}$-semi-continuous (totally $\xi_{\Im}$-α-continuous) map*, we have $\mathcal{F}^{-1}$(L, M) is *$\mathcal{T}_{\Im}$-semi-clopen ($\mathcal{T}_{\Im}$-α-clopen)* in (Z, $\mathcal{T}$). We know that *$\mathcal{T}_{\Im}$-semi-clopen ($\mathcal{T}_{\Im}$-α-clopen) set* is *$\mathcal{T}_{\Im}$-semi-open ($\mathcal{T}_{\Im}$-α-open) set,* therefore $\mathcal{F}^{-1}$(L, M) is *$\mathcal{T}_{\Im}$-semi-open ($\mathcal{T}_{\Im}$-α-open) set* in (Z, $\mathcal{T}$). Thus $\mathcal{F}: (Z, \mathcal{T}) \to \Upsilon_1 \times \Upsilon_2$ is *$\xi_{\Im}$-semi-continuous ($\xi_{\Im}$-α-continuous) map.*

The converse is explained in Example 3.3.

Example 3.3: Suppose Z = {1, 2, 3}, Υ_1 = {a_1, a_2} and Υ_2 = {b_1, b_2}. Then clearly $\mathcal{T}$ = {∅, {1, 2}, {2, 3}, Z} is G_T, $\Im$ = {∅, {2, 3}} is an ideal on Z and ξ = {(∅, ∅), ({a_1}, {b_1}), ({a_2}, {Υ_2}), (Υ_1, Υ_2)} is ξ_T from Υ_1 to Υ_2. Now consider the mapping $\mathcal{F}: (Z, \mathcal{T}) \to \Upsilon_1 \times \Upsilon_2$ defined as $\mathcal{F}(1) = (a_1, b_1) = \mathcal{F}(3)$. Thus $\mathcal{F}^{-1}$(L, M) for every *ξ-open set* (L, M) in (Υ_1, Υ_2, ξ) is *$\mathcal{T}_{\Im}$-semi-open ($\xi_{\Im}$-α-open) set* in (Z, $\mathcal{T}$). Hence $\mathcal{F}: (Z, \mathcal{T}) \to \Upsilon_1 \times \Upsilon_2$ is *$\xi_{\Im}$-semi-continuous map* but not *totally $\xi_{\Im}$-semi-continuous (totally $\xi_{\Im}$-α-continuous) map* because $\mathcal{F}^{-1}(\{a_1\}, \{b_1\}) = \{1, 3\}$ where {1, 3} is not $\mathcal{T}_{\Im}$-semi-clopen ($\xi_{\Im}$-α-clopen) set in (Z, $\mathcal{T}$).

Proposition 3.3: *Totally $\xi_{\Im}$-α-continuous map $\Rightarrow \not\Leftarrow$ totally $\xi_{\Im}$-semi-continuous.*

Proof Suppose (L, M) is *ξ-open set* in (Υ_1, Υ_2, ξ). Then as $\mathcal{F}: (Z, \mathcal{T}) \to \Upsilon_1 \times \Upsilon_2$ is *totally $\xi_{\Im}$-α-continuous map*, we have $\mathcal{F}^{-1}$(L, M) *is $\mathcal{T}_{\Im}$-α-clopen set* in (Z, $\mathcal{T}$). We know that every *$\mathcal{T}_{\Im}$-α-clopen set* is *$\mathcal{T}_{\Im}$-semi-clopen,* therefore $\mathcal{F}^{-1}$(L, M) is *$\mathcal{T}_{\Im}$-semi-clopen set* in (Z, $\mathcal{T}$). Thus $\mathcal{F}$ is *totally $\xi_{\Im}$-semi-continuous map.*

The converse is explained in Example 3.4.

Example 3.4: Let Z={1, 2, 3, 4}, Υ_1={a_1, a_2} and Υ_2={b_1, b_2}. Then $\mathcal{T}$ = {∅, {1}, {2}, {3}, {1, 2}, {1, 3}, {1, 4}, {2, 3}, {3, 4}, {1, 2, 3}, {1, 3, 4}, {1, 2, 4}, {2, 3, 4}, Z} is G_T, $\Im$ = {∅, {2}} is an ideal on Z and ξ = {(∅, ∅), ({a_1}, {b_1}), ({a_2}, {b_2}), ({a_1}, {Υ_2}), (Υ_1, Υ_2)} is ξ_T from Υ_1 to Υ_2. Now consider the mapping $\mathcal{F} : (Z, \mathcal{T}) \to \Upsilon_1 \times \Upsilon_2$ defined as $\mathcal{F}(2) = (a_2, b_2)$ and $\mathcal{F}(1) = \mathcal{F}(3) = (a_1, b_1) = \mathcal{F}(4)$. Thus $\mathcal{F}^{-1}$(L, M) for every *ξ-open set* (L, M) in (Υ_1, Υ_2, ξ) is *$\mathcal{T}_\Im$-semi-clopen set* but not *$\mathcal{T}_\Im$-α-clopen set* in (Z, $\mathcal{T}$). Hence $\mathcal{F} : (Z, \mathcal{T}) \to \Upsilon_1 \times \Upsilon_2$ is *totally $\xi_\Im$-semi-continuous* but *totally $\xi_\Im$-α-continuous map.*

Definition 3.3: Suppose (Z, $\mathcal{T}$) is G_TS and (Υ_1, Υ_2, ξ, $\Im$) is Iξ_T*S*. Then the mapping $\mathcal{F} : (Z, \mathcal{T}) \to \Upsilon_1 \times \Upsilon_2$ is known as ***$\xi_\Im$-pre-continuous ($\xi_\Im$-β-continuous) map*** if $\mathcal{F}^{-1}$(L, M) is *$\mathcal{T}_\Im$-pre-open ($\mathcal{T}_\Im$-β-open) set* in (Z, $\mathcal{T}$) for every *ξ-open set* (L, M) in (Υ_1, Υ_2, ξ).

Definition 3.4: Suppose (Z, $\mathcal{T}$) is G_TS and (Υ_1, Υ_2, ξ, $\Im$) is Iξ_T*S*. Then the mapping $\mathcal{F} : (Z, \mathcal{T}) \to \Upsilon_1 \times \Upsilon_2$ is known as ***totally $\xi_\Im$-pre-continuous (totally $\xi_\Im$-β-continuous)*** map if $\mathcal{F}^{-1}$(L, M) is *$\mathcal{T}_\Im$-pre-clopen ($\mathcal{T}_\Im$-β-clopen) set* in (Z, $\mathcal{T}$) for every *ξ-open set* (L, M) in (Υ_1, Υ_2, ξ).

Example 3.5: Suppose Z = {1, 2, 3}, Υ_1 = {a_1, a_2} and Υ_2 = {b_1, b_2}. Then clearly $\mathcal{T}$ = {∅, {1, 2}, {2, 3}, Z} is G_TS, $\Im$ = {∅, {2}, {2, 3}} is an ideal on Z and ξ = {(∅, ∅), ({a_2}, {b_2}), ({a_1}, {Υ_2}), (Υ_1, Υ_2)} is ξ_T from Υ_1 to Υ_2. Now consider the mapping $\mathcal{F} : (Z, \mathcal{T}) \to \Upsilon_1 \times \Upsilon_2$ defined as $\mathcal{F}(1) = (a_1, \Upsilon_2) = \mathcal{F}(2)$ and $\mathcal{F}(3) = (a_2, b_2)$. Thus $\mathcal{F}^{-1}$(L, M) for every *ξ-open set* (L, M) in (Υ_1, Υ_2, ξ) is *$\mathcal{T}_\Im$-pre-clopen ($\mathcal{T}_\Im$-β-clopen) set* in (Z, $\mathcal{T}$). Hence $\mathcal{F} : (Z, \mathcal{T}) \to \Upsilon_1 \times \Upsilon_2$ is *totally $\xi_\Im$-pre-continuous (totally $\xi_\Im$-β-continuous) map.*

Proposition 3.4: *ξ-continuous (ξ-semi-continuous, ξ-α-continuous, ξ-pre-continuous, ξ-β-continuous) map ⇒ ⇍ totally $\xi_\Im$-pre-continuous (totally $\xi_\Im$-β-continuous) map.*

Proof The proof follows from the definitions, while the converse is explained in Example 3.6.

Example 3.6: The mapping $\mathcal{F} : (Z, \mathcal{T}) \to \Upsilon_1 \times \Upsilon_2$ in Example 3.5, is *totally $\xi_\Im$-pre-continuous (totally $\xi_\Im$-β-continuous) map* but not *ξ-continuous (ξ-semi-continuous, ξ-α-continuous, ξ-pre-continuous, ξ-β-continuous) map.*

Proposition 3.5: *Totally $\xi_\Im$-β-continuous map ⇒ ⇍ totally $\xi_\Im$-pre-continuous map.*

Proof Suppose (L, M) is ξ-open set in (Υ_1, Υ_2, ξ). Then as $\mathcal{F} : (Z, \mathcal{T}) \to \Upsilon_1 \times \Upsilon_2$ is *totally $\xi_\Im$-β-continuous map*, we have $\mathcal{F}^{-1}$(L, M) is *$\mathcal{T}_\Im$-β-clopen set* in (Z, $\mathcal{T}$). We know that every *$\mathcal{T}_\Im$-β-clopen set* is *$\mathcal{T}_\Im$-pre-clopen*, therefore $\mathcal{F}^{-1}$(L, M) is *$\mathcal{T}_\Im$-pre-clopen set* in (Z, $\mathcal{T}$). Thus $\mathcal{F} : (Z, \mathcal{T}) \to \Upsilon_1 \times \Upsilon_2$ is *totally $\xi_\Im$-pre-continuous map.*

The converse is explained in Example 3.7.

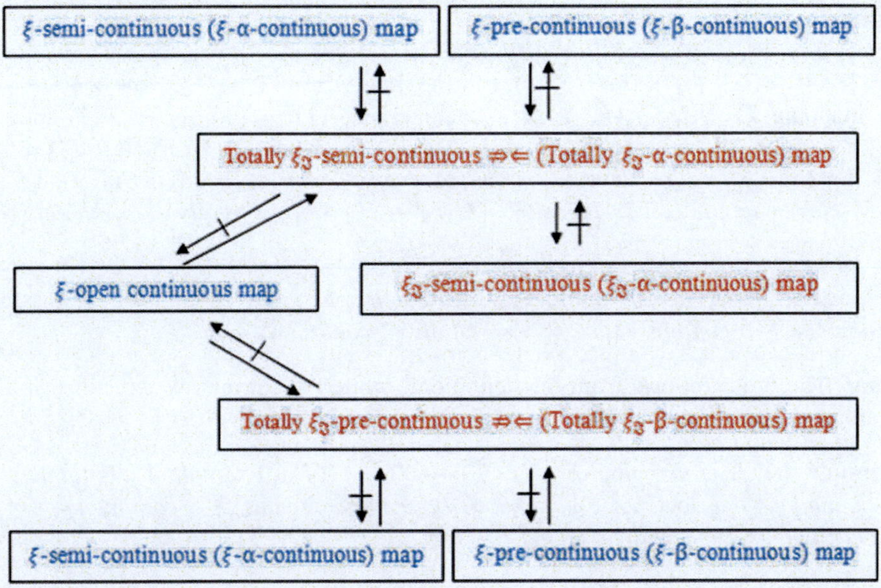

Fig. 1 Relationships of totally $\xi_{\Im}$−continuous maps with some other maps

Example 3.7: Suppose in Example 3.4, $\Im = \{\emptyset, \{2\}, \{2, 3\}\}$, then the mapping $\mathcal{F}$: $(Z, \mathcal{T}) \rightarrow \Upsilon_1 \times \Upsilon_2$ is *totally $\xi_{\Im}$-pre-continuous map* but not *totally $\xi_{\Im}$-β-continuous map*.

Remark 3.1: The following implications as shown in Fig. 1 are the direct consequences of the definitions of various totally generalized maps that we discussed in this section.

4 Totally $\xi^*_{\Im}$-Continuous Maps

In this section, we established the relationships between totally $\xi^*_{\Im}$-continuous maps in Iξ_TS and some other maps.

Definition 4.1: Suppose $(Z, \mathcal{T})$ is G_TS and $(\Upsilon_1, \Upsilon_2, \xi, \Im)$ is an Iξ_TS. Then the mapping $\mathcal{F} : (Z, \mathcal{T}) \rightarrow \Upsilon_1 \times \Upsilon_2$ is known as ***$\xi^*_{\Im}$-continuous map*** if $\mathcal{F}^{-1}$(L, M) is *$\mathcal{T}^*_{\Im}$-open set* in $(Z, \mathcal{T})$ for every *ξ-open set* (L, M) in $(\Upsilon_1, \Upsilon_2, \xi)$.

Definition 4.2: Suppose $(Z, \mathcal{T})$ is G_T and $(\Upsilon_1, \Upsilon_2, \xi, \Im)$ is Iξ_TS. Then the mapping $\mathcal{F} : (Z, \mathcal{T}) \rightarrow \Upsilon_1 \times \Upsilon_2$ is known as ***totally $\xi^*_{\Im}$-continuous map*** if $\mathcal{F}^{-1}$(L, M) is *$\mathcal{T}^*_{\Im}$-Clopen* in $(Z, \mathcal{T})$ for every *ξ-open set* (L, M) in $(\Upsilon_1, \Upsilon_2, \xi)$.

Example 4.1: Suppose $Z = \{1, 2, 3\}$, $\Upsilon_1 = \{a_1, a_2\}$ and $\Upsilon_2 = \{b_1, b_2\}$. Then clearly $\mathcal{T} = \{\emptyset, \{1, 2\}, \{2, 3\}, Z\}$ is G_T, $\Im = \{\emptyset, \{2\}, \{3\}, \{2, 3\}\}$ is an ideal on Z and $\xi = \{(\emptyset, \emptyset), (\{a_1\}, \{b_1\}), (\{a_2\}, \{\Upsilon_2\}), (\Upsilon_1, \Upsilon_2)\}$ is ξ_T from Υ_1 to Υ_2. Now consider the mapping $\mathcal{F} : (Z, \mathcal{T}) \to \Upsilon_1 \times \Upsilon_2$ defined as $\mathcal{F}(1) = (a_1, b_1) = \mathcal{F}(3)$ and $\mathcal{F}(2) = (a_2, \Upsilon_2)$. Therefore $\mathcal{F}^{-1}(\emptyset, \emptyset) = \{\emptyset\}$, $\mathcal{F}^{-1}(a_1, b_1) = \{1, 3\}$, $\mathcal{F}^{-1}(a_2, \Upsilon_2) = \{2\}$, $\mathcal{F}^{-1}(\Upsilon_1, \Upsilon_2) = \{Z\}$. Thus $\mathcal{F}^{-1}(L, M)$ for every *ξ-open set* (L, M) in $(\Upsilon_1, \Upsilon_2, \xi)$ is $\mathcal{T}^*_\Im$*-clopen set* in $(Z, \mathcal{T})$. Hence $\mathcal{F} : (Z, \mathcal{T}) \to \Upsilon_1 \times \Upsilon_2$ is *totally* $\xi^*_\Im$*-continuous map.*

Proposition 4.1: *ξ-continuous (ξ-semi-continuous, ξ-α-continuous, ξ-pre-continuous, ξ-β-continuous) map* $\Rightarrow \not\Leftarrow$ *totally* $\xi^*_\Im$*-continuous map.*

Proof The proof follows from the definitions, while the converse is explained in Example 4.2.

Example 4.2: The mapping $\mathcal{F} : (Z, \mathcal{T}) \to \Upsilon_1 \times \Upsilon_2$ in Example 4.1 is *totally* $\xi^*_\Im$*-continuous map* but not *ξ-continuous (ξ-semi-continuous, ξ-α-continuous, ξ-pre-continuous, ξ-β-continuous) map.*

Proposition 4.2: *Totally* $\xi^*_\Im$*-continuous map* $\Rightarrow \not\Leftarrow$ $\xi_\Im$*-semi-continuous* ($\xi_\Im$*-α-continuous) map.*

Proof Suppose (L, M) is ξ-open set in $(\Upsilon_1, \Upsilon_2, \xi)$. Then as $\mathcal{F} : (Z, \mathcal{T}) \to \Upsilon_1 \times \Upsilon_2$ is *totally* $\xi^*_\Im$*-continuous map*, we have $\mathcal{F}^{-1}(L, M)$ is $\mathcal{T}^*_\Im$*-clopen set* in $(Z, \mathcal{T})$. We know that every $\mathcal{T}^*_\Im$*-clopen set* is $\mathcal{T}_\Im$*-semi-open* ($\mathcal{T}_\Im$*-α-open) set*, therefore $\mathcal{F}^{-1}(L, M)$ is $\mathcal{T}_\Im$*-semi-open* ($\mathcal{T}_\Im$*-α-open) set* in $(Z, \mathcal{T})$. Thus $\mathcal{F} : (Z, \mathcal{T}) \to \Upsilon_1 \times \Upsilon_2$ is $\xi_\Im$*-semi-continuous* ($\xi_\Im$*-α-continuous) map.*

The converse is explained in Example 4.3.

Example 4.3: Suppose $Z=\{1, 2, 3, 4\}$, $\Upsilon_1=\{a_1, a_2\}$and $\Upsilon_2=\{b_1, b_2\}$. Then clearly $\mathcal{T} = \{\emptyset,\{1\},\{2\},\{3\},\{1,2\},\{1,3\},\{1,4\},\{2,3\},\{1,2,3\},\{1,2,4\},\{1, 3, 4\}, \{2, 3, 4\}, Z\}$ is G_T, $\Im=\{\emptyset, \{2, 3\}\}$ is an ideal on Z and $\xi=\{(\emptyset, \emptyset), (\{a_1\}, \{b_1\}), (\{a_2\}, \{b_2\}), (\{a_2\}, \{\Upsilon_2\}), (\Upsilon_1, \Upsilon_2)\}$ is ξ_T from Υ_1 to Υ_2. Now consider the mapping $\mathcal{F} : (Z, \mathcal{T}) \to \Upsilon_1 \times \Upsilon_2$ defined as $\mathcal{F}(3) = (a_1, b_1) = \mathcal{F}(4)$. Thus $\mathcal{F}^{-1}(L, M)$ for every *ξ-open set* (L, M) in $(\Upsilon_1, \Upsilon_2, \xi)$ is $\mathcal{T}_\Im$*-semi-open* ($\mathcal{T}_\Im$*-α-open) set* but not $\mathcal{T}^*_\Im$*-clopen set* in $(Z, \mathcal{T})$. Hence $\mathcal{F} : (Z, \mathcal{T}) \to \Upsilon_1 \times \Upsilon_2$ is $\xi_\Im$*-semi-continuous* ($\xi_\Im$*-α-continuous) map* but not *totally* $\xi^*_\Im$*-continuous map.*

Proposition 4.3: *Totally* $\xi^*_\Im$*-continuous map* $\Rightarrow \not\Leftarrow$ $\xi_\Im$*-pre-continuous* ($\xi_\Im$*-β-continuous) map.*

Proof Suppose (L, M) is ξ-open set in $(\Upsilon_1, \Upsilon_2, \xi)$. Then as $\mathcal{F} : (Z, \mathcal{T}) \to \Upsilon_1 \times \Upsilon_2$ is *totally* $\xi^*_\Im$*-continuous map*, we have $\mathcal{F}^{-1}(L, M)$ is $\mathcal{T}^*_\Im$*-clopen set* in $(Z, \mathcal{T})$. We know that every $\mathcal{T}^*_\Im$*-clopen set* is $\mathcal{T}_\Im$*-pre-open* ($\mathcal{T}_\Im$*-β-open) set*, therefore $\mathcal{F}^{-1}(L, M)$ is $\mathcal{T}_\Im$*-pre-open* ($\mathcal{T}_\Im$*-β-open) set* in $(Z, \mathcal{T})$. Thus $\mathcal{F} : (Z, \mathcal{T}) \to \Upsilon_1 \times \Upsilon_2$ is $\xi_\Im$*-pre-continuous* ($\xi_\Im$*-β-continuous) map.*

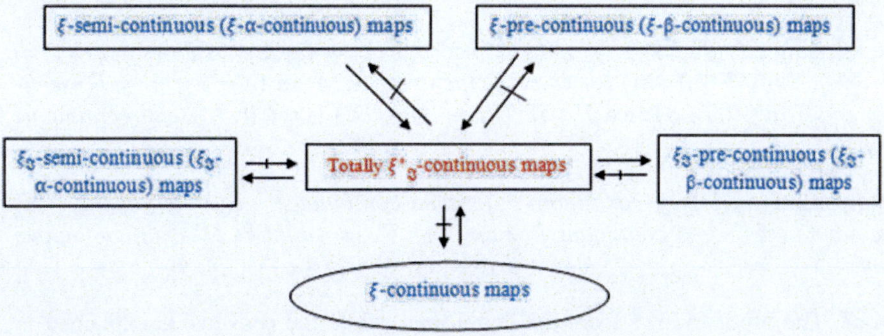

Fig. 2 Relationships of totally $\xi^*_{\Im}$—continuous maps with some other maps

The converse is explained in Example 4.4.

Example 4.4: Suppose Z={1, 2, 3, 4}, Υ_1={a_1, a_2} and Υ_2={b_1, b_2}. Then clearly $\mathcal{T}$={∅, {1}, {3}, {1, 3}, {1, 4}, {2, 3}, {1, 2, 3}, {1, 3, 4}, {1, 2, 4}, {2, 3, 4}, Z} is G_T, $\Im$ = {∅, {1}} is an ideal on Z and ξ = {(∅, ∅), ({a_1}, {b_1}), ({a_2}, {b_2}), ({a_2}, {Υ_2}), (Υ_1, Υ_2)} is ξ_T from Υ_1 to Υ_2. Now consider the mapping $\mathcal{F} : (Z, \mathcal{T}) \rightarrow \Upsilon_1 \times \Upsilon_2$ defined as $\mathcal{F}(3) = (a_1, b_1) = \mathcal{F}(4)$. Thus $\mathcal{F}^{-1}$(L, M) for every ξ*-open set* (L, M) in (Υ_1, Υ_2, ξ) is $\mathcal{T}_{\Im}$*-pre-open* ($\mathcal{T}_{\Im}$*-β-open) set* but not $\mathcal{T}^*_{\Im}$*-clopen set* in (Z, $\mathcal{T}$). Hence $\mathcal{F} : (Z, \mathcal{T}) \rightarrow \Upsilon_1 \times \Upsilon_2$ is $\xi_{\Im}$*-pre-continuous* ($\xi_{\Im}$*-β-continuous) map* but not *totally* $\xi^*_{\Im}$*-continuous map.*

Remark 4.1: The following implications as shown in Fig. 2 are the direct consequences of the definitions of various totally generalized maps that we discussed in this section.

5 Totally $\xi^*_{\Im}$-Semi-continuous Maps

In this section, we established the relationship between totally $\xi^*_{\Im}$-semi-continuous maps in Iξ_TS and some other maps.

Definition 5.1: Suppose (Z, $\mathcal{T}$) is G_TS and (Υ_1, Υ_2, ξ, $\Im$) is Iξ_TS. Then the mapping $\mathcal{F} : (Z, \mathcal{T}) \rightarrow \Upsilon_1 \times \Upsilon_2$ is known as ***$\xi^*_{\Im}$-semi-continuous map*** if $\mathcal{F}^{-1}$(L, M) is $\mathcal{T}^*_{\Im}$*-semi-open set* in (Z, $\mathcal{T}$) for every ξ*-open set* (L, M) in (Υ_1, Υ_2, ξ).

Definition 5.2: Suppose (Z, $\mathcal{T}$) is G_TS and (Υ_1, Υ_2, ξ, $\Im$) is Iξ_TS. Then the mapping $\mathcal{F} : (Z, \mathcal{T}) \rightarrow \Upsilon_1 \times \Upsilon_2$ is known as ***totally $\xi^*_{\Im}$-semi-continuous map*** if $\mathcal{F}^{-1}$(L, M) is $\mathcal{T}^*_{\Im}$*-semi-clopen set* in (Z, $\mathcal{T}$) for every ξ*-open set* (L, M) in (Υ_1, Υ_2, ξ).

Example 5.1: Suppose Z = {1, 2, 3}, Υ_1 = {a_1, a_2} and Υ_2 = {b_1, b_2}. Then clearly $\mathcal{T}$ = {∅, {1, 2}, {2, 3}, Z} is G_T, $\Im$ = {∅, {2, 3}, Z} is an ideal on Z and

$\xi = \{(\emptyset, \emptyset), (\{a_1\}, \{b_1\}), (\{a_2\}, \{\Upsilon_2\}), (\Upsilon_1, \Upsilon_2)\}$ is ξ_T from Υ_1 to Υ_2. Now consider the mapping $\mathcal{F} : (Z, \mathcal{T}) \to \Upsilon_1 \times \Upsilon_2$ defined as $\mathcal{F}(2) = (a_1, b_1) = \mathcal{F}(3)$ and $\mathcal{F}(1) = (a_2, \Upsilon_2)$. Thus $\mathcal{F}^{-1}(L, M)$ for every ξ*-open set* (L, M) in $(\Upsilon_1, \Upsilon_2, \xi)$ is $\mathcal{T}^*_{\Im}$*-semi-clopen set* in $(Z, \mathcal{T})$. Hence $\mathcal{F} : (Z, \mathcal{T}) \to \Upsilon_1 \times \Upsilon_2$ is *totally* $\xi^*_{\Im}$*-semi-continuous map.*

Proposition 5.1: *ξ-continuous (ξ-semi-continuous, ξ-α-continuous, ξ-pre-continuous, ξ-β-continuous) map $\Rightarrow$ $\not\Leftarrow$ totally $\xi^*_{\Im}$-semi-continuous map.*

Proof The proof follows from the definitions, while the converse is explained in Example 5.2.

Example 5.2: The mapping $\mathcal{F} : (Z, \mathcal{T}) \to \Upsilon_1 \times \Upsilon_2$ in Example 5.1 is *totally $\xi^*_{\Im}$-semi-continuous map* but not *ξ-continuous (ξ-semi-continuous, ξ-α-continuous, ξ-pre-continuous, ξ-β-continuous) map.*

Proposition 5.2: *Totally $\xi^*_{\Im}$-semi-continuous map $\Rightarrow$ $\not\Leftarrow$ $\xi_{\Im}$-semi-continuous ($\xi_{\Im}$-α-continuous) map.*

Proof Suppose (L, M) is ξ*-open set* in $(\Upsilon_1, \Upsilon_2, \xi)$. Then as $\mathcal{F} : (Z, \mathcal{T}) \to \Upsilon_1 \times \Upsilon_2$ is *totally $\xi^*_{\Im}$-semi-continuous map*, we have $\mathcal{F}^{-1}(L, M)$ is $\mathcal{T}^*_{\Im}$*-semi-clopen set* in $(Z, \mathcal{T})$. We know that every $\mathcal{T}^*_{\Im}$*-semi-clopen set* is $\mathcal{T}_{\Im}$*-semi-open ($\mathcal{T}_{\Im}$-α-open) set,* therefore $\mathcal{F}^{-1}(L, M)$ is $\mathcal{T}_{\Im}$*-semi-open ($\mathcal{T}_{\Im}$-α-open) set* in $(Z, \mathcal{T})$. Thus $\mathcal{F} : (Z, \mathcal{T}) \to \Upsilon_1 \times \Upsilon_2$ is *$\xi_{\Im}$-semi-continuous ($\xi_{\Im}$-α-continuous) map.*

The converse is explained in Example 5.3.

Example 5.3: The mapping $\mathcal{F} : (Z, \mathcal{T}) \to \Upsilon_1 \times \Upsilon_2$ in Example 4.3 is *$\xi_{\Im}$-semi-continuous ($\xi_{\Im}$-α-continuous) map* but not *totally $\xi^*_{\Im}$-semi-continuous map.*

Definition 5.3: Suppose $(Z, \mathcal{T})$ is G_TS and $(\Upsilon_1, \Upsilon_2, \xi, \Im)$ is $I\xi_TS$. Then the mapping $\mathcal{F} : (Z, \mathcal{T}) \to \Upsilon_1 \times \Upsilon_2$ is known as ***totally $\xi^*_{\Im}$-α-continuous map*** if $\mathcal{F}^{-1}(L, M)$ is $\mathcal{T}^*_{\Im}$*-α-clopen set* in $(Z, \mathcal{T})$ for every ξ*-open set* (L, M) in $(\Upsilon_1, \Upsilon_2, \xi)$.

Example 5.4: The mapping $\mathcal{F} : (Z, \mathcal{T}) \to \Upsilon_1 \times \Upsilon_2$ in Example 5.1 is totally $\xi^*_{\Im}$-α-continuous map.

Proposition 5.3: *ξ-continuous (ξ-semi-continuous, ξ-α-continuous, ξ-pre-continuous, ξ-β-continuous) map $\Rightarrow$ $\not\Leftarrow$ totally $\xi^*_{\Im}$-α-continuous map.*

Proof The proof follows from the definitions, while the converse is explained in Example 5.2.

Proposition 5.4: *Totally $\xi^*_{\Im}$-α-continuous map $\Rightarrow$ $\not\Leftarrow$ $\xi_{\Im}$-semi-continuous ($\xi_{\Im}$-α-continuous) map.*

Proof Suppose (L, M) is *ξ-open set* in $(\Upsilon_1, \Upsilon_2, \xi)$. Then as $\mathcal{F} : (Z, \mathcal{T}) \rightarrow \Upsilon_1 \times \Upsilon_2$ is *totally $\xi_{\Im}^*$-α-continuous map*, we have $\mathcal{F}^{-1}$(L, M) is *$\mathcal{T}_{\Im}^*$-α-clopen set* in $(Z, \mathcal{T})$. We know that every *$\mathcal{T}_{\Im}^*$-α-clopen set* is *$\mathcal{T}_{\Im}$-semi-open ($\mathcal{T}_{\Im}$-α-open) set*, therefore $\mathcal{F}^{-1}$(L, M) is *$\mathcal{T}_{\Im}$-semi-open ($\mathcal{T}_{\Im}$-α-open) set* in $(Z, \mathcal{T})$. Thus $\mathcal{F} : (Z, \mathcal{T}) \rightarrow \Upsilon_1 \times \Upsilon_2$ is *$\xi_{\Im}$-semi-continuous ($\xi_{\Im}$-α-continuous) map.*

The converse is explained in Example 5.3, where we found $\mathcal{F} : (Z, \mathcal{T}) \rightarrow \Upsilon_1 \times \Upsilon_2$ is *$\xi_{\Im}$-semi-continuous ($\xi_{\Im}$-α-continuous) map* but not *totally $\xi_{\Im}^*$-α-continuous map.*

Proposition 5.5: *Totally $\xi_{\Im}^*$-semi-continuous map $\Rightarrow$ $\not\Leftarrow$ totally $\xi_{\Im}^*$-α-continuous map.*

Proof The proof follows from the definitions, while the converse is explained in Example 5.5.

Example 5.5: Suppose Z={1, 2, 3, 4}, Υ_1={a_1, a_2} and Υ_2={b_1, b_2}. Then clearly $\mathcal{T}$={∅, {1}, {1, 3}, {1, 4}, {3, 4}, {1, 2, 3}, {1, 2, 4}, {1, 3, 4}, Z} is G_T, $\Im$ = {∅, {1}} is an ideal on Z and ξ = {(∅, ∅), ({a_1}, {b_1}), ({a_2}, {b_2}), ({a_2}, {Υ_2}), (Υ_1, Υ_2) is ξ_T from Υ_1 to Υ_2. Now consider the mapping $\mathcal{F} : (Z, \mathcal{T}) \rightarrow \Upsilon_1 \times \Upsilon_2$ defined as $\mathcal{F}(2) = \mathcal{F}(3) = (a_1, b_1) = \mathcal{F}(4)$. Thus $\mathcal{F}^{-1}$(L, M) for every *ξ-open set* (L, M) in $(\Upsilon_1, \Upsilon_2, \xi)$ is *$\mathcal{T}_{\Im}^*$-α-clopen set* but not *$\mathcal{T}_{\Im}^*$-semi-clopen set* in $(Z, \mathcal{T})$. Hence $\mathcal{F} : (Z, \mathcal{T}) \rightarrow \Upsilon_1 \times \Upsilon_2$ is *totally $\xi_{\Im}^*$-α-continuous map* but not *totally $\xi_{\Im}^*$-semi-continuous map.*

Definition 5.4: Suppose $(Z, \mathcal{T})$ is G_TS and $(\Upsilon_1, \Upsilon_2, \xi, \Im)$ is $I\xi_TS$. Then the mapping $\mathcal{F} : (Z, \mathcal{T}) \rightarrow \Upsilon_1 \times \Upsilon_2$ is known as ***$\xi_{\Im}^*$-pre-continuous map*** if $\mathcal{F}^{-1}$(L, M) is *$\mathcal{T}_{\Im}^*$-pre-open set* in $(Z, \mathcal{T})$ for every *ξ-open set* (L, M) in $(\Upsilon_1, \Upsilon_2, \xi)$.

Definition 5.5: Suppose $(Z, \mathcal{T})$ is G_TS and $(\Upsilon_1, \Upsilon_2, \xi, \Im)$ is $I\xi_TS$. Then the mapping $\mathcal{F} : (Z, \mathcal{T}) \rightarrow \Upsilon_1 \times \Upsilon_2$ is known as ***totally $\xi_{\Im}^*$-pre-continuous map*** if $\mathcal{F}^{-1}$(L, M) is *$\mathcal{T}_{\Im}^*$-pre-clopen set* in $(Z, \mathcal{T})$ for every *ξ-open set* (L, M) in $(\Upsilon_1, \Upsilon_2, \xi)$.

Example 5.6: Suppose Z = {1, 2, 3,}, Υ_1 = {a_1, a_2} and Υ_2 = {b_1, b_2}. Then clearly $\mathcal{T}$ = {∅, {1, 2}, {2, 3}, Z} is G_T, $\Im$ = {∅, {2}, {1, 2}} is an ideal on Z and ξ = {(∅, ∅), ({a_2}, {b_2}), ({a_1}, {Υ_2}), (Υ_1, Υ_2)} is ξ_T from Υ_1 to Υ_2. Now consider the mapping $\mathcal{F} : (Z, \mathcal{T}) \rightarrow \Upsilon_1 \times \Upsilon_2$ defined as $\mathcal{F}(1) = (a_2, b_2)$ and $\mathcal{F}(2) = (a_1, \Upsilon_2) = \mathcal{F}(3)$. Thus $\mathcal{F}^{-1}$(L, M) for every *ξ-open set* (L, M) in $(\Upsilon_1, \Upsilon_2, \xi)$ is *$\mathcal{T}_{\Im}^*$-pre-clopen set* in $(Z, \mathcal{T})$. Hence $\mathcal{F} : (Z, \mathcal{T}) \rightarrow \Upsilon_1 \times \Upsilon_2$ is *totally $\xi_{\Im}^*$-pre-continuous map.*

Proposition 5.6: *ξ-continuous (ξ-semi-continuous, ξ-α-continuous, ξ-pre-continuous, ξ-β-continuous) map $\Rightarrow$ $\not\Leftarrow$ totally $\xi_{\Im}^*$-pre-continuous map.*

Proof The proof follows from the definitions, while the converse is explained in Example 5.7.

Example 5.7: The mapping $\mathcal{F} : (Z, \mathcal{T}) \rightarrow \Upsilon_1 \times \Upsilon_2$ in Example 5.6 is *totally* $\xi^*_{\Im}$-*pre-continuous map set* but not ξ-*continuous* (ξ-*semi-continuous*, ξ-α-*continuous*, ξ-*pre-continuous*, ξ-β-*continuous*) *map*.

Proposition 5.7: *Totally* $\xi^*_{\Im}$-*pre-continuous map* $\Rightarrow \not\Leftarrow$ $\xi_{\Im}$-*semi-continuous* ($\xi_{\Im}$-α-*continuous*) *map*.

Proof Suppose (L, M) is ξ-*open set* in $(\Upsilon_1, \Upsilon_2, \xi)$. Then as $\mathcal{F} : (Z, \mathcal{T}) \rightarrow \Upsilon_1 \times \Upsilon_2$ is *totally* $\xi^*_{\Im}$-*pre-continuous map*, we have $\mathcal{F}^{-1}(L, M)$ is $\mathcal{T}^*_{\Im}$-*pre-clopen set* in $(Z, \mathcal{T})$. We know that every $\mathcal{T}^*_{\Im}$-*pre-clopen set* is $\mathcal{T}_{\Im}$-*semi-open* ($\mathcal{T}_{\Im}$-α-*open*) *set*, therefore $\mathcal{F}^{-1}(L, M)$ is $\mathcal{T}_{\Im}$-*semi-open* ($\mathcal{T}_{\Im}$-α-*open*) *set* in $(Z, \mathcal{T})$. Thus $\mathcal{F} : (Z, \mathcal{T}) \rightarrow \Upsilon_1 \times \Upsilon_2$ is $\xi_{\Im}$-*semi-continuous* ($\xi_{\Im}$-α-*continuous*) *map*.

The converse is explained in Example 5.8.

Example 5.8: Suppose $Z = \{1, 2, 3, 4\}$, $\Upsilon_1 = \{a_1, a_2\}$ and $\Upsilon_2 = \{b_1, b_2\}$. Then clearly $\mathcal{T} = \{\emptyset, \{1, 3\}, \{1, 4\}, \{1, 2, 3\}, \{1, 2, 4\}, \{1, 3, 4\}, Z\}$ is G_T, $\Im = \{\emptyset, \{2, 3\}\}$ is an ideal on Z and $\xi = \{(\emptyset, \emptyset), (\{a_1\}, \{b_1\}), (\{a_2\}, \{b_2\}), (\{a_2\}, \{\Upsilon_2\}), (\Upsilon_1, \Upsilon_2)\}$ is ξ_T from Υ_1 to Υ_2. Now consider the mapping $\mathcal{F} : (Z, \mathcal{T}) \rightarrow \Upsilon_1 \times \Upsilon_2$ defined by $\mathcal{F}(2) = (a_1, b_1) = \mathcal{F}(4)$. Thus $\mathcal{F}^{-1}(L, M)$ for every ξ-*open set* (L, M) in $(\Upsilon_1, \Upsilon_2, \xi)$ is $\mathcal{T}_{\Im}$-*semi-open* ($\mathcal{T}_{\Im}$-α-*open*) *set* but not $\mathcal{T}^*_{\Im}$-*pre-open set* in $(Z, \mathcal{T})$. Hence $\mathcal{F} : (Z, \mathcal{T}) \rightarrow \Upsilon_1 \times \Upsilon_2$ is $\xi_{\Im}$-*semi-continuous* ($\xi_{\Im}$-α-*continuous*) *map* but not *totally* $\xi^*_{\Im}$-*pre-continuous map*.

Definition 5.6: Suppose $(Z, \mathcal{T})$ is G_TS and $(\Upsilon_1, \Upsilon_2, \xi, \Im)$ is $I\xi_TS$. Then the mapping $\mathcal{F} : (Z, \mathcal{T}) \rightarrow \Upsilon_1 \times \Upsilon_2$ is known as *totally* $\xi^*_{\Im}$-β-*continuous map* if $\mathcal{F}^{-1}(L, M)$ is $\mathcal{T}^*_{\Im}$-β-*clopen set* in $(Z, \mathcal{T})$ for every ξ-*open set* (L, M) in $(\Upsilon_1, \Upsilon_2, \xi)$.

Example 5.9: The mapping $\mathcal{F} : (Z, \mathcal{T}) \rightarrow \Upsilon_1 \times \Upsilon_2$ in Example 5.6 is *totally* $\xi^*_{\Im}$-β-*continuous map*.

Proposition 5.8: ξ-*continuous* (ξ-*semi-continuous*, ξ-α-*continuous*, ξ-*pre-continuous*, ξ-β-*continuous*) *map* $\Rightarrow \not\Leftarrow$ $\xi^*_{\Im}$-β-*continuous map*.

Proof The proof follows from the definitions, while the converse is explained in Example 5.10.

Example 5.10: The mapping $\mathcal{F} : (Z, \mathcal{T}) \rightarrow \Upsilon_1 \times \Upsilon_2$ in Example 5.6 is *totally* $\xi^*_{\Im}$-β-*continuous map* but not ξ-*continuous* (ξ-*semi-continuous*, ξ-α-*continuous*, ξ-*pre-continuous*, ξ-β-*continuous*) *map*.

Proposition 5.9: *Totally* $\xi^*_{\Im}$-β-*continuous map* $\Rightarrow \not\Leftarrow$ $\xi_{\Im}$-*semi-continuous* ($\xi_{\Im}$-α-*continuous*) *map*.

Proof Suppose (L, M) is ξ-*open set* in $(\Upsilon_1, \Upsilon_2, \xi)$. Then as $\mathcal{F} : (Z, \mathcal{T}) \rightarrow \Upsilon_1 \times \Upsilon_2$ is *totally* $\xi^*_{\Im}$-β-*continuous map*, we have $\mathcal{F}^{-1}(L, M)$ is $\mathcal{T}^*_{\Im}$-β-*clopen set* in $(Z, \mathcal{T})$. We know that every $\mathcal{T}^*_{\Im}$-β-*clopen set* is $\mathcal{T}_{\Im}$-*semi-open* ($\mathcal{T}_{\Im}$-α-*open*) *set*, therefore $\mathcal{F}^{-1}(L, M)$ is $\mathcal{T}_{\Im}$-*semi-open* ($\mathcal{T}_{\Im}$-α-*open*) *set* in $(Z, \mathcal{T})$. Thus $\mathcal{F} : (Z, \mathcal{T}) \rightarrow \Upsilon_1 \times \Upsilon_2$ is $\xi_{\Im}$-*semi-continuous* ($\xi_{\Im}$-α-*continuous*) *map*.

The converse is explained in Example 5.11.

Example 5.11: The mapping $\mathcal{F} : (Z, \mathcal{T}) \to \Upsilon_1 \times \Upsilon_2$ in Example 5.8 is *$\xi_{\Im}$-semi-continuous ($\xi_{\Im}$-α-continuous) map* but not *totally $\xi_{\Im}^*$-β-continuous map.*

Remark 5.1: *Totally $\xi_{\Im}^*$-continuous maps $\nLeftrightarrow$ totally $\xi_{\Im}^*$-semi-continuous maps.*

This is explained in Example 5.12 and Example 5.13.

Example 5.12: Suppose $Z = \{1, 2, 3, 4\}$, $\Upsilon_1 = \{a_1, a_2\}$ and $\Upsilon_2 = \{b_1, b_2\}$. Then clearly $\mathcal{T} = \{\emptyset, \{1\}, \{1, 3\}, \{1, 4\}, \{3, 4\}, \{1, 2, 3\}, \{1, 2, 4\}, \{1, 3, 4\}, Z\}$ is G_T, $\Im = \{\emptyset, \{1\}\}$ is an ideal on Z and $\xi = \{(\emptyset, \emptyset), (\{a_2\}, \{b_2\}), (\{a_1\}, \{\Upsilon_2\}), (\Upsilon_1, \Upsilon_2)\}$ is ξ_T from Υ_1 to Υ_2. Now consider the mapping $\mathcal{F} : (Z, \mathcal{T}) \to \Upsilon_1 \times \Upsilon_2$ defined by $\mathcal{F}(1) = (a_1, \Upsilon_2)$ and $\mathcal{F}(2) = \mathcal{F}(3) = (a_2, b_2) = \mathcal{F}(4)$. Thus $\mathcal{F}^{-1}(L, M)$ for every *ξ-open set* (L, M) in $(\Upsilon_1, \Upsilon_2, \xi)$ is *$\mathcal{T}_{\Im}^*$-clopen set* but not *$\mathcal{T}_{\Im}^*$-semi-clopen set* in $(Z, \mathcal{T})$. Hence $\mathcal{F} : (Z, \mathcal{T}) \to \Upsilon_1 \times \Upsilon_2$ is *totally $\xi_{\Im}^*$-continuous map* but not *totally $\xi_{\Im}^*$-semi-continuous map.*

Example 5.13: Suppose $Z = \{1, 2, 3\}$, $\Upsilon_1 = \{a_1, a_2\}$ and $\Upsilon_2 = \{b_1, b_2\}$. Then $\mathcal{T} = \{\emptyset, \{1, 3\}, \{2, 3\}, Z\}$ is G_T, $\Im = \{\emptyset, \{1, 3\}\}$ is an ideal on Z and $\xi = \{(\emptyset, \emptyset), (\{a_2\}, \{b_2\}), (\{a_1\}, \{\Upsilon_2\}), (\Upsilon_1, \Upsilon_2)\}$ is ξ_T from Υ_1 to Υ_2. Now consider the mapping $\mathcal{F} : (Z, \mathcal{T}) \to \Upsilon_1 \times \Upsilon_2$ defined as $\mathcal{F}(1) = (a_1, \Upsilon_2)$ and $\mathcal{F}(2) = (a_2, b_2) = \mathcal{F}(3)$. Thus $\mathcal{F}^{-1}(L, M)$ for every *ξ-open set* (L, M) in $(\Upsilon_1, \Upsilon_2, \xi)$ is *$\mathcal{T}_{\Im}^*$-semi-clopen set* in $(Z, \mathcal{T})$. Hence $\mathcal{F} : (Z, \mathcal{T}) \to \Upsilon_1 \times \Upsilon_2$ is *totally $\xi_{\Im}^*$-semi-continuous map* but not *totally $\xi_{\Im}^*$-continuous map.*

Remark 5.2: *Totally $\xi_{\Im}^*$-continuous maps $\nLeftrightarrow$ totally $\xi_{\Im}^*$-α-continuous maps.*

This is explained in Example 5.14 and Example 5.15.

Example 5.14: The mapping $\mathcal{F} : (Z, \mathcal{T}) \to \Upsilon_1 \times \Upsilon_2$ in Example 5.12 is *totally $\xi_{\Im}^*$-continuous map* but not *totally $\xi_{\Im}^*$-α-continuous map.*

Example 5.15: The mapping $\mathcal{F} : (Z, \mathcal{T}) \to \Upsilon_1 \times \Upsilon_2$ in Example 5.13 is *totally $\xi_{\Im}^*$-α-continuous map* but not *totally $\xi_{\Im}^*$-continuous map.*

Remark 5.3: *Totally $\xi_{\Im}^*$-semi-continuous maps $\nLeftrightarrow$ totally $\xi_{\Im}^*$-pre-continuous maps.*

This is explained in Example 5.16 and Example 5.17.

Example 5.16: Suppose $Z = \{1, 2, 3, 4\}$, $\Upsilon_1 = \{a_1, a_2\}$ and $\Upsilon_2 = \{b_1, b_2\}$. Then clearly $\mathcal{T} = \{\emptyset, \{1\}, \{1, 3\}, \{1, 4\}, \{3, 4\}, \{1, 2, 3\}, \{1, 2, 4\}, \{1, 3, 4\}, Z\}$ is G_T, $\Im = \{\emptyset, \{1, 2, 4\}\}$ is an ideal on Z and $\xi = \{(\emptyset, \emptyset), (\{a_2\}, \{b_2\}), (\{a_1\}, \{\Upsilon_2\}), (\Upsilon_1, \Upsilon_2)\}$ is ξ_T from Υ_1 to Υ_2. Now consider the mapping $\mathcal{F} : (Z, \mathcal{T}) \to \Upsilon_1 \times \Upsilon_2$ defined as $\mathcal{F}(1) = (a_1, \Upsilon_2) = \mathcal{F}(3)$ and $\mathcal{F}(2) = (a_2, b_2) = \mathcal{F}(4)$. Thus $\mathcal{F}^{-1}(L, M)$ for every *ξ-open set* (L, M) in $(\Upsilon_1, \Upsilon_2, \xi)$ is *$\mathcal{T}_{\Im}^*$-semi-clopen set* but not *$\mathcal{T}_{\Im}^*$-pre-clopen set* in $(Z, \mathcal{T})$. Hence $\mathcal{F} : (Z, \mathcal{T}) \to \Upsilon_1 \times \Upsilon_2$ is *totally $\xi_{\Im}^*$-semi-continuous map* but not *totally $\xi_{\Im}^*$-pre-continuous map.*

Example 5.17: In Example 5.16, let I $= \{\emptyset, \{3\}\}$ on ideal on Z and define $\mathcal{F} : (Z, \mathcal{T}) \rightarrow \Upsilon_1 \times \Upsilon_2$ defined as $\mathcal{F}(1) = (a_2, b_2) = \mathcal{F}(2)$ and $\mathcal{F}(3) = (a_1, \Upsilon_2) = \mathcal{F}(4)$. Thus $\mathcal{F}^{-1}(L, M)$ for every ξ*-open set* (L, M) in $(\Upsilon_1, \Upsilon_2, \xi)$ is $\mathcal{T}_{\Im}^{*}$*-pre-clopen set* but not $\mathcal{T}_{\Im}^{*}$*-semi-clopen set* in $(Z, \mathcal{T})$. Hence $\mathcal{F} : (Z, \mathcal{T}) \rightarrow \Upsilon_1 \times \Upsilon_2$ is *totally* $\xi_{\Im}^{*}$*-pre-continuous map* but not *totally* $\xi_{\Im}^{*}$*-semi-continuous map.*

Remark 5.4: *Totally* $\xi_{\Im}^{*}$*-continuous map* $\Rightarrow$ $\nLeftarrow$ *totally* $\xi_{\Im}^{*}$*-pre-continuous* ($\xi_{\Im}^{*}$*-*β*-continuous) map.*

The proof is quite easy while the converse is explained in Example 5.18.

Example 5.18: Suppose $Z = \{1, 2, 3, 4\}$, $\Upsilon_1 = \{a_1, a_2\}$ and $\Upsilon_2 = \{b_1, b_2\}$. Then clearly $\mathcal{T} = \{\emptyset, \{1, 3\}, \{1, 4\}, \{1, 2, 3\}, \{1, 2, 4\}, \{1, 3, 4\}, Z\}$ is G_T, $\Im = \{\emptyset, \{2\}\}$ is an ideal on Z and $\xi = \{(\emptyset, \emptyset), (\{a_1\}, \{b_1\}), (\{a_2\}, \{b_2\}), (\{a_2\}, \{\Upsilon_2\}), (\Upsilon_1, \Upsilon_2)\}$ is ξ_T from Υ_1 to Υ_2. Now consider the mapping $\mathcal{F} : (Z, \mathcal{T}) \rightarrow \Upsilon_1 \times \Upsilon_2$ defined as $\mathcal{F}(1) = (a_1, b_1)$ and $\mathcal{F}(2) = \mathcal{F}(3) = \mathcal{F}(4) = (a_2, \Upsilon_2)$ Thus $\mathcal{F}^{-1}(L, M)$ for every ξ*-open set* (L, M) in $(\Upsilon_1, \Upsilon_2, \xi)$ is $\mathcal{T}_{\Im}$*-pre-clopen* ($\mathcal{T}_{\Im}$*-*β*-clopen) set* but not $\mathcal{T}_{\Im}^{*}$*-clopen set* in $(Z, \mathcal{T})$. Hence $\mathcal{F} : (Z, \mathcal{T}) \rightarrow \Upsilon_1 \times \Upsilon_2$ is *totally* $\xi_{\Im}^{*}$*-pre-continuous* ($\xi_{\Im}^{*}$*-*β*-continuous) map* but not *totally* $\xi_{\Im}^{*}$*-continuous map.*

Remark 5.5: The direct consequences of the definitions of various generalized maps that we discussed in this section are given in Fig. 3.

6 Conclusion

The very useful concept of $\xi_{\Im}$-semi-continuous, $\xi_{\Im}^{*}$-continuous and $\xi_{\Im}^{*}$-semi-continuous maps has been introduced and the relationships between these maps and several other maps like ξ-continuous, ξ-semi-continuous, ξ-pre-continuous, ξ-α-continuous, ξ-β-continuous, $\xi_{\Im}$-semi-continuous, $\xi_{\Im}$-pre-continuous, $\xi_{\Im}$-α-continuous, $\xi_{\Im}$-β-continuous and $\xi_{\Im}^{*}$-continuous maps, etc. are established in this paper. In the present direction, we have categorized maps in ideal generalized binary topological spaces and investigated the behaviour of presented maps by utilizing ideal.

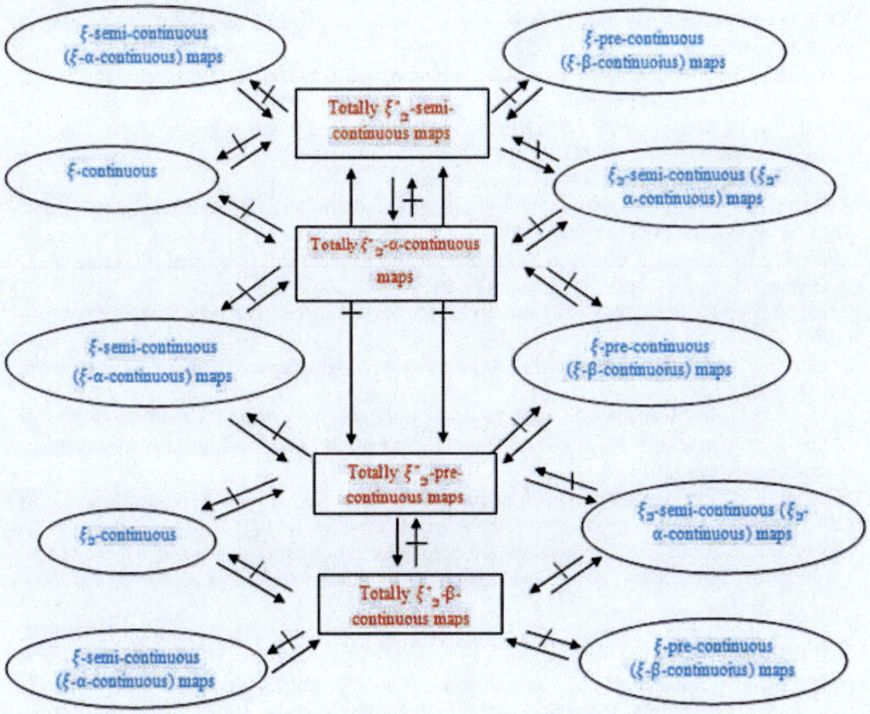

Fig. 3 Relationships of totally $\xi^*_{\Im}$-semi-continuous maps with some other maps

References

1. Ahengar, N.A., Maitra, J.K.: On g-binary continuity. J. Emerg. Technol. Inov. Res. **7**, 240–244 (2018)
2. Chen, C.C., Conejero, J.A., Kostic, M., Murillo-Arcila., M.: Dynamics on binary relations over topological spaces. Sym. **10**, 211 (2018). https://doi.org/10.3390/sym10060211
3. Csaszar, A.: Generalized topology, generalized continuity. Acta Math. Hungar **96**, 351–357 (2002)
4. Devi, R., Balachandran K., Maki, H.: Semi-generalized closed maps and generalized semi-closed maps. Mem. Fac. Sci. Kochi Univ. Ser. A Math. **14**, 41–54 (1993)
5. Dontchev, J.: On pre-I-open sets and a decomposition of I-continuity. Bayan Math. L **2** (1996)
6. Egenhofer, M.J.: Reasoning about binary topological relations. In: Symposium on Spatial Databases SSD 1991. Advances in Spatial Databases, pp. 141–160 (1991)
7. Gevorgyan, P.S.: Groups of binary operations and binary G-spaces. Topol. Appl. **201**, 18–28 (2016)
8. Hatir, E., Noiri, T.: On β–I-open sets and decomposition of almost-I-continuity. Bull. Malays. Math. Sci. **29**, 119–124 (2006)
9. Hatir, E., Noiri, T.: Decompositions of continuity and complete continuity. Acta Math. Hung. **4**, 281–287 (2006)
10. Jafari, S., Rajesh, N.: Generalized closed sets with respect to an ideal. Eur. J. Pure Appl. Math. **2**, 147–151 (2011)
11. Jankovic, D., Hamlett, T.R.: Compactible extensions of ideals. Bull. Mat. Ital. **6-B**, 453–465 (1992)

12. Khalimsky, E.D., Kopperman, R., Meyer, P.R.: Computer graphics and connected topologies an finite ordered sets. Topol. Appl. **36**, 1–17 (1990)
13. Kong, T.Y., Kopperman, R., Meyer, P.R.: A topological approach to digital topology. Amer. Math. Mon. **98**, 901–917 (1991)
14. Kovalesky, V., Kopperman, R.: Some topology-based imaged processing algorithms. Ann. NY Acad. Sci. **728**, 174–182 (1994)
15. Kuratowski, K., Topologie, I.: Warszawa (1930)
16. Kırbas, H., Aslım, G.: Decompositions of continuity and some weak forms of continuity. Chaos, Solitons Fractals **41**, 1684–1690 (2009)
17. Landi, G.: An Introduction to Non Commutative Spaces and Their Geometrics (Lecture Notes in Physics). Springer-Verlag, New York (1997)
18. Levine, N.: A decomposition of continuity in topological spaces. Am. Math. Mon. **68**, 44–46 (1961)
19. Levine, N.: Semi open sets and semi continuity in topological spaces. Amer. Math. Mon. **70**, 36–41 (1963)
20. Levine, N.: Generalized closed sets in topology. Rend. Cir. Mat. Palermo **2**, 89–96 (1970)
21. Maitra, J.K., Tripathi, H.K.: Local function in generalized ideal topological spaces. VISLESANA **1**, 191–195 (2014)
22. Michael, F.: On semi-open sets with respect to an ideal. Eur. J. Pure Appl. Math. **6**, 53–58 (2013)
23. Moore, E.L.F., Peters, T.J.: Computational topology for geometric design and molecular design. In: Ferguson, D.R., Peters, T.J. (eds.) Mathematics in Industry Challenges and Frontiers 2003. SIAM (2005)
24. NithyananthaJothi, S., Thangavelu, P.: On binary topological spaces. Pac. Asian J. Math. **2**, 133–138 (2011)
25. Njastad, O.: On some classes of nearly open sets. Pac. J. Math. **15**, 961–970 (1965)
26. Pawlak, Z.: Rough sets: theoretical aspects of reasoning about data. In: System Theory, Knowledge Engineering and Problem Solving, vol. 9. Dordrecht, Kluwer (1991)
27. Rodyna, A.H., Deena, A.-K.: Types of generalized open sets with ideal. Int. J. Comput. Appl. **4**, 0975-8887 (2013)
28. Son, M.J., Park, J.H., Lim, K.M.: Weakly clopen functions. Chaos, Solitons Fractals **33**, 1746–1755 (2007)
29. Svozil, K.: Quantum field theory on fractal space–time: a new regularization method. J. Phys. A Math. Gen. **20**, 3861–3875 (1987)

Application of Soft Set Theory for Dimensionality Reduction Approach in Machine Learning

P. D. Lanjewar and B. F. Momin

1 Introduction

Nowadays uncertainty is presented in decision-making process. It has been found that the problem of uncertainty is too complicated to capture within a single framework. There are so many difficulties in engineering, social science, medical science, and economics, which include imprecise or uncertain data. For this purpose, a number of different theories like probability theory, fuzzy sets, interval mathematics, and rough sets have been developed that handle uncertain data, but these theories have their own problem. Such as in the theory of probability, there is some problem in defining the probability density function and it is applicable only for stochastically stable systems. In fuzzy set theory, there is difficulty in defining the membership function for different cases. Interval mathematics is unreliable, not adequate and defective information is presented. In rough set theory, there is no equivalence relation between decision attribute values and conditional attribute values. So that these traditional methods are not useful because of various types of difficulties presented above. The main reason for these problems is the insufficiency of the parameterization concept in the above theories. In traditional mathematics, there is a need to develop a mathematical model for an object with the exact solution. But usually, it is too complex to define the precise answer to mathematical model. Noting such a problem the novel idea of Soft Set theory that is SST is being developed by Russian Scientist D. Molodtsov in 1999 [1] that handles uncertain data means unknown and missing data to solve the problem of decision-making in data mining and machine learning areas. The researchers have

P. D. Lanjewar (✉) · B. F. Momin
Department of Computer Science & Engineering, Walchand College of Engineering, Sangli, Maharashtra 416415, India
e-mail: priyanka.lanjewar@walchandsangli.ac.in

B. F. Momin
e-mail: bashirahamad.momin@walchandsangli.ac.in

P. Srivastava et al. (eds.), *Mathematical, Computational Intelligence and Engineering Approaches for Tourism, Agriculture and Healthcare*, Lecture Notes in Networks and Systems 214, https://doi.org/10.1007/978-981-16-3807-7_17

been focusing on the topic called "Soft Set theory" as parameter reduction theory that is based on parameterization property for the selection of parameters to give the optimal solution. So the SST theory is free from existing problems and provides an approximate solution to the problem and no need to define a specific solution. There is no restriction on the estimated depiction of an object which makes the soft set theory more convenient and easily applicable in practice. It is used to recognize improper data and construct the reduction of parameters for decision-making to give optimal selection. Maji et al. [2] has shown the use of SST by giving the example of the selection of house for optimal selection process using SST theory concept for the first time and give the definition of the soft set as follows:

$$F : E \to P(U), \tag{1}$$

where *P (U)* is the power set of *U* and *(F, E)* is a soft set if *F* is the mapping of *E* it means that, set *U* is the subset of the parameterized soft set.

2 Related Works

SST is a parameter reduction tool that uses the concept of parameterization property to take the optimal decision which is now explored as dimensionality reduction for big data analysis. There are various parameter reduction methods based on SST that give optimal solution over data as follows.

The first example and the set of rules for selecting a house using SST are presented in [2] as follows:

Consider, set U = $\{x_1, x_2, x_3, x_4, x_5, x_6\}$ as a combination of houses and E = {costly, attractive, timber, low-priced, green-atmospheres, modern, good-condition, bad- condition} is the set of attributes or parameters.

For this example, soft set (F, E) defines the 'characteristics of house', specified by

(F, E) = {costly = Ø,

Attractive houses = $\{x_1, x_2, x_3, x_4, x_5, x_6\}$,
Timber houses = $\{x_1, x_2, x_6\}$,
Modern houses = $\{x_1, x_2, x_6\}$,
Houses in bad-condition = $\{x_2, x_4, x_5\}$,
Low-priced houses = $\{x_1, x_2, x_3, x_4, x_5, x_6\}$,
Houses in good-condition = $\{x_1, x_3, x_6\}$,
Houses in green-atmospheres = $\{x_1, x_2, x_3, x_4, x_6\}$}.

If someone is attentive to buy a house considering these constraints as: 'attractive', 'timber', 'low-priced', 'green-atmospheres', 'good-condition' that represent as a subset P = {attractive(y_1), timber(y_2), low-priced (y_3), green-atmospheres(y_4), in good- condition(y_5)} of the set E.

Table 1 Soft set (F, E)

U	y_1	y_2	y_3	y_4	y_5
x_1	1	1	1	1	1
x_2	1	1	1	1	0
x_3	1	0	1	1	1
x_4	1	0	1	1	0
x_5	1	0	1	0	0
x_6	1	1	1	1	1

According to this above information, tabular presentation of the soft set is given in Table 1.

The algorithm for the optimal selection of houses is as follows.

1. Input the soft set (F, E).
2. Input the set P of choice parameters which is a subset of E.
3. Identify all reduct-soft sets of (F, P).
4. Select one reduct-soft set say (F, Q) of (F, P).
5. Obtain k, for which $Ck = max\ c_i$.

The above algorithm first takes the soft set (F, E) as input and then sets the choice of parameters. As parameters y_1 and y_3 have the same value for each house so according to that calculate the reduct-soft set from the original soft set (F, E) as there are two reduct-soft sets (F, P) = $\{y_1, y_2, y_4, y_5\}$ and $\{y_2, y_3, y_4, y_5\}$ by choosing one reduct set as (F, Q) = $\{y_1, y_2, y_4, y_5\}$ as shown in Table 2. And after that calculate the choice value for reduct-soft set as follows:

$$c_i = \sum\nolimits_j x_{ij} \tag{2}$$

where c_i is the choice value is given by the summation of each x_{ij} entry in Table 1 and the selection of optimal value is made according to the maximum value of c_i as shown in Table 2. As choice values for houses x_1 and x_6 are maximum so the optimum selection will be x_1 and x_6 for the selection of houses as in Table 2. But in

Table 2 Reduct-soft set (F, Q)

U	y_1	y_2	y_4	y_5	Choice value = c_i
x_1	1	1	1	1	4
x_2	1	1	1	0	3
x_3	1	0	1	1	3
x_4	1	0	1	0	2
x_5	1	0	0	0	1
x_6	1	1	1	1	4

Table 3 Parametrization reduction

U	y_2	y_5	Choice value = c_i
x_1	1	1	2
x_2	1	0	1
x_3	0	1	1
x_4	0	0	0
x_5	0	0	0
x_6	1	1	2

[2] first, the reduction of the input set is calculated after calculating the choice value that leads toward a problem that is the reduction of the soft set are common which results in inconsistent parameter reduction for the selection of house.

This is improved by reducing parameters of soft sets as presented by Degang Chen in [3] using parameterization reduction property of soft set. The correct parameter reduction of the above algorithm is pointed out and the other way of parameter reduction is provided as follows.

If parameters y_1, y_3, y_4, are deleted from Table 1, then the optimal object remains the same. If one of the parameters is omitted from $\{y_2, y_5\}$ then definitely optimal object will be changed. It means that if parameter $\{y_2, y_4\}$ is deleted then optimal object selection will be $\{x_1, x_3, x_6\}$. Thus the final parameter reduction of the above example is $\{y_2, y_5\}$. So here by using only two parameters y_2 and y_5, optimal object selection of house will be x_1 and x_6 as exposed in below Table 3.

Kong et al. [4] have presented the first normal parameter reduction (NPR) algorithm for soft set in which the problem of suboptimal choice is solved by adding the parameter set of the soft set. This method computes parameter importance degree and decision partition to give the optimal solution for decision-making. But this method is not easy to understand and needs more computation. Herawan et al. [5] have shown dimensionality reduction using multi-valued information which is to be transformed into multi-soft sets. AND, OR operations are used to reduce the soft set attributes. So the difficulty of [4] is solved by Ma et al. [6] in which NENPR is developed called a new efficient NPR technique, without considering parameter important degree parameter and partition of decision. Ma et al. [7] discussed soft set algorithm depends on parameterization value reduction that considers the least parameter for optimal selection. Kumar and Rengasamy [8] also presented better decisions for soft set theory using parameterization reduction by converting sample data into binary data without any loss of real data. Danjuma et al. [9] proposed ANPR alternative approach to NPR that overcome the difficulty of existing NPR algorithms in terms of computational complexity but the last choice not always consider and also gives comparisons of NPR, NENPR, and ANPR.

The above all soft set based NPR algorithm mainly focuses on dummy data but when they performed on the original dataset then they faced problems. To recover such a problem Ma [10] proposed a weighted normal parameter reduction algorithm for decision-making. This is the first algorithm that is implemented for real-time mobile phone dataset for 15 parameters and 28 objects, each parameter has its own weights. Finally, optimal and suboptimal choice is shown for real-time decision-making process. The review of various dimensionality reduction techniques under rough set theory is presented by Thangavel in [11]. Hybrid of fuzzy, rough set, and neural network methods for decision-making under missing data for various applications are also presented. Another dimensionality reduction method for complex massive data has been proposed by Zhe in [12] based on a rough set. Two methods of reduction are proposed one is for an inconsistent complete decision system and the other is for an incomplete decision system and experimented for various UCI datasets [13]. Feature or attribute reduction method based on rough set is also presented by Wann et al. [14] that reduces the dimensionality of feature space. A soft computing method for dimensionality reduction is presented by Pant et al. [15] based on SST for multi-valued information systems based on operations AND and OR. Another hybrid approach for dimensionality reduction based on fuzzy SST is suggested by Sharma et al. [16] in which soft set is converted in the fuzzy soft set then calculate grade membership function and the threshold value of fuzzy soft set to obtained the reduced data.

3 Methodology

As soft set theory is the latest reduction theory as mentioned in the literature survey, it provides various parameter reduction algorithms, but all the NPR algorithms are performed for the Boolean type dummy data. There is no NPR algorithm that is performed on real-time datasets. In decision-making process, if the number of parameters is increasing then it becomes difficult to provide the optimal solution. This paper explores the notion of dimensionality reduction method using the theory of soft set for big data analysis. As dimensionality reduction is the data mining technique that reduces the number of dimensions or features or parameters. To work out the problem of decision-making in big data analysis, a dimensionality reduction approach depend on SST is proposed. The normal dimensionality reduction soft set algorithm (NDRSS) is proposed here to give the optimal solution over big data. The flow of research work is shown in Fig. 1. At first input data is taken from the dataset, some preprocessing on input data is done and then soft set based dimensionality reduction is applied to give the optimal output (Fig. 2).

Degree of importance of e_i calculated as

$$\mathrm{r_{ei}} = \frac{1}{|U|}(\alpha 1, \mathrm{ei} + \alpha 2\mathrm{ei} + \ldots + \alpha s, \mathrm{ei}) \tag{3}$$

and cardinality of a set is defined as

$$\alpha_{k,e_i} = \begin{cases} \left|E_{f_k} - \overline{E - e_{if_{z'}}}\right|, \\ \left|E_{f_k}\right|, \quad \text{otherwise.} \end{cases} \tag{4}$$

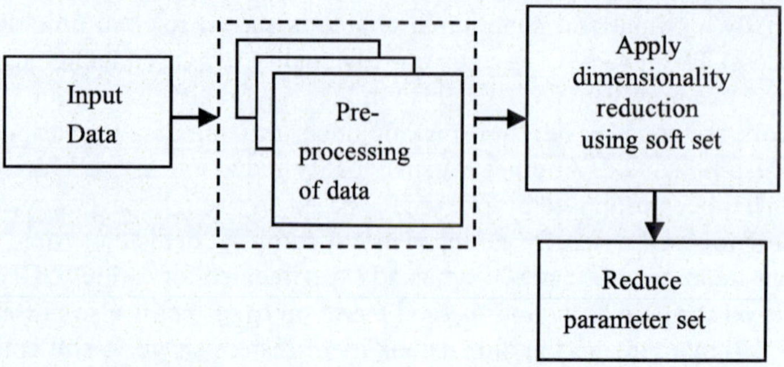

Fig. 1 Research flow of proposed work

Pseudo Code of NDRSS Algorithm:

Step 1: Take the input from dataset.
Step 2: Covert it into soft set (F, E)
Step 3: Degree of importance is calculated
Step 4: Choose highest cardinality
Step 5: Compute reduct data which is NDRSS

Fig. 2 NDRSS algorithm

4 Result and Discussion

The implementation of the above dimensionality reduction soft set algorithm is performed on a machine learning platform using Python version of 3.7.1. The algorithm first takes the digit dataset of the MNIST dataset and then performed soft set operations to reduce the dimensionality of data. MNIST dataset contains 60,000 images for training and 10,000 images for testing. Each image is of 28*28 that becomes 784*1 vectors as high dimension. The digit dataset of MNIST contains 0 to 9 handwritten digits as shown in Fig. 3. Implementation takes the first 60,000 samples to train which is run for 50 epochs with batch size 256 and then validate on 10,000 samples. Figure 4 shows the comparison of two algorithms that is PCA and NDRSS in which the output of NDRSS shows the more reduction of digit dataset as compared to the PCA algorithm. The output of PCA and NDRSS contains clusters of dots of different colors that represent the images of 0 to 9 digits in which each cluster represents a single-digit number. So by observing the output it clearly shows that the proposed algorithm gives more reduction on high-dimensional data of digit dataset.

Fig. 3 Original data of digit datasets

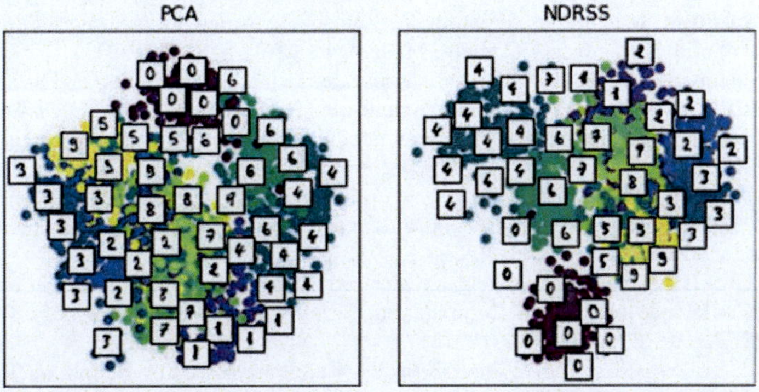

Fig. 4 Output of PCA and NDRSS algorithm

The performance of the above algorithm is calculated with regard to complexity and accuracy over existing algorithms.

5 Conclusion

The SST is the latest mathematical philosophy that solves the problem of uncertain data in decision-making process of big data analysis. Various soft set based NPR methods are presented in the literature survey of this paper. The SST gives a better parameter reduction than the rough set and fuzzy set. In the literature survey, no such method in the direction of the dimensionality of data of real-time dataset using soft set. Thus this paper proposed such a normal dimensionality reduction soft set algorithm that gives the reduction data for optimal selection in decision-making. The result shows how the proposed method gives more reduction than PCA over the MNIST dataset. In future, there is scope for other mining methods like classification and clustering and association rule mining based on soft set.

References

1. Molodtsov, D.: Soft set theory—first results. Comput. Math. Appl. **37**(4–5), 19–31 (1999)
2. Maji, P.K., Roy, A.R., Biswas, R.: An application of soft sets in a decision making problem. Comput. Math. Appl. **44**(8–9), 1077–1083 (2002)
3. Chen, D., Tsang, E.C.C., Yeung, D.S., Wang, X.: The parameterization reduction of soft sets and its applications. Int. J. Comput. Math. with Appl. **49**, 757–763 (2005)
4. Kong, Z., Liqun, G., Lifu, W., Steven, L.: The normal parameter reduction of soft sets and its algorithm. Comput. Math. Appl. **56**, 3029–3037 (2008)
5. Herawan, T., Rose, A.N.M., Deris, M.M.: Soft Set Theoretic Approach for Dimensionality, pp. 171–178 (2009)

6. Ma, X., Sulaiman, N., Qin, H., Herawan, T., Zain, J.M.: A new efficient normal parameter reduction algorithm of soft sets. Comput. Math. Appl. **62**(2), 588–598 (2011)
7. Ma, X., Sulaiman, N., Qin, H.: Parameterization value reduction of soft Sets and its algorithm. In: 2011 IEEE Colloquium on Humanities, Science and Engineering. IEEE, pp. 261–264 (2011)
8. Kumar, D.A., Rengasamy, R.: Parameterization reduction using soft set theory for better decision making. In: 2013 International Conference on Pattern Recognition, Informatics and Mobile Engineering. IEEE, pp. 365–367 (2013)
9. Danjuma, S., Ismail, M.A., Herawan, T.: An alternative approach to normal parameter reduction algorithm for soft set theory. IEEE Access **5**, 4732–4746 (2017)
10. Ma, X., Qin, H.: Weighted normal parameter reduction for soft set. In: 2018 14th International Conference on Natural Computation, Fuzzy Systems and Knowledge Discovery (ICNC-FSKD). IEEE, pp. 712–716 (2018)
11. Thangavel, K., Pethalakshmi, A.: Dimensionality Reduction Based on Rough Set Theory: A Review, vol. 9, pp. 1–12 (2009)
12. Zhe, D.: A Dimensionality reduction based on rough set theory for complex massive data. In: 2015 8th International Congress on Image and Signal Processing (CISP). IEEE, pp. 1520–1528 (2015)
13. UCI Machine Learning Repository. https://archive.ics.uci.edu/ml/datasets.php
14. Wang, S.Q., Gao, C.Y., Luo, C., Zheng, G.M., Zhou, Y.N.: Research on feature selection/attribute reduction method based on rough set theory. Procedia Comput. Sci. **154**, 194–198 (2019)
15. Pant, J., Juyal, A., Bahuguna, S.: Soft set, a soft computing approach for dimensionality reduction. Int. J. Innov. Sci. Eng. Technol. **2**(4), 728–735 (2015)
16. Sharma, O., Gupta, P., Sheoran, A.: Dimensionality reduction using fuzzy soft set theory. Int. J. Stat. Reliab. Eng. 154–158 (2018)

QoS Evaluation for Wireless Networks Using *K*-Nearest Neighbor (k-NN) Algorithm

Siddharth Goutam and Srija Unnikrishnan

1 Introduction

Several mobile and wireless technologies will have co-existence in Next Generation Wireless Networks (NGWNs), which will be providing high data rate to the mobile users. Mobile users will be roaming across the various heterogeneous wireless networks with expectation of seamless mobility and high data rate coupled with good Quality of Service (QoS) [1]. For realization of these metrics efficient Vertical Handover (VHO) is an essential requirement. Vertical Handover Decision Algorithm (VHDA) should also be designed in a way that QoS is assured and does not get degraded [2].

QoS of the network is mainly depends upon the network parameters like latency, jitter and packet loss.

$QoS\ of\ Network = f^{n}(latency, jitter, packetloss)$ [3]

$$QoS\ of\ Network \propto \frac{1}{Latency} \tag{1}$$

$$QoS\ of\ Network \propto \frac{1}{Jitter} \tag{2}$$

$$QoS\ of\ Network \propto \frac{1}{Packet\ Loss} \tag{3}$$

Improvement in QoS can be achieved by reduction in Latency, Jitter and Packet Loss. The decisions pertaining to incorrect handovers can be avoided by using classification algorithms like k-NN Algorithm [4].

S. Goutam (✉) · S. Unnikrishnan
Fr. Conceicao Rodrigues College of Engineering, Mumbai 400050, India

P. Srivastava et al. (eds.), *Mathematical, Computational Intelligence and Engineering Approaches for Tourism, Agriculture and Healthcare*, Lecture Notes in Networks and Systems 214, https://doi.org/10.1007/978-981-16-3807-7_18

This research paper is organized as follows: Sect. 2 captures the literature survey and the theoretical background of k-NN classification is captured in Sect. 3. The description of the system model and implementation of algorithm is captured in Sect. 4. The Simulation Results are captured in Sect. 5 followed by Statistical Analysis in Sect. 6. Section 7 captures the conclusion.

2 Literature Review

In this research paper, we have explored the use of k-NN Algorithm for determination of QoS of the wireless networks. Through literature review, we have performed study and analysis of the research performed by research scholars and academicians on the applications of k-NN Algorithm. In the paper presented by Imandoust, economic forecasting is implemented by using k-NN classification method. Important factors like financial distress on stakeholders have been predicted using machine learning techniques [5]. In the research paper presented by Zhang, short term traffic flow is predicted using the improved K-nearest neighbor model. The main factors considered in the prediction of short term traffic flow for a route are previous data, search mechanism and prediction plan [6]. In the research paper presented by Lin, K-NN based local linear wavelet neural network model is used for estimation of traffic volumes for short duration of time [7].

3 Theoretical Background

The k-NN algorithm is simple and a popularly used algorithm for classification and regression applications.

The Euclidean distance between two points (S1, S2) in an n-dimensional space is given by

$$distance(S_1, S_2) = \sqrt{\sum_{i=1}^{n} (s_{1i} - s_{2i})^2} \tag{4}$$

Where

$S1 = (s11, s12, \ldots\ldots\ldots\ldots s1n)$

$S2 = (s21, s22, \ldots\ldots\ldots\ldots s2n)$

n—Number of attributes of data set D

Each point in the training data set is kept in the n dimensional space pattern.

k-NN classification algorithm finds the space pattern for the k points in training data set for a new test point. And these k points are referred to as the nearest neighbors of the new point [8]. K-NN is used for classification as well as regression analysis [5–7].

3.1 K-NN for Classification

It is the simplest classification technique. The execution and effectiveness of the k-NN as the classifier is dependent on the following two factors;

i. Value of k—decided based on the implementation and speed
ii. Distance metric [5].

The performance analysis of the classification algorithm is done using following factors.

i. *Confusion Matrix* is defined as the grouping of actual and predicted class by the classifiers.

ii. $$Precision = \frac{\text{TruePositives}}{(\text{True Positives} + \text{FalsePositives})} \tag{5}$$

iii. $$Recall = \frac{\text{TruePositives}}{(\text{True Positives})} \tag{6}$$

iv. $$F - measure = \frac{2}{\left(\frac{1}{\text{Precision}} + \frac{1}{\text{Recall}}\right)} \tag{7}$$

v. $$Accuracy = \frac{\text{TP} + \text{TN}}{(\text{P} + \text{N})} \tag{8}$$

vi. *F1 score* is the weighted average of value of recall and value of precision [9].

3.2 K-NN for Regression

The performance of the regression algorithms is analyzed on the basis of the following parameters-

i. *R-Square Value* (R^2) is the measure to check the closeness of the data fitment to the line of regression. R-Square Value is also referred as the coefficient of determination.

ii. *Adjusted R-Square* (Ra^2) value is mostly lesser than R Square value. Adjusted R-Square can be termed as a modified form of the R-square [10].

$$R^2 = \frac{SCE_p}{SCE_{tot}} \tag{9}$$

$$R_a^2 = \frac{\frac{SCE_p}{p}}{\frac{SCE_{tot}}{(n-1)}} \tag{10}$$

where

SCE_p—sum of the squares pertaining to regression

SCE_{tot}—total sum value of squares

n—Total number of observations

p—Number of the parameters for estimation

e_i—error for ith sample

iii. $$Mean\ Absolute\ Error = \frac{1}{n}\sum_{i=1}^{n}|e_i| \tag{11}$$

iv. $$Root\ Mean\ Square\ Error = \sqrt{\frac{1}{n}\sum_{i=1}^{n}e_i^2} \tag{12}$$

Root mean square error denotes the spread in the residuals (difference in values i.e. predicted value and actual value) [11].

4 System Model and Implementation

Figure 1 gives the block diagram of the system model for QoS evaluation [12]. As

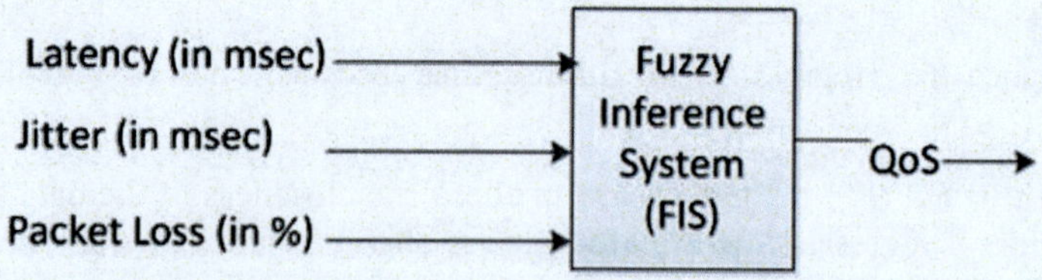

Fig. 1 Block diagram for FIS for QoS

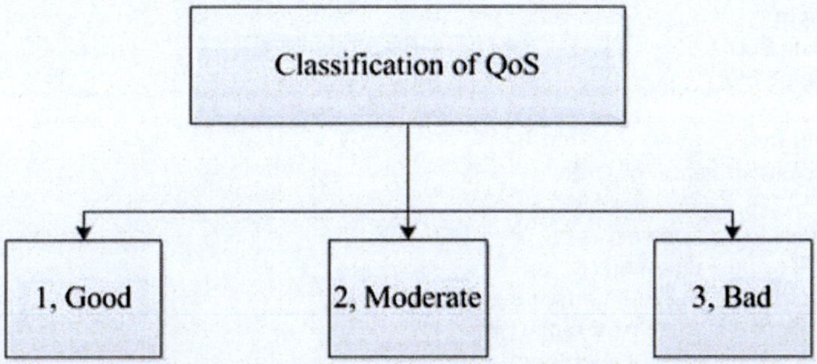

Fig. 2 Classification of QoS

Table. 1 Universe of discourse of parameters in dataset

Parameter	Unit of measurement	Minimum value	Maximum value
Latency	Msecs	0	6
Packet loss	%	0	6
Jitter	Msecs	0	6
QoS	(No units)	0	8

shown in the figure, the critical input parameters for predicting QoS are latency, jitter and packet loss of the wireless network. The FIS gives QoS value as the output, which is classified Good, Moderate and Bad.

Figure 2 shows the classification of the QoS for wireless networks.

The Universe of discourse for the input parameters which are under consideration is captured in Table 1.

Figure 3 gives the k-NN Classification Algorithm for QoS.

In a similar way, we have k-NN Regression Algorithm for QoS which gives the output basis the average value of the neighbors.

Thus, K-NN algorithm can be used to evaluate the QoS of wireless networks. The presence of overlapping heterogeneous wireless networks provide the choice to the user to get connected to the Best Network with the desired QoS. From the network operator perspective, since handoff involves high use of network resources, with good QoS evaluation, the frequency of handoff can be reduced.

5 Simulation and Results

We have presented a case study here where the dataset used for simulation of the algorithm are practically measured values of the network parameters.

The simulation environment is Python & WEKA (Waikato Environment for Knowledge Analysis) workbench.

Input –
Data Set D
Value of k

Output-
Classification of QoS

Steps to be followed –
1, Read the dataset (D).
2, Split the dataset into training and test dataset
3, Define k value over range (3 to 11)
k value has to be odd for classification
4, Calculate distance of newly arrived test point using the distance metric
5, Classify the point using class label and check its nearest neighbors
6, Find the classification of the test sample
7, End

Fig. 3 K-NN classification algorithm for QoS

5.1 K-NN Classification

Figure 4 shows the output of the k-NN classifier.

Figure 5 captures the confusion matrix for k = 5. Figure 6 captures the accuracy score of k-NN classifier for our dataset with variation in values of k.

Fig. 4 Output of k-NN classifier (k = 5)

Classification of QoS using K- Nearest Neighbor

M: Moderate QoS
N: Poor QoS
Y: Good QoS

{0: 'M', 1:'N', 2:'Y'}

Classification report :

Class	Precision	Recall	F1-Score
0	1.000	0.991	0.996
1	0.500	1.000	0.667
2	1.000	1.000	1.000
Weighted Average	0.996	0.991	0.993

Input Latency in msec : 12
Input Packet Loss in % : 0.8
Input Jitter in msec : 148

QoS of Wireless Network is Moderate

		Predicted Class of QoS		
		Good	Moderate	Poor
Actual Class of QoS	Good	1	0	0
	Moderate	0	114	1
	Poor	0	0	1

Fig. 5 Confusion matrix for k = 5

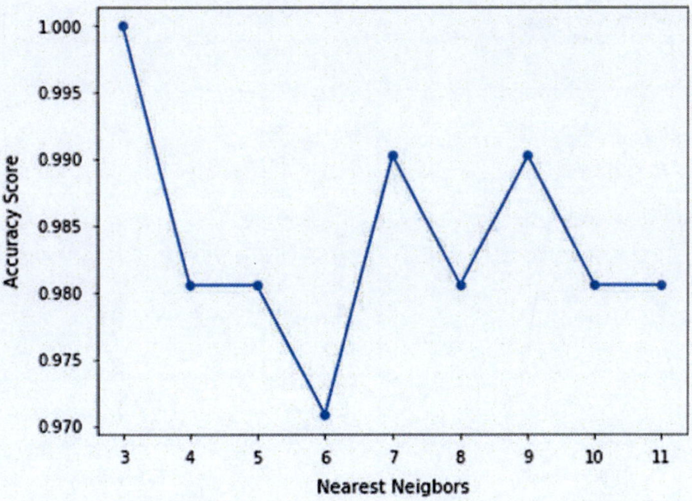

Fig. 6 Accuracy score with variation in value of k

Observations from Fig. 6: The graph shows the number of nearest neighbors, k vs the accuracy score. The value of k refers to the number of neighbors that needs to be considered in order to predict the value of the actual data point. Accuracy score is one of the metric to check the accuracy of classifications. For accurate prediction, we need to choose the k value, giving the highest accuracy score.

5.2 K-NN Regression

Figure 7 captures the output of k-NN Regressor with k = 5.

Observations from Fig. 7: If the value of the inputs parameters are within the values of universe of discourse, output is provided by Fuzzy Inference System but if the values of input parameters lie outside the range of universe of discourse, then the value of QoS can be obtained by using machine learning techniques like k-NN algorithm.

```
****Quality Service Analysis using K-Nearest Neighbor Regressor****

Model Behaviour Stats
R Square:  97.2441209812
Ajusted R Square:  97.2197326713
Mean Squared Error:  0.928779645116
Root Mean Squared Error:  9.63732143864
Input Latency (Ping) in msec     12
Input Packet Loss in %  0.8
Input Jitter in msec     148
Quality of Service value of Network is:  [ 3.93136461]
```

Fig. 7 Output of k-NN regressor

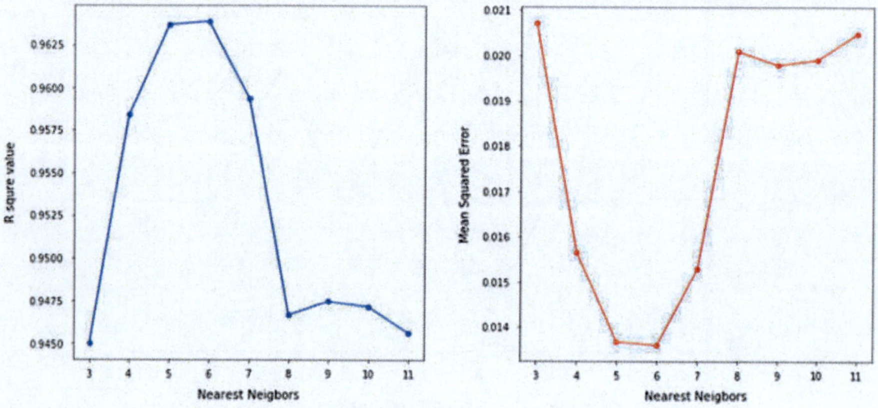

Fig. 8 Performance analysis of k-NN regressor

For the case study given in Fig. 7, the algorithm gives the output as QoS of the network being Moderate, based on the input parameters latency, jitter and packet loss.

Figure 8 captures the performance analysis of the k-NN Regressor. It captures the R square value and the mean square error as a function of k.

Observations from Fig. 8: Fig. 8 captures the performance analysis of k-NN Regressor and has two parts. First part captures variation in R square value with change in nearest neighbors (value of k). R square value is an indication how close the predictions are to the actual values ie: R-square is an accuracy measuring metric to check the accuracy of the predicted regression values. We need to choose the k value having the highest R Square.

In the second part of Fig. 8, mean square error is plotted against the values of k. As is clearly visible from the graph, it is the inverse of the first part of Fig. 8, as we are plotting error against the k value. We choose the k where the Mean Squared Error is the least.

k-NN algorithm output can also be represented as a continuous QoS value instead of classifying as Good, Moderate and Poor.

6 Analysis of Results

We have performed QoS analysis of the wireless network using k-NN classification and regression. For k = 5, in k-NN classifier we get the output that QoS is moderate while in k-NN Regressor we get the value of QoS as 3.93 on a scale of 0–8, as given in Table 1.

6.1 Analysis of Results of k-NN Classification

From Fig. 4, we observe that precision and recall values are high. The precision shows the measure of exactness while recall is referred to as the measure of completeness.

With high scores of precision and recall, F1 score is also high. The accuracy of model is very high (around 99%) as the test data sets are correctly classified after training.

6.2 Analysis of Results of k-NN Regression

R-square describes the degree of goodness-of –fit of data for the regression models. In our simulation environment, for k = 5, we have received the value of R-square as 97.2%. This denotes correct fitment of the data, based on similar results [10].

Low values of the means square error denotes that the estimation is correct [11].

7 Conclusion and Future Scope

This research paper gives the classification and evaluation of QoS for wireless networks based on k-NN algorithm. The main parameters which impact QoS practically have been considered in the system model. The proposed algorithm may be utilized in the network design and site optimization by telecom network operators. QoS evaluation is very necessary for making good handover decisions. Future scope includes extending the algorithm for evaluation of networks for vertical handover.

References

1. Nkansah-Gyekye, Y., Agbinya, J.I.: A vertical handoff decision algorithm for next generation wireless networks. In: Third International Conference on Broadband Communications, Information Technology & Biomedical Applications. IEEE Computer Society (2008)

2. Yan, X., Mani, N., Sekercioglu, Y.A.: A traveling distance prediction based method to minimize unnecessary handovers from cellular networks to WLANs. IEEE Commun. Lett. **12**(1), 14–16 (2008)
3. Goutam, S., Unnikrishnan, S.: Analysis and comparison of decision tree algorithms for vertical handover in wireless networks. Grenze Int. J. Eng. Technol. **5**, 83–89 (2019)
4. Ma, B., Wang, D., Cheng, S., Xie, X.: Modeling and analysis for vertical handoff based on the decision tree in a heterogeneous vehicle network. IEEE Access **5**, 8812–8824 (2017)
5. Han, J., Kamber, M., Pei, J.: Data mining concepts and techniques, 3rd edn. Elsevier Morgan Kaufmann Publishers, MA (2011)
6. Imandoust, S.B., Bolandraftar, M.: Application of K-Nearest Neighbor (K-NN) approach for predicting economic events: theoretical background. Int. J. Eng. Res. Appl. **3**(5), 605–610 (2013)
7. Zhang, L., Liu, Q., Yang, W., Wei, N., Dong, D.: An improved K-Nearest Neighbor model for short-term traffic flow prediction. In: Elsevier Procedia—Social and Behavioral Sciences (2013)
8. Lin, L., Li, Y., Sadek, A.: A k nearest neighbor based local linear wavelet neural network model for on-line short-term traffic volume prediction. In: International Conference of Transportation Professionals, Elsevier Procedia (2013)
9. Goutam, S., Unnikrishnan, S., Prabavathy, S., Kudu, N., Goutam, A.: Assessment and prediction of quality of service of wireless networks using support vector machines. Grenze Int. J. Eng. Technol. **5**, 119–126 (2019)
10. Akossou, A.Y., Palm, R.: Impact of data structure on the estimators R-square and adjusted R-square in linear regression. Int. J. Methematics Comput. CESER Publ. **20**(3), 84–93 (2013)
11. Chai, T., Draxler, R.R.: Root Mean Square Error (RMSE) or Mean Absolute Error (MAE) arguments against avoiding RMSE in the literature. Geosci. Model. 1247–1250 (2014)
12. Goutam, S., Unnikrishnan, S.: QoS based vertical handover decision algorithm using fuzzy logic. In: International Conference on Nascent Technologies in Engineering, IEEE, Mumbai (2019)

On the Generalized Fractional Calculus Operators Involving the H-Function of Several Variables

Utpal Kumar Saha

1 Introduction

On account of the importance and usefulness of the fractional integral operators in the theory of fractional calculus and other allied topics, these operators have been studied from time to time by many authors notably Erdélyi [1], Riemann–Liouville [2], Gupta and Soni [3], Kiryakova [4], Kober [5], Lowndes [6], Purohit et al. [7], Raina [8], Saigo et al. [9], Saigo and Raina [10], Saigo et al. [11], Saxena and Kumbhat [12], Srivastava et al. [13], Srivastava and Saxena [14], etc. Baleanu and Agarwal [15] investigated generalized operators of the fractional integration involving Appell's function $F_3(.)$ due to Marichev-Saigo-Maeda. Saiful et al. [16] studied Marichev–Saigo–Maeda fractional integration operators involving generalized Bessel functions. Sunil Kumar Sharma and his colleague Shekhawat [17] established image formulas for M-series, involving some operators of fractional integrals in terms of Fox's H-function. Recently, Suthar et al. [18] studied generalized multivariable Mittag-Leffler function via generalized fractional calculus operators involving Fox's H-function as a kernel. Inspired and motivated by the above researchers, we introduced two fractional calculus operators involving the H-function of several variables in the kernel. In this paper, our main aim is to study two key results via our operators in the form of two theorems.

U. K. Saha (✉)
Department of Basic and Applied Science, National Institute of Technology Arunachal Pradesh, Jote, Arunachal Pradesh, India
e-mail: utpal@nitap.ac.in

P. Srivastava et al. (eds.), *Mathematical, Computational Intelligence and Engineering Approaches for Tourism, Agriculture and Healthcare*, Lecture Notes in Networks and Systems 214, https://doi.org/10.1007/978-981-16-3807-7_19

2 Definitions

(a) The H-function of one and several variables: Charles Fox [19] introduced a function which is more general than Meijer's G-function, and this function is well known in the literature of special functions as Fox's H-function or simply the H-function. This function is defined and represented by means of the following Mellin–Barnes-type contour integral:

$$H[z] = H^{m,n}_{p,q}[z] = H^{m\,,\,n}_{p\,,\,q}\left[z \,\middle|\, \begin{matrix} (a_j, e_j)_{1,p} \\ (b_j, f_j)_{1,q} \end{matrix} \right] = \frac{1}{2\pi i} \int_L \theta(s)\, z^s\, ds \tag{1}$$

where $i = \sqrt{-1}$,

$$\theta(s) = \frac{\prod_{j=1}^{m} \Gamma(b_j - f_j s) \prod_{j=1}^{n} \Gamma(1 - a_j + e_j s)}{\prod_{j=m+1}^{q} \Gamma(1 - b_j + f_j s) \prod_{j=n+1}^{p} \Gamma(a_j - e_j s)}. \tag{2}$$

Owing to the popularity of the H-function, details regarding the various parametric conditions and existence of the contour L, we refer to the monograph by Srivastava et al. [20].

The H-function of r variables of Srivastava and Panda [21] occurring in this paper is represented as follows: (see also Srivastava et al. [20]; pp. 251–254)

$$H[z_1, \cdots, z_r] = H^{0\,,\,n\,:\,m_1\,,n_1\,;\,\cdots\,;\,m_r\,,n_r}_{p\,,\,q\,:\,p_1\,,q_1\,;\,\cdots\,;\,p_r\,,q_r}$$
$$\left[\begin{matrix} z_1 \\ \vdots \\ z_r \end{matrix} \,\middle|\, \begin{matrix} (a_j; \alpha'_j, \cdots, \alpha^{(r)}_j)_{1,p} : (c'_j, \gamma'_j)_{1,p_1}; \cdots ; (c^{(r)}_j, \gamma^{(r)}_j)_{1,p_r} \\ (b_j; \beta'_j, \cdots, \beta^{(r)}_j)_{1,q} : (d'_j, \delta'_j)_{1,q_1}; \cdots ; (d^{(r)}_j, \delta^{(r)}_j)_{1,q_r} \end{matrix} \right]$$
$$= \frac{1}{(2\pi\omega)^r} \int_{L_1} \cdots \int_{L_r} \varphi_1(\xi_1) \cdots \varphi_r(\xi_r)\, \psi(\xi_1, \cdots, \xi_r)\, z_1^{\xi_1} \cdots z_r^{\xi_r}\, d\xi_1 \cdots d\xi_r \tag{3}$$

where $\omega = \sqrt{-1}$,

$$\phi_i(\xi_i) = \frac{\prod_{j=1}^{m_i} \Gamma(d^{(i)}_j - \delta^{(i)}_j \xi_i) \prod_{j=1}^{n_i} \Gamma(1 - c^{(i)}_j + \gamma^{(i)}_j \xi_i)}{\prod_{j=m_i+1}^{q_i} \Gamma(1 - d^{(i)}_j + \delta^{(i)}_j \xi_i) \prod_{j=n_i+1}^{p_i} \Gamma(c^{(i)}_j - \gamma^{(i)}_j \xi_i)} \tag{4}$$

$\forall\, i \in \{1, \cdots, r\}$, and

$$\psi(\xi_1, \cdots, \xi_r) = \frac{\prod_{j=1}^{n} \Gamma(1 - a_j + \sum_{i=1}^{r} \alpha_j^{(i)} \xi_i)}{\prod_{j=n+1}^{p} \Gamma(a_j - \sum_{i=1}^{r} \alpha_j^{(i)} \xi_i) \prod_{j=1}^{q} \Gamma(1 - b_j + \sum_{i=1}^{r} \beta_j^{(i)} \xi_i)} \tag{5}$$

For conditions of existence, etc. of the H-function of several variables, we refer to Srivastava et al. [20], pp. 252–253, Eqs. C.4–C.8.

(b) The fractional integral operators: We now introduce the two fractional integral operators studied in this paper as follows:

$$\begin{aligned} E\,[f(x)] &= E_{0,x}^{\mu,\eta}[f(x)] \\ &= \frac{cx^{-c\mu-\eta}}{\Gamma(\mu)} \int_0^x (x^c - t^c)^{\mu-1} t^\eta H \\ &\left[z_1 \left(\frac{t^c}{x^c}\right)^{u_1} \left(1 - \frac{t^c}{x^c}\right)^{v_1}, \cdots, z_r \left(\frac{t^c}{x^c}\right)^{u_r} \left(1 - \frac{t^c}{x^c}\right)^{v_r}\right] f(t)dt \end{aligned} \tag{6}$$

$$\begin{aligned} K\,[f(x)] &= K_{x,\infty}^{\mu,\eta}[f(x)] \\ &= \frac{cx^{\eta}}{\Gamma(\mu)} \int_x^\infty (t^c - x^c)^{\mu-1} t^{-c\mu-\eta} H \\ &\left[z_1 \left(\frac{x^c}{t^c}\right)^{u_1} \left(1 - \frac{x^c}{t^c}\right)^{v_1}, \cdots, z_r \left(\frac{x^c}{t^c}\right)^{u_r} \left(1 - \frac{x^c}{t^c}\right)^{v_r}\right] f(t)dt \end{aligned} \tag{7}$$

The conditions of the validity of the operator defined by (6) are as follows:

(i) The quantities $u_1, v_1, \cdots, u_r, v_r, c$ are all positive (some of them may decrease to zero provided that the resulting operator has a meaning),

$$\text{(ii) } \operatorname{Re}(\mu) + \sum_{i=1}^{r} v_i \min_{1 \le j \le M_i} \operatorname{Re}\left[\frac{d_j^{(i)}}{\delta_j^{(i)}}\right] + 1 > 0,$$

$$\text{(iii) } \operatorname{Re}(\eta + \gamma) + c \sum_{i=1}^{r} u_i \min_{1 \le j \le M_i} \operatorname{Re}\left[\frac{d_j^{(i)}}{\delta_j^{(i)}}\right] + 1 > 0,$$

where

$$f(x) = \begin{cases} o\,(|x\,|)^{\gamma}, & |x| \to 0 \\ o\,(|x|)^{\delta}, & |x| \to \infty \end{cases}$$

and the operator defined by (7) exists if

$$\text{Re}(\eta - \delta) + c \sum_{i=1}^{r} u_i \min_{1 \le j \le M_i} \text{Re}\left[\frac{d_j^{(i)}}{\delta_j^{(i)}}\right] > 0,$$

together with the sets of conditions (i) and (ii) specified for the existence of the operator (6) are satisfied.

Remark If we set $c = 1,\ r = 1,\ u_1 = v_1 = 0$ and $z_1 \to 1$, these operators reduce to the integral operators defined by Erdélyi [1] and Kober [5], respectively.

3 The Main Results

Theorem 1 *For the fractional integral given by (6), we have*

$$E_{0,x}^{\mu,\eta}\left\{x^{\lambda} H_{p,q}^{m,n}\left[b x^{\beta} \,\middle|\, \begin{matrix}(e_j, E_j)_{1,p}\\(f_j, F_j)_{1,q}\end{matrix}\right]\right\}$$
$$= \frac{x^{\lambda-c+1}}{\Gamma(\mu)} H_{P+2,Q+1\,:\,P_1,Q_1\,;\,\cdots\,;\,P_r,Q_r\,;\,p,q}^{0,N+2\,:\,M_1,N_1\,;\,\cdots\,;\,M_r,N_r\,;\,m,n}$$
$$\left[\begin{matrix} z_1\\ \vdots\\ z_r\\ b x^{\beta}\end{matrix} \,\middle|\, \begin{matrix}(1-\frac{\eta+\lambda+1}{c}; u_1, \cdots, u_r, \beta/c), (1-\mu, v_1, \cdots, v_r, 0),\\ (1-\mu-\frac{\eta+\lambda+1}{c}; u_1+v_1, \cdots, u_r+v_r, \beta/c),\end{matrix}\right. \tag{8}$$
$$\left.\begin{matrix}(a_j; \alpha'_j, \cdots, \alpha_j^{(r)}, 0)_{1,P} : (c'_j, \gamma'_j)_{1,P_1}; \cdots; (c_j^{(r)}, \gamma_j^{(r)})_{1,P_r}; (e_j, E_j)_{1,p}\\ (b_j; \beta'_j, \cdots, \beta_j^{(r)}, 0)_{1,Q} : (d'_j, \delta'_j)_{1,Q_1}; \cdots; (d_j^{(r)}, \delta_j^{(r)})_{1,Q_r}; (f_j, F_j)_{1,q}\end{matrix}\right]$$

where the quantities $\beta, u_1, v_1, \cdots, u_r, v_r, c$ are all positive (some of them may, however, decrease to zero provided that the resulting operator has a meaning),

$$\text{Re}(\eta + \gamma) + c \sum_{i=1}^{r} u_i \min_{1 \le j \le M_i} \text{Re}\left[\frac{d_j^{(i)}}{\delta_j^{(i)}}\right] + \beta \min_{1 \le j \le m} \text{Re}\left[\frac{f_j}{F_j}\right] + 1 > 0$$

and

$$\text{Re}(\mu) + \sum_{i=1}^{r} v_i \min_{1 \le j \le M_i} \text{Re}\left[\frac{d_j^{(i)}}{\delta_j^{(i)}}\right] + 1 > 0.$$

Theorem 2 *For the fractional integral (7), we have*

$$K_{x,\infty}^{\mu,\eta}\left\{x^{\lambda} H_{p,q}^{m,n}\left[b\,x^{-\beta}\,\middle|\,\begin{matrix}(e_j,E_j)_{1,p}\\(f_j,F_j)_{1,q}\end{matrix}\right]\right\}$$

$$= \frac{x^{\lambda-c+1}}{\Gamma(\mu)} H^{0,N+2\ :M_1,N_1\ ;\ \cdots\ ;\ M_r,N_r\ ;\ m,n}_{P+2,Q+1\ :\ P_1,Q_1\ ;\ \cdots\ ;\ P_r,Q_r\ ;\ p,q}$$

$$\left[\begin{matrix} z_1\vdots z_r b x^{-\beta}(-\frac{\eta-\lambda-1}{c};u_1,\cdots,u_r,\beta/\beta c-c),(1-\mu,v_1,\cdots,v_r,0),\\ (-\mu-\frac{\eta-\lambda-1}{c};u_1+v_1,\cdots,u_r+v_r,\beta/\beta c-c),\end{matrix}\right]$$

$$\left.\begin{matrix}(a_j;\alpha_j',\cdots,\alpha_j^{(r)},0)_{1,P}:(c_j',\gamma_j')_{1,P_1};\cdots;(c_j^{(r)},\gamma_j^{(r)})_{1,P_r};(e_j,E_j)_{1,p}\\ \\ (b_j;\beta_j',\cdots,\beta_j^{(r)},0)_{1,Q}:(d_j',\delta_j')_{1,Q_1};\cdots;(d_j^{(r)},\delta_j^{(r)})_{1,Q_r};(f_j,F_j)_{1,q}\end{matrix}\right] \tag{9}$$

provided

$$\mathrm{Re}(\eta+\gamma)+c\sum_{i=1}^{r}u_i\min_{1\le j\le M_i}\mathrm{Re}\left[\frac{d_j^{(i)}}{\delta_j^{(i)}}\right]-\beta\min_{1\le j\le m}\mathrm{Re}\left[\frac{f_j}{F_j}\right]+1>0$$

and

$$\mathrm{Re}(\mu)+\sum_{i=1}^{r}v_i\min_{1\le j\le M_i}\mathrm{Re}\left[\frac{d_j^{(i)}}{\delta_j^{(i)}}\right]+1>0$$

4 Proofs of the Theorems

Proofs of Theorems 1 and 2 are based upon (1) and (3) and on the beta integral

$$B(\alpha,\beta)=\int_0^1 t^{\alpha-1}(1-t)^{\beta-1}dt=\frac{\Gamma(\alpha)\,\Gamma(\beta)}{\Gamma(\alpha+\beta)}. \tag{10}$$

Denoting, for convenience, the expression on the left-hand side of (8) by Ω and applying the definition of $E\,[f(x)]$ in it, we obtain

$$\Omega \equiv E_{0,x}^{\mu,\eta} x^{\lambda} H_{p,q}^{m,n} b x^{\beta}(e_j,E_j)_{1,p}(f_j,F_j)_{1,q}$$

$$= \frac{c x^{-c\mu-\eta}}{\Gamma(\mu)}\int_0^x t^{\eta+\lambda}(x^c-t^c)^{\mu-1}H$$

$$\left[z_1\left(\frac{t^c}{x^c}\right)^{u_1}\left(1-\frac{t^c}{x^c}\right)^{v_1},\cdots,z_r\left(\frac{t^c}{x^c}\right)^{u_r}\left(1-\frac{t^c}{x^c}\right)^{v_r}\right]H_{p,q}^{m,n}\left[bt^{\beta}\right]dt$$

Substituting the values of the H-functions of one and several variables from (1) and (3) and then changing the order of integrations in the resulting expression, we get

$$\begin{aligned}\Omega =&\frac{cx^{-c\mu-\eta}}{(2\pi\omega)^{r+1}\Gamma(\mu)}\int\limits_{L}\int\limits_{L_1}\cdots\int\limits_{L_r}\phi(s)\,\phi_1(s_1)\cdots\phi_r(s_r)\,\psi(s_1,\cdots,s_r)\,z_1^{s_1}\cdots z_r^{s_r}\\ &\times\ b^s x^{-c(u_1s_1+\cdots+u_rs_r)-c(v_1s_1+\cdots+v_rs_r)}\\ &\left\{\int\limits_0^x t^{c(u_1s_1+\cdots+u_rs_r)+\beta s+\eta+\lambda}(x^c-t^c)^{\mu-1+v_1s_1+\cdots+v_rs_r}\,dt\right\}ds\,ds_1\cdots ds_r\\ =&\frac{cx^{-\eta-c}}{(2\pi\omega)^{r+1}\Gamma(\mu)}\int\limits_{L}\int\limits_{L_1}\cdots\int\limits_{L_r}\varphi(s)\,\varphi_1(s_1)\cdots\varphi_r(s_r)\,\psi(s_1,\cdots,s_r)\,z_1^{s_1}\cdots z_r^{s_r}\\ &\times\ b^s x^{-c(u_1s_1+\cdots+u_rs_r)}\\ &\left\{\int\limits_0^x t^{c(u_1s_1+\cdots+u_rs_r)+\beta s+\eta+\lambda}\left(1-\frac{t^c}{x^c}\right)^{\mu-1+v_1s_1+\cdots+v_rs_r}dt\right\}ds\,ds_1\cdots ds_r\end{aligned}$$

On substituting $\frac{t^c}{x^c}=y$ in the inner integral, the above expression leads to

$$\begin{aligned}\Omega =&\frac{x^{\lambda-c+1}}{(2\pi\omega)^{r+1}\Gamma(\mu)}\int\limits_{L}\int\limits_{L_1}\cdots\int\limits_{L_r}\phi(s)\,\phi_1(s_1)\cdots\phi_r(s_r)\,\psi(s_1,\cdots,s_r)\,z_1^{s_1}\cdots z_r^{s_r}\\ &\times\ b^s x^{\beta S}\left\{\int\limits_0^1 y^{\frac{c(u_1s_1+\cdots+u_rs_r)+\beta s+\eta+\lambda+1}{c}-1}(1-y)^{(\mu+v_1s_1+\cdots+v_rs_r)\,-1}dy\right\}\\ &ds\,ds_1\cdots ds_r\end{aligned}$$

where the functions $\phi(s)$, $\phi_i(s_i)$ and $\psi(s_1,\cdots,s_r)$ are obtainable from (2), (4) and (5), respectively. Finally, evaluating the inner beta integral by using (10) and then interpreting the $(r+1)$ contour integrals with the help of (3), we arrive at

$$\begin{aligned}&E_{0,x}^{\mu,\eta}\left\{x^{\lambda}H_{p,q}^{m,n}\left[bx^{\beta}\left|\begin{matrix}(e_j,E_j)_{1,p}\\(f_j,F_j)_{1,q}\end{matrix}\right.\right]\right\}\\ &=x^{\lambda}H_{P+1,Q+1:P_1Q_1;p,q}^{0,\ N+1:M_1N_1;m,n}\left[\begin{matrix}z\\bx^{\beta}\end{matrix}\left|\begin{matrix}(-\eta-\lambda,0,\beta),\\(-\mu-\eta-\lambda,0,\beta),\end{matrix}\right.\right.\\ &\qquad\left.\begin{matrix}(a_j;\alpha_j,0)_{1,P}:(c_j,\gamma_j)_{1,P_1};(e_j,E_j)_{1,p}\\(b_j;\beta_j,0)_{1,Q}:(d_j,\delta_j)_{1,Q_1};(f_j,F_j)_{1,q}\end{matrix}\right]\end{aligned}$$

this completes the proof of Theorem 1.

In a similar fashion, Theorem 2 can be proved.

5 Applications

We choose to record here the following consequences of our theorems: If we set $m = q = \beta = F_1 = 1$, $n = p = f_1 = 0$ and by the relation [20], pp. 18, Eq. 2.6.2:

$$H_{0,1}^{1,0}\left[x\left|\begin{matrix}\text{---}\\(b,\beta)\end{matrix}\right.\right] = \beta^{-1}x^{b/\beta}\exp(-x^{1/\beta}) \tag{11}$$

Theorem 1 leads to the following formula:

$$\begin{aligned}&E_{0,x}^{\mu,\eta}\left\{x^{\lambda}H_{p,q}^{m,n}\left[bx^{\beta}\left|\begin{matrix}(e_j,E_j)_{1,p}\\(f_j,F_j)_{1,q}\end{matrix}\right.\right]\right\}\\&= x^{\lambda}H_{P+1,Q+1:P_1Q_1;p,q}^{0,\ N+1:M_1N_1;m,n}\left[\begin{matrix}z\\bx^{\beta}\end{matrix}\left|\begin{matrix}(-\eta-\lambda,0,\beta),\\(-\mu-\eta-\lambda,0,\beta),\end{matrix}\right.\right.\\&\qquad\left.\begin{matrix}(a_j;\alpha_j,0)_{1,P}:(c_j,\gamma_j)_{1,P_1};(e_j,E_j)_{1,p}\\(b_j;\beta_j,0)_{1,Q}:(d_j,\delta_j)_{1,Q_1};(f_j,F_j)_{1,q}\end{matrix}\right]\end{aligned} \tag{12}$$

provided

$$\mathrm{Re}(\eta+\gamma)+c\sum_{i=1}^{r}u_i\min_{1\le j\le M_i}\mathrm{Re}\left[\frac{d_j^{(i)}}{\delta_j^{(i)}}\right]+1>0$$

$$\mathrm{Re}(\mu)+\sum_{i=1}^{r}v_i\min_{1\le j\le M_i}\mathrm{Re}\left[\frac{d_j^{(i)}}{\delta_j^{(i)}}\right]+1>0.$$

Similarly, Theorem 2 yields the following formula:

$$\begin{aligned}&E_{0,x}^{\mu,\eta}\left\{x^{\lambda}H_{p,q}^{m,n}\left[bx^{\beta}\left|\begin{matrix}(e_j,E_j)_{1,p}\\(f_j,F_j)_{1,q}\end{matrix}\right.\right]\right\}\\&= x^{\lambda}H_{P+1,Q+1:P_1Q_1;p,q}^{0,\ N+1:M_1N_1;m,n}\left[\begin{matrix}z\\bx^{\beta}\end{matrix}\left|\begin{matrix}(-\eta-\lambda,0,\beta),\\(-\mu-\eta-\lambda,0,\beta),\end{matrix}\right.\right.\\&\qquad\left.\begin{matrix}(a_j;\alpha_j,0)_{1,P}:(c_j,\gamma_j)_{1,P_1};(e_j,E_j)_{1,p}\\(b_j;\beta_j,0)_{1,Q}:(d_j,\delta_j)_{1,Q_1};(f_j,F_j)_{1,q}\end{matrix}\right]\end{aligned} \tag{13}$$

together with the conditions given with (12) are satisfied.

6 Particular Cases

(i) For $c = 1,\ r = 1,\ u_1 = v_1 = 0$ and $z_1 = z$, (8) reduces to the result

$$E_{0,x}^{\mu,\eta}\left\{x^{\lambda} H_{p,q}^{m,n}\left[bx^{\beta}\left| \begin{matrix}(e_j, E_j)_{1,p}\\ (f_j, F_j)_{1,q}\end{matrix}\right.\right]\right\}$$

$$= x^{\lambda} H_{P+1,Q+1:P_1Q_1;p,q}^{0,\ N+1:M_1N_1;m,n}\left[\begin{matrix}z\\ bx^{\beta}\end{matrix}\left| \begin{matrix}(-\eta-\lambda, 0, \beta),\\ (-\mu-\eta-\lambda, 0, \beta),\end{matrix}\right.\right.$$
$$\left.\begin{matrix}(a_j; \alpha_j, 0)_{1,P} : (c_j, \gamma_j)_{1,P_1}; (e_j, E_j)_{1,p}\\ (b_j; \beta_j, 0)_{1,Q} : (d_j, \delta_j)_{1,Q_1}; (f_j, F_j)_{1,q}\end{matrix}\right] \tag{14}$$

(ii) Again for $c = 1\ r = 1,\ u_1 = v_1 = 0,\ z_1 = z$ and $\beta > 0$, (9) reduces to the result

$$K_{x,\infty}^{\mu,\eta}\left\{x^{\lambda} H_{p,q}^{m,n}\left[bx^{-\beta}\left| \begin{matrix}(e_j, E_j)_{1,p}\\ (f_j, F_j)_{1,q}\end{matrix}\right.\right]\right\}$$

$$= x^{\lambda} H_{P+1,Q+1:P_1Q_1;p,q}^{0,N+1\ \ :M_1N_1;m,n}\left[\begin{matrix}z\\ bx^{-\beta}\end{matrix}\left| \begin{matrix}(-\eta+\lambda+1, 0, \beta),\\ (-\mu-\eta+\lambda+1, 0, \beta),\end{matrix}\right.\right.$$
$$\left.\begin{matrix}(a_j; \alpha_j, 0)_{1,P} : (c_j, \gamma_j)_{1,P_1}; (e_j, E_j)_{1,p}\\ (b_j; \beta_j, 0)_{1,Q} : (d_j, \delta_j)_{1,Q_1}; (f_j, F_j)_{1,q}\end{matrix}\right] \tag{15}$$

Acknowledgements This work is supported under the project "SG-TEQIP-III/BAS/UKS" of National Institute of Technology, Arunachal Pradesh, India.

References

1. Erdelyi, A.: On fractional integration and its application to the theory of Hankel transforms. Quart. J. Math. Oxford Ser. **11**, 293–303 (1940)
2. Erdelyi, A., Magnus, W., Oberhettinger, F., Tricomi, F.G.: Tables of integral transforms. Vol. II, Ch. 13, McGraw Hill Book Company, New York, Toronto, London (1954)
3. Gupta, K.C., Soni, R.C.: On composition of some general fractional integral operators. Proc. Ind. Acad. Sci. (Math. Sci.) **104**(2), 339–349 (1994)
4. Kiryakova, V.: Generalized fractional calculus and applications. Longman Scientific & Technical, Harlow and John Wiley & Sons, Inc., New York (1994)
5. Kober, H.: On fractional integrals and derivatives. Quart. J. Math. Oxford Ser. **11**, 193–211 (1940)
6. Lowndes, J.S.: A generalization of the Erdélyi-Kober operators. Proc. Edinburgh Math. Soc. **17**, 139–148 (1970)

7. Purohit, S.D., Suthar, D.L., Kalla, S.L.: Marichev-Saigo-Maeda fractional integration operators of the Bessel functions. Le Matematiche **LXVII** Fasc. I, 21–32 (2012). https://doi.org/10.4418/2012.67.1.2
8. Raina, R.K.: A note on multidimensional modified fractional calculus operators. Proc. Ind. Acad. Sci. (Math. Sci.) **106**(2), 155–162 (1996)
9. Saigo, M., Kant, S., Koul, C.L.: On generalized fractional integrals involving the H-function associated with a general class of polynomials. Fukuoka Univ. Sci. Rep. **20**(1), 15–21 (1990)
10. Saigo, M., Raina, R.K.: Fractional calculus operators associated with the H-function of several variables. Anal. Geom. Groups Riemann Legacy Vol. 527–538 (1993)
11. Saigo, M., Saxena, R.K., Ram, J.: On the fractional calculus operator associated with the H-function. Ganita Sandesh **6**(1), 36–47 (1992)
12. Saxena, R.K., Kumbhat, R.K.: Integral operators involving H-function. Ind. J. Pure Appl. Math. **5**(1), 1–6 (1974)
13. Srivastava, H.M., Goyal, S.P., Jain, R.M.: Fractional integral operators involving a general class of polynomials. J. Math. Anal. Appl. **148**(1), 87–100 (1990)
14. Srivastava, H.M., Saxena, R.K.: Operators of fractional integration and their applications. Appl. Math. Comput. **118**, 1–52 (2001)
15. Baleanu, D., Agarwal, P.: On generalized fractional integral operators and the generalized Gauss hypergeometric functions. Abstr. Appl. Anal. **2014**, 1–5 (2014). Article ID 630840, https://doi.org/10.1155/2014/630840
16. Mondal, S.R., Nisar, K.S.: Marichev-Saigo-Maeda fractional integration operators involving generalized Bessel functions. Math. Prob. Eng. **2014**, 1–11 (2014). Article ID 274093. https://doi.org/10.1155/2014/274093
17. Sharma, S.K., Shekhawat, A.S.: A unified presentation of generalized fractional integral operators and H-function. Glob. J. Pure Appl. Math. **13**(2), 393–403 (2017). http://www.ripublication.com
18. Suthar, D.L., Andualem, M., Debalkie, B.: A study on generalized multivariable Mittag-Leffler function via generalized fractional calculus operators. J. Math. **2019**, 1–7 (2019). Article ID 9864737. https://doi.org/10.1155/2019/9864737
19. Fox. C.: The G-and H-functions as symmetrical Fourier kernels. Trans. Am. Math. Soc. **98**, 395–429 (1961)
20. Srivastava, H.M., Gupta, K.C., Goyal, S.P.: The H-Functions of One and Two Variables with Applications. South Asian Publishers, New Delhi, Madras (1982)
21. Srivastava, H.M., Panda, R.: Expansion theorems for the H-function of several complex variables. J. Reine Angew. Math. **288**, 129–145 (1976)

Fuzzy Logic for Enhancing the Evaluation and Monitoring of Agile Artifacts in Cardiovascular Disease

Alka Gulati, Kanak Saxena, and Umesh Banodha

1 Introduction

The study of the field of Software Engineering arose because of the necessity to mediate the shared development of huge software systems by studying, researching, and developing artifacts. Though the traditional aspects of Software Engineering are mandatorily needed as a base to develop new techniques and tools, still they are not effective enough to support the collaborative aspects of big scale software development, hence, we need absolutely new and smart developments to efficiently curb by providing smart solutions for our ongoing issues and changes [1]. Artifacts are sequences of quantified units generated or used in a development process. Technical artifacts are objects which don't exist naturally. They are physical objects with functions [2].

Cardiovascular disease affects millions of people all over the world every year. Although much progress has been done in medical science, early detection of this disease with very few factors is a challenge for prevention. Agile methodology helps programmers in dealing with dynamic, changing, and uncertain type of requirements quickly. Agile along with software engineering can be beneficial for the development and maintenance of healthcare systems. To deal with the complexity and to come up with effective innovation, "Agile" can be used. One of the basic principles of Agile is to address the requirements and adopting changes of stakeholders. Agile's limit is not only in correlating health care system with the latest information technology but it is used beyond it. The input sources for the software being developed for the healthcare industry are numerous dissimilar groups. To build user-friendly software, healthcare software engineers gather requirements from all actors involved like

A. Gulati (✉)
Barkatullah University, Bhopal, Madhya Pradesh, India

K. Saxena · U. Banodha
Department of CS and IT, Samrat Ashok Technological Institute, Vidisha, Madhya Pradesh, India

P. Srivastava et al. (eds.), *Mathematical, Computational Intelligence and Engineering Approaches for Tourism, Agriculture and Healthcare*, Lecture Notes in Networks and Systems 214, https://doi.org/10.1007/978-981-16-3807-7_20

hospital administrators, patients, doctors, nurses, officials, and medical technicians. Agile is an iterative, team-based approach to development. While development of project Agile consider voices of all involved actors. The iterative development cycles of agile methodology are highly beneficial for the software development of healthcare systems [3].

In this paper, Sect. 2 describes related work, Sect. 3 describes the healthcare domain, Sect. 4 describes the agile methodology, Sect. 5 describes the proposed work, and Sect. 6 describes the case study.

2 Related Work

Crotty et al. [4] have found that healthcare organizations have adopted changes and innovations effectively in patient care, digital health technology by using Agile methodology. They recognized that it is necessary to override health care design to incorporate the values of innovated technology trends. The agile approach can easily integrate the redesign of health care and innovated latest trends of technology. New agile health practices will be enabled if the agile methodology is embraced by healthcare-associated organizations. With this, the latest technologies can be adopted effectively for the betterment and care of patients.

Turk et al. [5] found that in a particular situation, application of agile process leads to a good understanding of agile methodology. They have studied the use of principles of the agile method in identifying essential assumptions. They also identified restrictions that may come up due to the identified assumptions and outlined how these restrictions can be resolved by associating various software development approaches and practices while using an agile development environment.

Beulah et al. [6] have predicted heart disease risk from various risk factors using ensemble classification. They have used a strategy called ensemble to progress the accurateness of prediction of risk of heart disease. They have improved the prediction accuracy and exhibited acceptable performance in finding heart disease risk.

Palaniappan and Awang [7] developed a framework model using data mining approaches like Bayes network, Neural Network and decision trees for intelligent heart disease prediction systems. They found that each data mining technique has its exclusive strength. The prediction system developed is flexible, easy to use, web based, consistent, and scalable. The parameters it uses for prediction are age, gender, blood pressure, and fasting blood sugar.

Chitra and Seenivasagam [8] gave a classification system that is unsupervised for predicting the possibility of a heart attack at the early phase using patient's health parameters. They have used the Fuzzy C-means classifier algorithm. They have shown that this algorithm gives a prediction of a heart attack in a patient in a more well-organized and gainful way than other existing algorithms.

Xiao [9] proposed a method that uses four steps. In the first step, the parameter's uncertainties are measured by attaching the conviction entropy. In the second step, the uncertainties of parameters are modulated. In the third step, suitable basic probability is assigned on the parameters of basic modulated uncertainty level. In the last step, Dempster's combination regulation is adopted to combine the independent parameters. They have validated the viability and efficacy of the method proposed by them busing a medical diagnosing application.

Celik and Yamak [10] have applied fuzzy soft sets and their arithmetic operations through Sanchez's findings for diagnosing patients' disease. They have given the case study to identify the disease of the patient. They have concluded that it is a simple and fruitful technique.

Agman et al. [11] redefined the fuzzy soft sets along with its operations. They have used fuzzy membership to explain its parameter approximate elements. They had shown a decision-making method for problems where uncertainty is high. They have shown an example to demonstrate the application of the method defined for f s-sets.

Yang et al. [12] proposed the perception of multi-fuzzy sets. This concept provides an innovative method to characterize different problems like color of pixels. They have combined Multi-fuzzy sets and Soft sets to form Multi-fuzzy soft sets. They had proven De Morgan's laws of Multi-fuzzy soft sets. They had demonstrated solving problems regarding decisions using Multi-fuzzy soft sets.

3 Healthcare Domain

According to the WHO, heart diseases are causing deaths in millions per year. Heart disease is a severe problem all over the world. Therefore, it is important to diagnose it but due to a large number of risk factors and uncertainty, it becomes hard for experts to diagnose it. In most of the cases, it is hard to predict the severity of cardiovascular disease results. It is not an easier task to evaluate cardiovascular disease in patients as it involves several factors. Experts require an exact tool that considers the factors and can find severity by using only a few parameters [13]. The use of computer technology nowadays for decision-making is being used in each field and the healthcare domain is not excluded. The decision regarding diagnostic depends upon practice, expertise and use of an appropriate method with powerful logical reasoning capability.

4 Agile Methodology

Agile leverages its iterative nature to define and control the project work. Agile is valuable for the healthcare domain as agile has moved beyond its origins in efficient software development among teams into productivity and customer responsiveness [14]. It offers concepts that can be implemented to optimize the processes.

By adopting the following key elements of agile, innovative ideas and changes arising due to uncertainties can be incorporated into cardiovascular system [15]. They are the following:

i. Correctness levels should be clearly specified. And quick solutions are provided.
ii. Impact of uncertainty is reduced by iteration or plan.
iii. Expert team efforts (build heterogeneous teams).
iv. Continuous improvement (non-judgmental retrospective).
v. Applicative prediction.

A powerful mathematical technique called fuzzy logic is used for handling uncertainties in data and helps in taking decisions while iterative process [16, 17]. Soft computing techniques are used to train the models and for which patients and their corresponding data are required. By using fuzzy logic, the minimal set of attributes can diagnose different types of strokes [18, 19]. The objective of this work is to incorporate Fuzzy logic in agile methodology to find the severity technically of heart disease with few technical artifacts.

5 Proposed Work

In this section, an algorithm using agile in association with fuzzy logic for evaluating the artifacts is proposed. The proposed algorithm works on technical artifacts using arithmetic operations on fuzzy sets. Let there be a set of m patients P = {p1, p2, p3, …, pm} with a set of n technical artifacts A = {s1, s2, s3, …, sn} related to a set of severity in heart disease.

5.1 Proposed Algorithm:

1. Correctness Function—Construct the correctness function using fuzzy logic in order to evaluate technical artifacts in terms of a patient and symptoms/artifacts matrix. Let it be denoted by M. It is defined as $\boldsymbol{F} : \boldsymbol{P} \to \boldsymbol{f}(\boldsymbol{A})$, where A is a set of artifacts and P is set of patients. Correctness function is defined as n * (n + 1) /2 for n artifacts.
2. Threshold Evaluation—Obtain the permissible low/high threshold symptom values for set A. Let it be denoted by N.
3. Retrospective calculation—In retrospective calculation every iteration results are identified. With the help of the correctness function and threshold evaluation, resultant matrix is generated. It shows the effect of one artifact on another.

$$R = M \times N$$

4. Evaluate the matrix obtained from step 3 in view of defuzzification. Normalize the combination of artifacts.
5. Evaluate the artifacts by calculating the max value obtained in matrix R in step 4. Find Pi in R for which Rip = max (Ril).
6. Repeat steps 3–5 in view of the expert perception. If the results hold within the permissible value the process will be stopped.

This proposed algorithm helps in evaluating and monitoring artifacts technically.

6 Case Study

This section shows a case study of the cardiovascular system. Datasets are taken from UCI. Cleveland datasets are used. The proposed algorithm is used for predicting the severity technically of heart disease of 250 patients. They have artifacts age, chest pain, resting blood pressure, cholesterol, fasting blood sugar, maximum heart rate achieved, resting ECG results, exercise-induced angina, ST depression, slope of the peak exercise ST segment, old peak, thalassemia. The standard normal values are used as mentioned by experts. These technical artifacts help in finding the severity of heart disease.

1. Firstly, create a correctness function by mentioning symptoms of patients.

$$M_{mn} = \begin{bmatrix} a_{11} & a_{12} & \ldots\ldots\ldots\ldots & a_{1n} \\ a_{21} & a_{22} & \ldots\ldots\ldots\ldots & a_{2n} \\ \ldots & \ldots & \ldots\ldots\ldots\ldots & \ldots \\ \ldots & \ldots & \ldots\ldots\ldots\ldots & \ldots \\ a_{m1} & a_{m2} & \ldots\ldots\ldots\ldots & a_{mn} \end{bmatrix}$$

2. Secondly, write the threshold values of all the symptoms.

$$N_{n1} = \begin{bmatrix} b_{11} \\ b_{21} \\ \ldots \\ \ldots \\ b_{n1} \end{bmatrix}$$

3. Next the patient severity matrix is evaluated.

$$R_{m1} = M_{mn} \times N_{n1}$$

4. Find the maximum value, which will represent the severity of heart disease.
5. Now normalize the artifacts and take the combination of artifacts.
6. Steps 3–5 are repeated by normalizing combinations of artifacts.

As a result of step 4, if all the technical artifacts are considered with their standard values then 5 patients P45, P115, P107, P120, and P72 are assessed the highest severity technically of heart disease, respectively, as they have higher values. Figure 1 shows the severity level of heart disease of patients and shows that patient P45 is severely prone to heart disease than other patients. This graph is made trivially on the patients' symptom values after applying threshold values. Thus it represents that doctors/experts at first glance can predict the severity and get it treated.

Heart disease is a critical disease and its prediction is difficult before a stroke. So by normalizing aim is to predict the severity of heart disease with less number of artifacts. If cholesterol is normalized then the severity is affected. As a result of step 5, patients P120, P234, P133, P237, and P40 have maximum values, respectively. Thus it can be used as an important factor to represent the severity of heart disease. That means cholesterol should be checked. Figure 2 shows that patient P120 is very

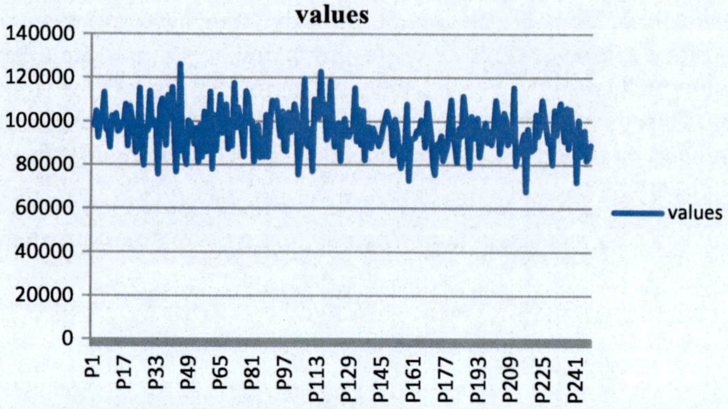

Fig. 1 Severity with max retrospective calculation

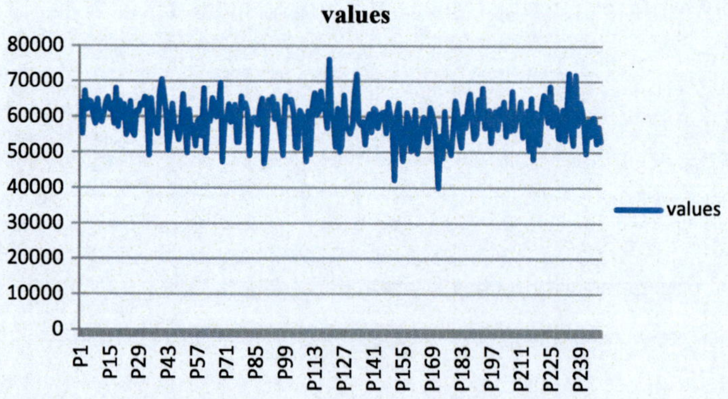

Fig. 2 Max value of retrospective calculation by normalizing chls

severe when chls is normalized and this shows that symptom cholesterol is very important in finding the severity of heart disease.

Similarly as a result of step 6, if the combination of cholesterol and chest pain is normalized (chls = cp = 0) then the severity is affected. It has similar results to when only cholesterol is normalized. That means a cholesterol check is enough.

When the combination of cholesterol and fasting blood sugar is normalized (chls = fbs = 0) then the severity is affected. The patients having maximum values respectively are P237, P193, P 148, P 234, and P50. Thus combination cholesterol and FBS are important symptoms in finding the severity of the heart disease of the patient.

If the combination of cholesterol, blood pressure, and fasting blood sugar is normalized (chls = bp = fbs = 0), then the patients having maximum values are, respectively, P120, P234, P228, P153, and P50. This shows that a combination of chls, bp, and fbs can't be missed while predicting the severity of heart disease. Figure 3 shows that patient P120 is very severe when chls, bp, fbs are normalized and this shows that symptoms chlos,bp,fbs together are very important in finding the severity of heart disease. The results are varying with step 4. Thus these three artifacts should be checked.

It is concluded from the above results that the patient with varying values of few artifacts can have heart disease. It is not necessary whose all artifact values are varying is more probable for heart disease. This concept can help at primary rural medical centers where resources are less and patients come with breathing or heart pain complain.

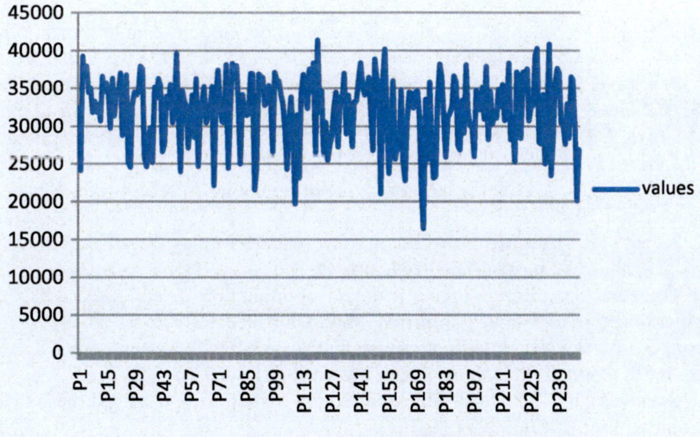

Fig. 3 Max value of retrospective calculation by normalizing chls, bp, fbs

7 Conclusion

Agile model is beneficial for healthcare organizations. Many of the Mathematical models also help in decision-making and adopting changes. Fuzzy logic is used to deal with uncertainty and evaluate artifacts. In the case study, algorithm is used in cardiovascular disease to find the severity of heart diseases. The paper describes the evaluation of artifacts during the development of the heart disease system with help of agile in association with fuzzy logic. This work represents the basis for software development in order to evaluate artifacts after their analysis. The patients are assigned values to their symptoms and the threshold values of their symptoms are used to find their severity for heart disease. Agile along with fuzzy logic helps in developing a system with uncertainty and changes which appear in different forms and affect decision-making. The proposed method is tested on heart data sets. The work focuses on the recognition of severity with very less information regarding symptoms of patients. This concept can be beneficial for handling patients at rural medical centers where specialized doctor for CVD is not available. With initial investigation and values of artifacts cholesterol, blood pressure, and fasting sugar they can predict the possibility and severity of heart disease. The limitations are varying range of technical artifacts are not considered.

References

1. Barthelmess, P., Anderson, K.M.: A view of software development environments based on activity theory. Computer Supported Cooperative Work (CSCW) **11**, 13–37 (2002)
2. Gulati, A., Saxena, K.: Performance prediction of technical artifacts in heart disease diagnosis. In: ICEMCI Jabalpur, 21–22 Dec'2019
3. Ma, Y., Saparova, L.D.: The role of iterative usability evaluation in agile development: a case study. UXMatters (2014)
4. Crotty, B.H., Somai, M., Carlile, N.: Adopting agile principles in health care. Health Affairs Blog, August 15, 2019
5. Turk, D., Robert, F., Rumpe, B.: Assumptions underlying agile software-development processes. J. Database Manage. (JDM) **1072**, 16(4)
6. Latha, C.B.C., Jeeva, S.C.: Improving the accuracy of prediction of heart disease risk based on ensemble classification techniques. Informat. Med. Unlocked **16**, 100203 (2019)
7. Palaniappan, S., Awang, R.: Intelligent heart disease prediction system using data mining techniques. Int. J. Comput. Sci. Netw. Secur. **8**(8), 108–115 (2008)
8. Chitra, R., Seenivasagam, V.: Heart attack prediction system using fuzzy C means classifier. IOSR J. Comput. Eng. **14**(2), 23–31 (2013)
9. Xiao, F.: A hybrid fuzzy soft sets decision making method in medical diagnosis
10. Çelik, Y., Yamak, S.: Fuzzy soft set theory applied to medical diagnosis using fuzzy arithmetic operations. J. Inequalities Appl. (2013)
11. Agman, N.C., Engino ˘ Glu, C., Citak, F.: Fuzzy soft set theory and its applications. Iran. J. Fuzzy Syst. **8**(3), 137–147 (2011)
12. Yang, Y., Tan, X., Meng, C.: The multi-fuzzy soft set and its application in decision making. Appl. Math. Model. **37**, 4915–2923 (2013)
13. Ali, L. et al.: An automated diagnostic system for heart disease prediction based on $\chi 2$ statistical model and optimally configured deep neural network. IEEE Access **7**, 34938–34945 (2019)

14. Joanna, K., Sharp, H., Blandford, A.: The importance of identity and vision to user experience designers on agile projects. In: AGILE '09. IEEE, New York (2009)
15. Holvitie, J. et al.: Technical debt and agile software development practices and processes: an industry practitioner survey. Inf. Softw. Technol. **96**, 141–160 (2018)
16. Iancu, I.: Heart disease diagnosis based on meditative fuzzy logic. Artif. Intell. Med. 51–60 (2018)
17. Suresh, M., Patri, R.: Agility assessment using fuzzy logic approach: a case of healthcare dispensary. BMC Health Serv. Res. **17**, 394 (2017)
18. Muthukumar, P., Sai Sundara Krishnan, G.: A similarity measure of intuitionistic fuzzy soft sets and its application in medical diagnosis. Appl. Soft Comput. **41**, 148–156 (2016)
19. Djatna, T., Hardhienata, M.K.D., Masruriyah, A.F.N.: An intuitionistic fuzzy diagnosis analytics for stroke disease. J. Big Data **5**, 35 (2018)

Development of Dust-Repellent Coating for Solar Panel and Evaluation of Energy Efficiency

Neha Bhatt, Abhilasha Mishra, Rekha Goswami, and Arun Pratap Singh Rathod

1 Introduction

Solar power is a clean renewable source of energy. Since the solar panel has some limitations, so the demand for solar panel is affected, it is because of very short-term durability. Along with irradiance, shadow effect, weather condition, there is one more major reason for the decrease in the output performance of the solar panel, which are tiny dust particles of carbon, soil, wood, plastic accumulation, which are carried by air and contamination by acidic/basic rainwater droplets. This layer by layer accumulation decreases the intensity of incident radiation on the glass top of the solar panel, which constantly decreases the efficiency of the solar panel. After a certain time, the panel does not work and the life of the panel ends [1–3].

The superhydrophobic coatings can be solution to this problem. Superhydrophobic surfaces have the tendency to clean their surfaces by rolling dust from the surface with rolling water droplets. They avoid dust and water accumulation on the panel. Superhydrophobicity on the surface is due to the presence of nanostructures and low surface energy of the surface. Nanostructures are also responsible for repelling dust particles from the surface. Lotus leaf is an example of a natural nanostructured, low-energy surface, which leads to superhydrophobicity of the leaf surface [4–8]. Surfaces having a contact angle less than 90° are called hydrophilic,

N. Bhatt · A. Mishra (✉)
Department of Chemistry, Graphic Era (Deemed To Be) University, Dehradun 248002, India

R. Goswami
Department of Environmental Science, Graphic Era (Deemed To Be) University, Dehradun 248002, India

Department of Environmental Science, Graphic Era Hill University, Dehradun 248002, India

A. P. S. Rathod
Department of Electronics and Communication, Graphic Era (Deemed To Be) University, Dehradun 248002, India

P. Srivastava et al. (eds.), *Mathematical, Computational Intelligence and Engineering Approaches for Tourism, Agriculture and Healthcare*, Lecture Notes in Networks and Systems 214, https://doi.org/10.1007/978-981-16-3807-7_21

which tend to attract water. Surfaces having a contact angle greater than 90° is called hydrophobic and surfaces that create a contact angle more than 150° are called superhydrophobic. These surfaces repel water [9].

Sol-gel technique was used to coat the glass surface to make it superhydrophobic. Sol-gel method is used for the polymerization of inorganic materials, for the hybrid of organic/inorganic materials, and nanocomposites of organic/inorganic materials [10]. The sol-gel technique is used widely by researchers in the field of coating technique, due to its versatility in synthesizing inorganic nanoparticles at easy conditions. Its reaction condition, morphology, transparency, and surface energy can be controlled by varying the temperature condition and chemical composition. Purity, homogeneity, controlled porosity, temperature stability, nanoscale structuring, physical rigidity, photochemical and thermal stability, desired optical transparency are the most amazing features of the sol-gel process. Sol has some problems in stability and adhesiveness. It provide dense and scratch resistance in fabrication [11–15]. SiO2 particles have good durability toward the environmental condition. The dense coatings have better scratch resistance capability. The high porosity of silica particles leads to low refractive index, but some disadvantages due to less adhesion tendency. Silica is also used to increase the roughness on the surface [12, 16]

Trialkoxy compounds are good cross-linkers, coupling agents and adhesives and modification can make it moisture curable also [17, 18]. Silicones, fluorocarbons, and long-chain fatty acids are used to enhance hydrophobicity. Long alkyl chain thiols, alkyl or fluorinated organic silanes, perfluorinated alkyl agents, long alkyl chain fatty acids, PDMS-based polymers, or their combinations are used for low surface energy modification. PDMS is low cost, transparent, environmental friendly, nontoxic, easy in fabrication, biocompatible, biodurable material, and also provide hydrophobic modification [13, 19, 20]. PDMS with high molecular weight incorporated with silica network and forms the bridge between silica particles and link them together. PDMS also act as a polymer binder. Two hydroxyl groups and many methyl groups make it hydrophobic. A condensation reaction occurs between hydroxyl groups of PDMS and hydroxyl groups on silica particles. PDMS provides high transmittance, low absorption of humidity. Heat treatment and aging are used to make porous robust and better coating structure. Binders increase adhesiveness, cross-linking and durability of coatings [10, 20, 21].

2 Material and Method

2.1 *Materials*

Methyl trimethoxysilane (MTMS) used as silica precursor, is used to prepare silica sol, was purchased from Renken. Ammonia was used as a hydrolyzing agent and also as a catalyst in hydrolysis reaction, purchased from Renken. Ethyl alcohol used as solvent was purchased from Merck. Polydimethylsiloxane (PDMS) used to increase the hydrophobicity of the coatings, was purchased from Renken. All the chemicals used were of AR grade. A multimeter was used to determine efficiency and purchased from the company Mastech. Solar panel from Ecco Company having specifications Pmax—3.0 W, Vmp—8.80 V, Imp—0.34A, Isc—0.38 A and Voc—10V was used. A 9W LED lamp source was used for the irradiating light on the solar panel.

2.2 *Preparation of Coating Material*

10 ml of methyltrimethoxy silane (MTMS) was taken as precursor to make silica sol and the hydrolysis was carried out by adding 10 ml of ethanol as a solvent. In the preparation of silica sol from silica precursor, the hydrolysis and polycondensation reactions are very slow. The reaction rate was increased by adding 2 ml of 1 molar ammonium hydroxide base. Base act as a catalyst and also increase the coating efficiency of the glass surface. After two hours of slow stirring with 300 rpm, of these mixtures at room temperature, obtained solutions were kept at room temperature for one week. This procedure provides a translucent low viscosity nanoparticle contained colloidal sol. The hydrophobic modification was done by adding polydimethylsiloxane (PDMS) in silica sol with a ratio of PDMS to MTMS was 3:7 and 4:6 and coded the sols as S1 and S2. PDMS incorporated between silica networks and increase the hydrophobic characteristic. For the completion of the reaction, the sols were kept at room temperature for another five days.

2.3 *Coating Method*

The sols were coated on 1mm thick and 3.5×1.5 cm^2 soda-lime glass by a spin coating method using spin coater model NXGP2 by Apex Instruments, India. Spin coating was done at 5000 rpm for 200 s by pouring 250 μl of sol at room temperature. Two layer coatings were done to increase the strength of the surface. The coated glass pieces were kept in a muffle furnace for one hour at 200 °C for calcinations. After calcinations, the glass surface shows better hydrophobicity and stability. The solar panels were coated by spray coating at a pressure of 50 lb/in^2 or 50 psi and the heat treatment was given by a hot air blow dryer.

2.4 *Surface Characterization*

The topography of the coated surfaces was determined by Atomic Force Microscopy, done by AFM instrument Park XE 70 at room temperature. Images showed the roughness and surface topography of the coated surfaces. Scanning Electron Microscopy analysis was also performed to see the surface morphological structure of coated surfaces by using JSM-6010LA In TouchScope™ multiple touch panel scanning electron microscopes. The prepared sample glasses were firstly sputtered with gold to gain better SEM images.

2.5 Contact Angle Measurements

The surface hydrophobicity is analyzed by pouring water drops on the surface. The degree of contact angle will determine the surface hydrophobicity or hydrophilicity. Surface hydrophobicity of the coated surfaces was characterized by the drop shape analyzer, DSA25S Kruss, Germany by pouring 2 μl of distilled water drop at 20 °C temperature on the surface.

2.6 Determination of Efficiency of Solar Panel

The coated and uncoated solar panels were connected by the appropriate circuit as shown in Fig. 1. and compared by making I-V curve, calculating power, efficiency, and fill factor. I-V curve, efficiency, fill factor of coated and uncoated panel surfaces were derived by using an LED lamp as light source, multimeter, and variable resistances to know the overall performance of panels.

Power of the solar cell is the product of current and voltage. Fill factor is the ratio of maximum power produced by solar cells divided by the product of Voc and Isc. Efficiency is defined as the ratio of energy output from the solar cell to input energy from the sun.

$$\text{Power} = \text{Vm} \times \text{Im}$$

$$\text{Fill factor} = \frac{\text{Vm} \times \text{Im}}{\text{Voc} \times \text{Isc}} = \frac{\text{Area of rectangle}}{\text{Area of the curve}}$$

$$\text{Efficiency} = \frac{\text{Vm} \times \text{Im}}{\text{E} \times \text{Area of the cell}}$$

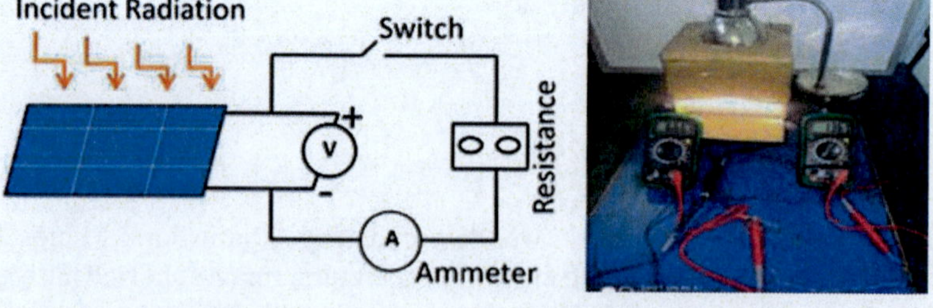

Fig. 1 Solar panel connection circuit diagram (left), connected solar panel (right)

where Vm and Im are the maximum voltage and current obtained from I-V curve. Voc and Isc are open-circuit voltage and short-circuit current, respectively. E is the power of incident light.

2.7 *Anti-Dust and Self-Cleaning Tendency*

Anti-dust and self-cleaning tendency were determined by pouring some dust on the coated substrate and putting a drop of water on it. The rolling pattern of drop on substrate and amount of dust collected by drop was observed to determine self-cleaning efficiency. Solar panels stability, dust repelling tendency of both surfaces were determined dusted substrate and compared. Dust repellence was checked by tilting the dusted substrate.

3 Result and Discussion

The coating material was prepared by sol-gel method. Prepared sol was a uniform colloidal solution shown in Fig. 2a. Superhydrophobic surfaces were prepared by coating S1 and S1 sol on soda-lime glass and solar panel, respectively, there was no effect of coating on transparency of glass so the glass and solar panel remain transparent after coating.

Fig. 2 **a** prepared sol **b** water droplet on coated solar panel **c** water droplet on coated (left) and uncoated (right) glass

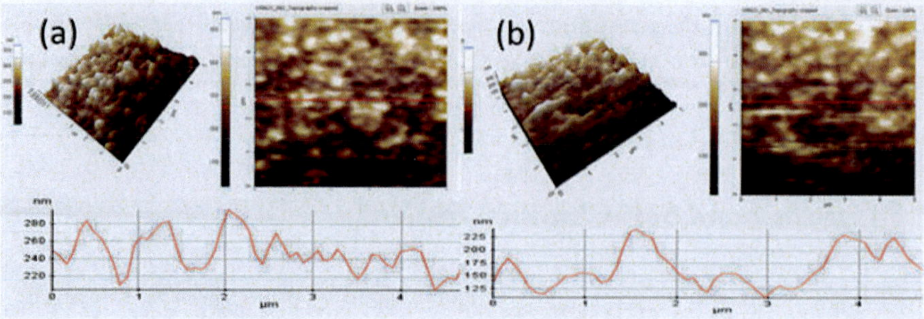

Fig. 3 Atomic force microscopy 3D image, 2D image, line profile of (left) Sol 1 coated surface, (right) Sol 2 coated surface

3.1 *Surface Characterization*

Atomic Force Microscopy

AFM images of prepared samples showed rough morphology over the coated surface. Surfaces were analyzed by the instrument Park XE 70. The root mean surface roughness was found between 247 nm and 174 nm for sol 1 and sol 2 coated glasses, respectively. Obtained images are shown in Fig.3.

Scanning Electron Microscopy

The surface morphology by SEM shows the presence of nanostructured particles present in the coated surface. Particles having a size in nano range between 15 and 30 nm is shown in Fig. 4 for both coatings. SEM images also showed uniform, homogenous surface with some cracks on the surface. SEM images concluded that the nanostructures are responsible for the superhydrophobicity of the surfaces.

3.2 *Contact Angle Analysis*

Surface wettability is used to check the hydrophobicity of any surface. The water contact angle is defined as the angle where a liquid-vapor interface meets a solid surface. 2 μl of water droplet was poured on each of the surfaces. The coated surfaces showed the superhydrophobic behavior as the contact angle increases to in the range of 145°–155° from $33 \pm 1°$, 0 in case of the plain uncoated sample as shown in Fig. 5.

The contact angle was also observed for acidic and basic solution. 1% HCl solution and 1% NaOH solution were prepared and contact angle was observed for the coated glass. There is no change observed for acidic and basic solution contact angle compared to water contact angle which shows the stability of coating in acidic and basic medium. The observed images are shown in Figs. 6 and 7. Observed contact angles on various surfaces are shown in Table 1.

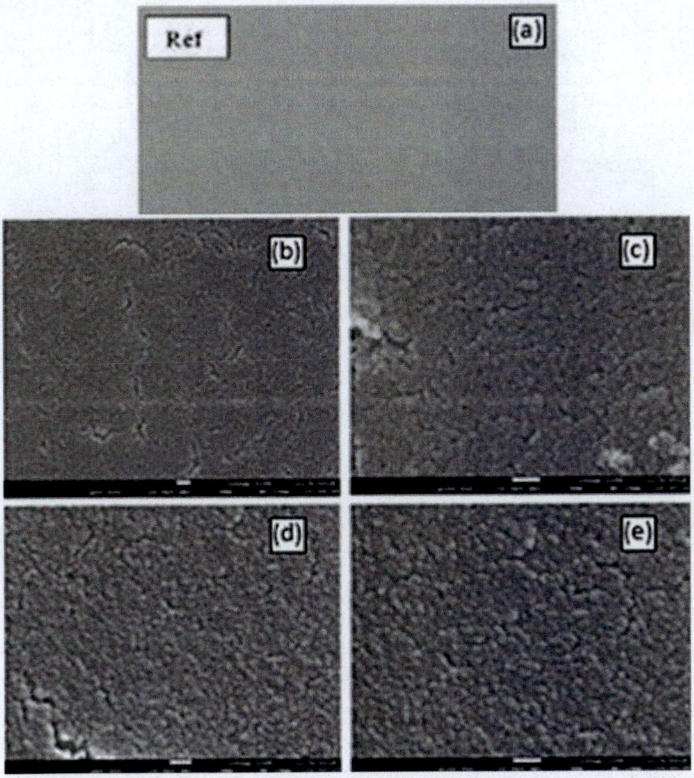

Fig. 4 **a** SEM image of plain uncoated glass sample, **b** S1 coated surface at 80000× magnification range, **c** S1 coated surface at 100000× magnification range, **d** S2 coated surface at 80000× magnification range, **e** S2 coated surface at 100000× magnification range

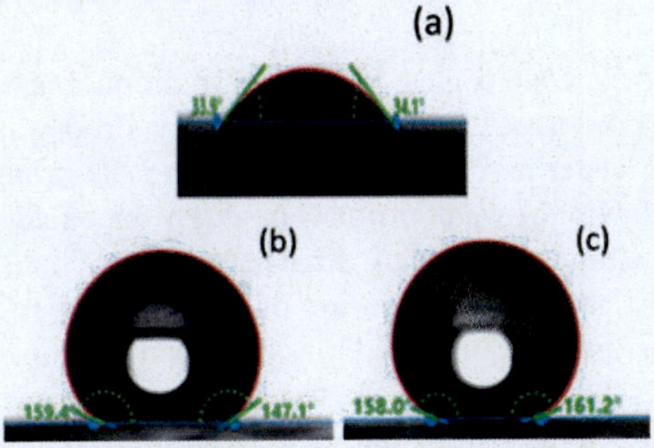

Fig. 5 Water contact angle **a** plain uncoated glass substrate, **b** S1 coated surface, **c** S2 coated surface

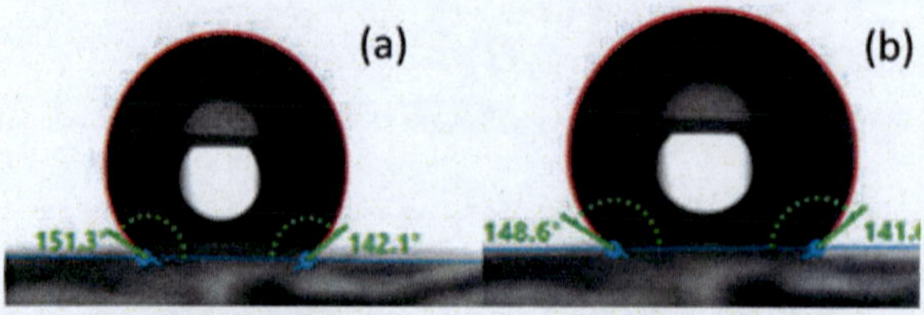

Fig. 6 Contact angle with 1 molar HCl solution, **a** S1 coated surface, **b** S2 coated surface

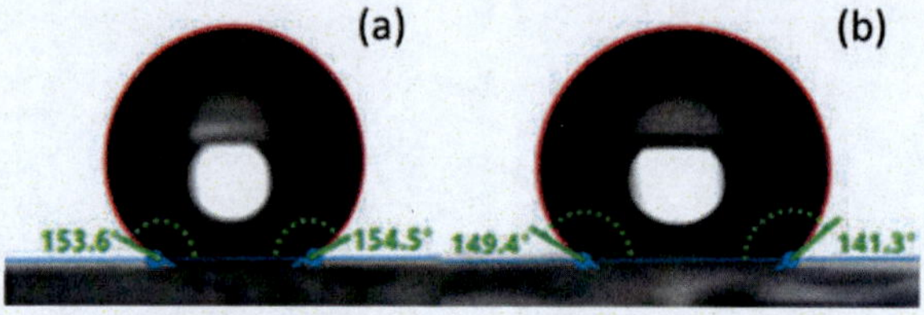

Fig. 7 Contact angle with 1 molar NaOH solution, **a** S1 coated surface, **b** S2 coated surface

Table 1 Observed contact angle on various surfaces

Drop type	Contact angle of soda lime glass uncoated surface	Contact angle by sol 1 coated surface	Contact angle by sol 2 coated surface
Water	34 ± 1°	159 ± 1°	158 ± 1°
1 M, HCl	–	151 ± 1°	146 ± 1°
1 M, NaOH	–	153 ± 1°	149 ± 1°

3.3 *Anti Dust and Self Cleaning Tendency*

As the coated glass surface is hydrophobic and hence having less affinity for dust as compared to water. When water drop was poured on the coated glass surface, the water carry dust particles with it as it has more affinity for dust particles. The behavior of water droplets on dusty uncoated and coated surfaces is shown in Fig. 8.

Comparisons were done between the coated and uncoated glass toward their anti-dust properties. Adhesion of dust is more on uncoated glass and repellency toward dust is more in coated glass compared to uncoated glass. The same amount of dust was poured on the coated and uncoated surface. On tilting the surfaces, coated glass removed all the dust while dust remained accumulated on uncoated glass. Observations are shown in Fig. 9.

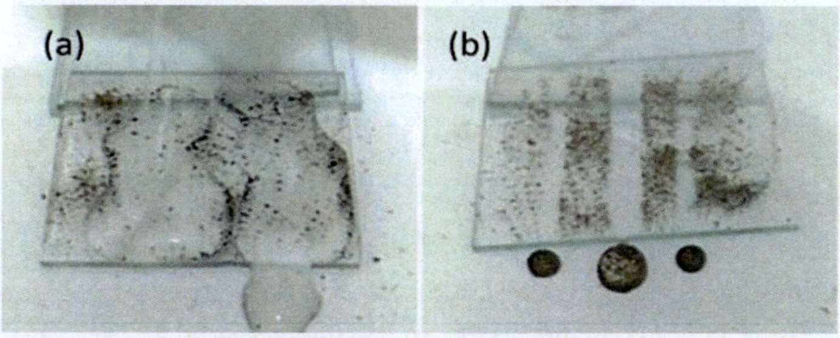

Fig. 8 Effect of water on dust over **a** uncoated and **b** coated surfaces

Fig. 9 **a** Poured dust in both coated and uncoated type of glass, **b** dust repelling observation on tilting glass

3.4 I-V Curve, Power, Fill Factor, and Efficiency Calculation

The I-V curve of the coated and uncoated surface were drawn after one week of exposure in open environmental conditions. Power, fill factor, and efficiency were calculated. Performances of coated solar panels showed better results compared to an uncoated surface. I-V curve for uncoated and coated sol 1 and sol 2 are shown in Fig. 10 and calculated power, efficiency, fill factor are represented in Table 2.

The increment in the power was 7.39% and 4.97%, fill factor increased by 11.36% and 9.09%, and the efficiency of the coated panel increased by 5.16% and 4.99% for sol1 and sol2 coated solar panel, respectively, compared to the uncoated solar panel.

4 Conclusion

The above analysis showed the prepared surfaces were nanostructured, low-energy superhydrophobic surfaces and having anti-dust properties. Superhydrophobic coated solar panels showed better output compared to uncoated glass panels, also increase the durability of panels by making them clean. The MTMS sol was stable for up to six months and every time it gives the same kind of superhydrophobic coating means the contact angles were the same at every interval of the coat. Although the

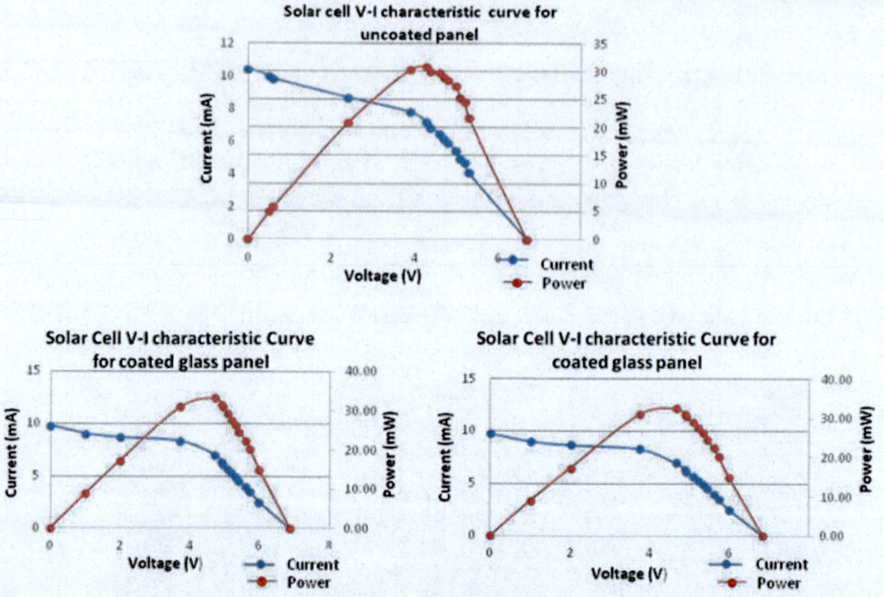

Fig. 10 Solar cell characteristic IV curve for uncoated (top), S1 (middle), and S2 (lower) sol-coated solar panel

Table 2 Observed results from solar panel analysis

Parameters	Uncoated surface	Coated surface (S1)	Coated surface (S2)	Percentage increase to coated glass compared to uncoated (%)	
				S1	S2
Maximum power obtained (mW)	30.96	33.25	32.50	7. 39	4.97
Fill factor	0.44	0.49	0.48	11.36	9.09
Efficiency	11.42	12.01	11.99	5.16	4.99

coated surface stability starts decreasing after three months. It could be a better alternative for enhancing solar panel efficiency by making some modifications. The silica precursor is an easy alternative to increase the efficiency of solar panels.

Acknowledgements This research work is financially supported by the Department of Science and Technology (DST)—Clean Energy Research Initiative DST/TMD/CERI/C28 (G).

References

1. Darwish, Z.A., Kazem, H.A., Sopian, K., Alghoul, M.A., Chaichan, M.T.: Impact of some environmental variables with dust on solar photovoltaic (PV) performance: review and research status. Int. J. Energy Environ. **7**(4), 152–159 (2013)
2. Elminir, H.K., Ghitas, A.E., Hamid, R.H., El-Hussainy, F., Beheary, M.M., Abdel-Moneim, K.M.: Effect of dust on the transparent cover of solar collectors. Energy Convers. Manage. **47**(18–19), 3192–3203 (2006)
3. Hirose, J., Takanohashi, H., Ogawa, S., Piliougine, R.M., Zorrilla, J., Carretero, J.: Evaluation of power-enhancement for photovoltaic modules. In: Proceedings of the 25th European Photovoltaic Solar Energy Conference and Exhibition 2010, pp. 4087–93
4. Gwon, H.J., Park, Y., Moon, C.W., Nahm, S., Yoon, S.J., Kim, S.Y., Jang H.W.: Superhydrophobic and antireflective nanograss-coated glass for high performance solar cells. Nano Res. **1**;**7**(5), 670–678 (2014)
5. Gholami, A., Alemrajabi, A.A., Saboonchi, A.: Experimental study of self-cleaning property of titanium dioxide and nanospray coatings in solar applications. Solar Energy **15**(157), 559–65 (2017)
6. Jang, G.G., Smith, D.B., Polizos, G., Collins, L., Keum, J.K., Lee, D.F.: Transparent superhydrophilic and superhydrophobic nanoparticle textured coatings: comparative study of anti-soiling performance. Nanoscale Adv. **1**(3), 1249–1260 (2019)
7. Sutha, S., Suresh, S., Raj, B., Ravi, K.R.: Transparent alumina based superhydrophobic self-cleaning coatings for solar cell cover glass applications. Solar Energy Mater. Solar Cells **1**(165), 128–137 (2017)
8. Vazirinasab, E., Jafari, R., Momen, G.: Application of superhydrophobic coatings as a corrosion barrier: a review. Surf. Coat. Technol. **15**(341), 40–56 (2018)
9. Falde, E.J., Yohe, S.T., Colson, Y.L., Grinstaff, M.W.: Superhydrophobic materials for biomedical applications. Biomaterials **1**(104), 87–103 (2016)
10. Wen, J., Wilkes, G.L.: Organic/inorganic hybrid network materials by the sol-gel approach. Chem. Mater. **14**;**8**(8), 1667–81(1996)
11. Satvekar, R.K., Phadatare, M.R., Karande, V.A., Patil, R.N., Tiwale, B.M., Pawar, S.H.: Influence of silane content on the optical properties of sol gel derived spin coated silica thin films. Int. J. Basic Appl. Sci. **1**, 468–476 (2012)
12. Li, W., Tan, X., Zhu, J., Xiang, P., Xiao, T., Tian, L., Yang, A., Wang, M., Chen, X.: Broadband antireflective and superhydrophobic coatings for solar cells. Mater. Today Energy **1**(12), 348–355 (2019)
13. Jankauskaitė, V., Narmontas, P., Lazauskas, A.: Control of polydimethylsiloxane surface hydrophobicity by plasma polymerized hexamethyldisilazane deposition. Coatings **9**(1), 36–45 (2019)
14. Terpilowski, K., Goncharuk, O.: Hydrophobic properties of hexamethyldisilazane modified nanostructured silica films on glass: effect of plasma pre-treatment of glass and polycondensation features. Mater. Res. Express **5**(1), 016409 (2018)
15. Kumar, D., Wu,. X, Fu, Q., Ho, J.W., Kanhere, P.D., Li, L., Chen, Z.: Hydrophobic sol–gel coatings based on polydimethylsiloxane for self-cleaning applications. Mater. Des. **5**(86), 855–62 (2015)
16. Wang, X., Shen, J.: Sol–gel derived durable antireflective coating for solar glass. J. Sol–Gel Sci. Technol. **53**(2), 322–327 (2010)
17. Jiang, H., Zheng, Z., Wang, X.: Kinetic study of methyltriethoxysilane (MTES) hydrolysis by FTIR spectroscopy under different temperatures and solvents. Vib. Spectrosc. **46**(1), 1–7 (2008)
18. Cai, C., Sang, N., Teng, S., Shen, Z., Guo, J., Zhao, X., Guo, Z.: Superhydrophobic surface fabricated by spraying hydrophobic R974 nanoparticles and the drag reduction in water. Surf. Coat. Technol. **15**(307), 366–73 (2016)
19. Xue, C.H., Jia, S.T., Zhang, J., Ma, J.Z.: Large-area fabrication of superhydrophobic surfaces for practical applications: an overview. Sci. Technol. Adv. Mater. **11**(3), 033002 (2010)

20. Ahangarani, S., Shanaghi, A.: Six-layer antireflection TiO 2/SiO 2 coatings for solar covers: role of Triton and PDMS additives. J. Mater. Sci.: Mater. Electron. **27**(7), 7141–7149 (2016)
21. Jafari, R., Farzaneh, M.: A simple method to create super-hydrophobic aluminium surfaces. Mater. Sci. Forum **706**(709), 2874–2879 (2012)

A Study of Nanoscale Zeolitic Imidazole Framework-8 in Removal of Humic Acid from Wastewater by Adsorption

Shikha Solanki, Malayaj Srivastava, Saransh Bhatnagar, and Surendra Sasi Kumar Jampa

1 Introduction

One of the major issue worldwide concerning drinking water treatment is the presence of natural organic matter (NOM), which is present everywhere in aquatic environments [1]. The significant component of NOM is humic acids (HA). Humic substances have humic acids as their principal component, which form the majority of the organic constituents of soil, coal, and peat. It is also a major organic constituent of ocean water and many dystrophic lakes. The formation of humic acid takes place by the biodegradation process of dead organic matter. It is not a single acid [1]; rather, It is a combination of acids containing phenolate and carboxylate groups, hence forming a complex mixture owing to which it behaves as a diabasic acid or, sometimes, as a tribasic acid [2].

It can easily dissolve in water and imparts dark brownish color. Humic acid has other problems such as converting into toxic compounds during chlorination and increases the concentration of iron and manganese in groundwater. Manganese and iron in drinking water have many health-related problems and can also stain the laundry, cause scaling on the plumbing, and make the water look, smell, or taste bad. Therefore, humic acid causes health issues for drinking water as well as scaling problems thus needs to be removed.

Humic acid is usually removed from water via filtration, flotation, and adsorption. Among the mentioned methods, adsorption is most employed because of its simple operation and low initial cost.

Recently, a new type of adsorbents has been developed and also studied, they are known as metal-organic frameworks (MOFs). Zeolitic imidazole frameworks (ZIFs) come under a special class of MOFs, which have a quality under an aqueous

S. Solanki · M. Srivastava (✉) · S. Bhatnagar · S. Sasi Kumar Jampa
Department of Chemical Engineering, Pandit Deendayal Petroleum University, Gandhinagar, Gujarat, India

P. Srivastava et al. (eds.), *Mathematical, Computational Intelligence and Engineering Approaches for Tourism, Agriculture and Healthcare*, Lecture Notes in Networks and Systems 214, https://doi.org/10.1007/978-981-16-3807-7_22

environment to maintain exceptional chemical stability [3]. The surface area of ZIF-8 is high and also by using a simple ethanol-washing process, they can be regenerated and hence can be subsequently reused for humic acid adsorption [3]. Hence owing to these features, for the removal of humic acid from water, ZIF-8 appeals to be a stable and efficient adsorbent for the same purpose.

2 Experimental

2.1 Materials

The chemicals used for the experiments are zinc nitrate hexahydrate ($Zn(NO_3)_2{\cdot}6H_2O$) and 2-Methylimidazole(2-MIM) was purchased from Sigma Aldrich. Whereas methanol, humic acid, hydrochloric acid and sodium hydroxide, all of these were obtained from Finar chemicals.

2.2 Synthesis of ZIF-8

The synthesis of ZIF-8 nanoparticles was done by the rapid precipitation method. Firstly, 2.34 g of $Zn(NO_3)_2{\cdot}6H_2O$ was dissolved in 16 ml of methanol to form solution I and simultaneously 44.70 g of 2-MIM was added to 80 ml of methanol to obtain solution II. Then these two solutions are mixed and stirred for 1 h in a magnetic stirrer at ambient temperature. Initially, the solution was transparent which turned milky with time. After 1 h, excess methanol is evaporated using rotavapour under vacuum conditions. Once the solution is reduced to 50 ml it is centrifuged at 13000 rpm for 10 min to collect the colloids. During this process, precipitation and washing with methanol are done 3 times to remove unreacted reactants and by-products. The white color product obtained was then dried at 80 °C for 1 h.

2.3 Humic Acid Adsorption Using ZIF-8

Batch-type experiments were done for the evaluation of humic acid adsorption on ZIF-8. The variation in adsorption was observed by varying the Solid/Liquid (S/L) ratio, temperature, time, and pH. The duplication and repetition of all the experiments were done atleast twice [3].

Firstly, we examined the effect of the S/L ratio on humic acid adsorption. For these experiments, we fixed the amount of ZIF-8, i.e., the weight of solid which was 0.025 g. The range of the S/L ratio (in g/L) was from 0.5 to 2.5 g/L. 50 PPM of the humic acid solution is taken with respective volume (according to S/L ratio and

0.025 g of solid) in 5 conical flasks (each for different S/L ratio). The conical flasks containing humic acid and ZIF-8 were kept in an orbital shaker at 40 °C for 1 h. Once this procedure is complete, the mixtures are kept in a centrifuge at 13000 rpm for 10 min, during which ZIF-8 with adsorbed humic acid settles at the bottom and UV-spectroscopy of the clear solution left is done at a wavenumber of 254 cm^{-1}.

For observing the variation in adsorption with time and temperature, at a fixed temperature, three conical flasks were taken in which humic acid and ZIF-8 was present at same S/L ratio and was stirred for different time, i.e. for 20, 40 and 60 min. Once the time is over in the orbital shaker, centrifugation and UV-spectroscopy are done.

This procedure was repeated for three different temperatures. The temperatures at which experiments were performed were 20, 30, 40, and 50 °C.

For studying the variation in adsorption of humic acid on ZIF-8 with pH, we performed the experiments at pH 3, 5, 7.5, and 10. The S/L ratio was the same for all the solutions. The experiments were performed at 20 °C for 1 h. After which centrifugation and UV-spectroscopy were done.

For calculating capacity and concentration at time 't', the following equation was being used (Eq. 1):

$$q_t = \frac{V(C_0 - C_t)}{m} \tag{1}$$

$y = 0.0357C_t$; where y is in abs.

Figure1 represents the absorbance of humic acid solution samples at various concentrations varying from 0 to 100 ppm. The absorbance is obtained using a UV Spectrometer.

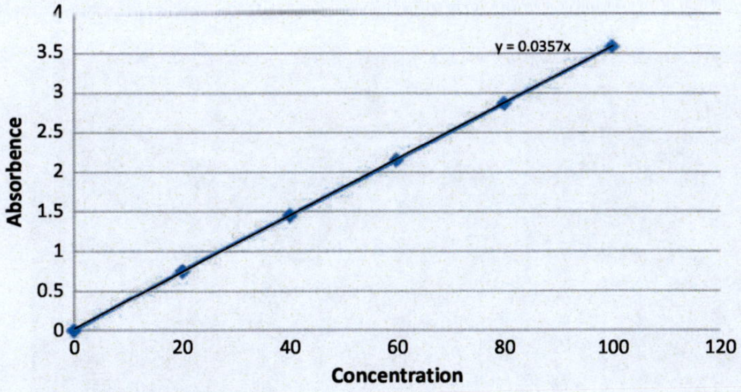

Fig. 1 Standard curve for humic acid

3 Results and Discussion

3.1 Characterization of ZIF-8

Sem Analysis

During our research, we used field emission scanning electron microscope JEOL JSM- 6700, JAPAN to study the morphology of as-synthesized ZIF-8. Figure 2 depicts the morphology of the as-synthesized ZIF-8 which visualizes a group of ZIF-8 particles bound together with a few other small ZIF-8 particles [1]. Each of the large and small particles reportedly exhibits a rhombic dodecahedra type of structure. It has been reported that along with synthesis time the morphology of ZIF-8 varies. As the synthesis time increases, shape changes from cubic to rhombic dodecahedron until the equilibrium is reached. Henceforth, the appearance of the rhombic dodecahedra shape of ZIF-8 in Fig. 2 clearly states that the as-synthesized ZIF-8 is at the equilibrium stage and the size of the ZIF-8 particle is in the range of 30–50 nm [3].

FTIR Analysis

In the spectrum of the humic acid-rich ZIF-8, several minor variations were observed. The variations found in our analysis are mainly due to the deposition (as a result of adsorption) of humic acid on the ZIF-8 surface. When we take the infrared spectrum of humic acid and ZIF-8-rich humic acid individually under consideration, it was found that the narrow band for humic acid, which was initially at 3,355 cm^{-1} shifted to 2931.22 cm^{-1}, which is due to the stretching vibration mode of C–H bond in the alkyl chain of humic acid. Strong bands were also exhibited by Humic acid at 1578.95 cm^{-1} and 1374.53 cm^{-1}, corresponding to the stretching vibrational mode of carbonyl groups of ketones/carboxylic acids and of C=C bond. It was also observed

Fig. 2 SEM image of the as-synthesized ZIF-8 nanoparticles

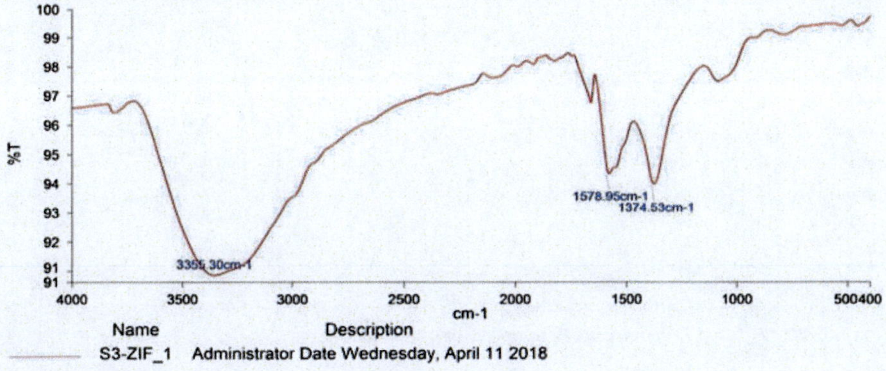

Fig. 3 FTIR spectra for pure humic acid

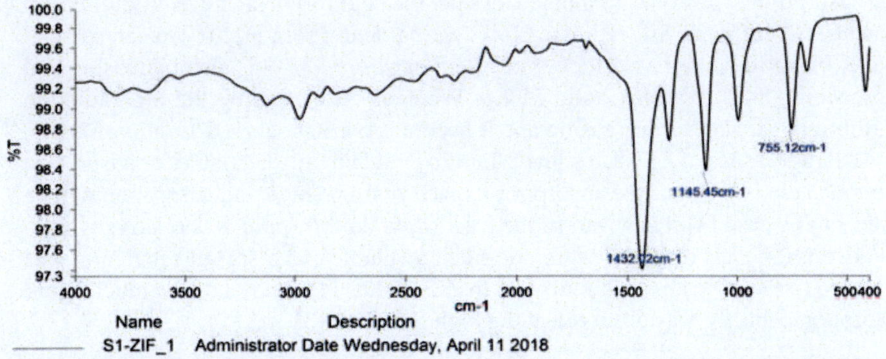

Fig. 4 FTIR Spectra for ZIF-8 adsorbent

that those bands were shifted to 1434.54 $cm^{-1}$and 1144.97 $cm^{-}1$, respectively, in the humic acid-rich ZIF-8. It advocates the presence of humic acid on the surface of ZIF-8. Owing to these results, humic acid adsorption on ZIF-8 was indeed confirmed. At 422 cm^{-1}, the association between nitrogen of 2-MIM and zinc can be observed which shows the existence of the Zn–N bonding (Figs. 3, 4 and 5).

3.2 *Effect of S/L Ratio on Adsorption Capacity of ZIF-8*

Adsorption capacity depends on the number of active adsorption sites and available specific surface area. Removal efficiency is determined by the tendency of adsorbate to load onto adsorbent compared to remaining in solution. To study the change in adsorption capacity of ZIF-8 with S/L ratio, we varied the ratio from 0.5 to 2.5 g/l and fixed the temperature at 20 °C and conducted the experiment for 90 min.

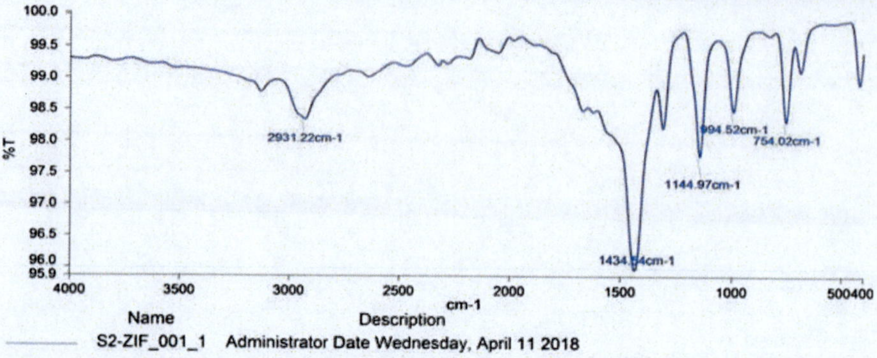

Fig. 5 FTIR Spectra for ZIF-8 rich humic acid

The adsorption capacity was calculated as per the equilibrium achieved. Equilibrium capacities at different S/L ratios at 20 °C were noted. From Fig. 6, we can see that as the S/L ratio is increased the removal efficiency of ZIF-8 is also increasing and approaching 84% at an S/L ratio of 2.5. Whereas on increasing the S/L ratio, the equilibrium capacity is decreasing and it becomes constant at an S/L ratio of 2.

Adsorption sites of ZIF-8 at a low S/L ratio could be fully available for adsorption to humic acid [4]. Hence the adsorption capacity will be high and adsorption will be faster in this case [4]. However, in the case when the S/L ratio was relatively high, the adsorption sites of ZIF-8 were not much available and a sort of restriction was observed in the exposure of humic acid to ZIF-8 sites [4]. Due to this, a much lower adsorption capacity was observed at the high S/L ratio [3].

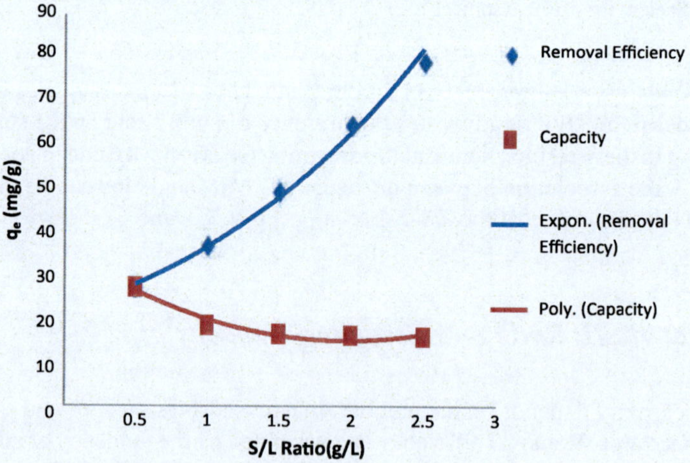

Fig. 6 Plot between q_e and removal efficiency with S/L ratio

3.3 Kinetic Analysis of Humic Acid Adsorption on ZIF-8

Determination of whether the adsorption process of Humic Acid follows pseudo-first order or second order should be done. The graphs were plotted according to the rate law equation for the respective order of the reaction. The pseudo-first-order rate law can be expressed in a linear form as follows (Eq. 2) [5]:

$$\ln(q_e - q_t) = \ln q_e - \frac{k_1 t}{2.303} \tag{2}$$

where k_1 (min^{-1}) is the pseudo-first-order rate constant (Fig. 7).

From the graph, we can see that the adsorption process of Humic acid on ZIF-8 does not follows a pseudo-first-order kinetics as there is no linear relation found as per the kinetic equation between the coordinates.

The pseudo-second-order rate law can be expressed in a linear form as follows (Eq. 3) [6]:

$$\frac{t}{q_t} = \frac{1}{k_2 q_e^2} + \frac{t}{q_e} \tag{3}$$

where k_2 ($g\ mg^{-1}\ min^{-1}$) denotes the pseudo-second-order rate constant [7] (Fig. 8).

From Fig. 8, we get a linear curve between t/q_t versus t hence proving that the adsorption of Humic Acid on ZIF-8 follows pseudo second-order kinetics.

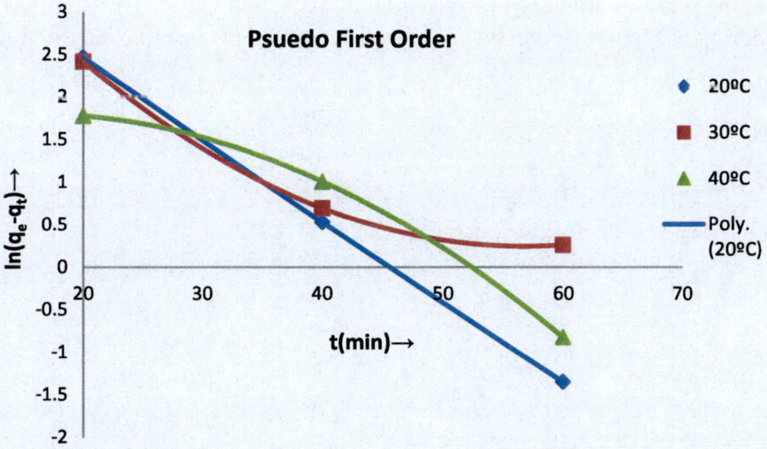

Fig. 7 Plot of $\ln(q_e - q_t)$ versus t

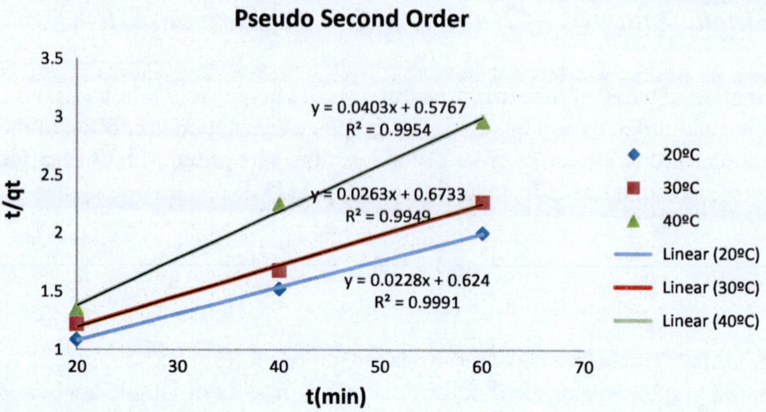

Fig. 8 Plot between t/q_t versus t

3.4 *Effect of pH on Adsorption of Humic Acid on ZIF-8*

We are discussing the removal of humic acid from water, but, the raw water or wastewater contains other molecules and compounds, which might change the pH of water [8]. So it is important to study the effect of pH in our adsorption studies. The pH range considered here is 3–12. The experiments were conducted at 20 °C for 1 h. The curve below shows the variation of q_e with pH (Fig. 9).

It can be seen that as the pH decreases from 7 to 3 the equilibrium capacity and removal efficiency increases. The equilibrium capacity increases from 20.89 to 29.36. Whereas, the removal efficiency increases from 62.92 to 88.46%. This signifies that the efficiency of adsorption can be very much increased in an acidic environment.

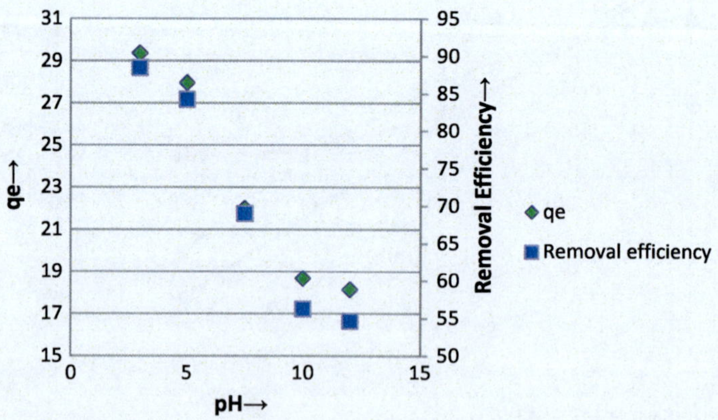

Fig. 9 Plot showing variation in equilibrium capacity and removal efficiency with pH

Moreover, when pH increases from 7 to 12, the equilibrium capacity and removal efficiency decreased to 18.14 and 54.64%, respectively. This indicates that higher pH is unfavourable for the adsorption of humic acid on ZIF-8.

4 Conclusions

In this study, ZIF-8 was firstly synthesized and adsorption of humic acid for its removal from water was observed subsequently. Factors affecting the adsorption were examined including S/L ratio, mixing time, temperature, and pH. It is observed that as the S/L ratio and time increase the removal efficiency increases [9]. On the other hand, as temperature and pH increase the removal efficiency decreases [9]. The adsorption kinetics and isotherm were determined to know that order of reaction, which came out to be of second order.

Acknowledgements We would like to give special thanks to the Department of Chemical Engineering, Pandit Deendayal Petroleum University (PDPU), and also the Department of Solar Energy (PDPU) for acquainting us with the required instruments, like FTIR and SEM analysis made it possible for us among many others to make our research work possible.

References

1. Elsheikh, A.F., Ahmad, U.K., Ramli, Z.: Investigations on humic acid removal from water using surfactant-modified zeolite as adsorbent in a fixed-bed reactor. Appl. Water Sci. **7**, 2843–2856 (2017)
2. Gastinel, L.N.: Principal component analysis in the era of «Omics» data. Principal Component Analysis–Multidisciplinary Applications (2012)
3. Lin, K.A., Chang, H.: Efficient adsorptive removal of humic acid from water using zeolitic imidazole framework-8 (ZIF-8). Water Air Soil Pollut. **226**, 10 (2015)
4. Lu, M., He, L., Cheng, Y.: Synthesis of Nd-doped ZSM-5 and its application to treating slightly polluted water. J. Water Reuse Desalin. **8**, 86–93 (2018)
5. Lagergren, S.: About the theory of so-called adsorption of soluble substances. Kungliga Svenska Vetenskapsakademiens Handlingar **24**, 1–39 (1898)
6. Ho, Y.S., Mckay, G.: Pseudo-second order model for sorption processes. Process Biochem. **34**, 451–465 (1999)
7. Aytar, D., Bozkurt, S., Erol, M.Ö., Çabuk, A.: Increased removal of reactive blue 72 and 13 acidic textile dyes by penicillium ochrochloron fungus isolated from acidic mine drainage. Desalin. Water Treat. **57**(41), 19333–19343 (2016)
8. Derakhshani, E., Naghizadeh, A.: Optimization of humic acid removal by adsorption onto bentonite and montmorillonite nanoparticles. J. Molec. Liquids **259**, 76–81 (2018)
9. Sarkar, S., Meikap, B.C., Chatterjee, S.G.: Modeling of removal of sulfur dioxide from flue gases in a horizontal cocurrent gas–liquid scrubber. Chem. Eng. J. **131**, 263–271 (2007)

Management of High-Volume Clashes During Clash Detection

Sagar Malsane, M. J. Merin, G. J. Gokul, P. Nithin Dev, and Smitha Treesa Babu

1 Introduction

Designing and preparation of construction drawings is one of the first steps in any construction project. Construction drawings are prepared by various departments like architectural, structural, MEPF, environmental, etc. from their independent offices. Therefore, when these drawings are integrated, chances of clashes between various building components are very high, e.g. structural component like a beam collides with an MEP component, a mechanical duct [1, 2]. The conventional method of resolving clashes is to overlap the various design drawings or Good for Construction (GFC) drawings and to manually coordinate between the different departments, which is a very hectic and time-consuming process. The efficiency of this process is also low and can even lead to conflicts between different departments. These clashes if not resolved before the construction can lead to reworks and idle time on site which results in time and cost overruns [3]. The Indian construction industry is plagued with many such challenges and hence holistic use of project management along with the use of technology is crucial for success [4].

The use of BIM as part of project management can lead any project to its optimum output. Building Information Modelling (BIMs) is a project management tool that can support owners, designers, and builders in their creation and coordination of the design of building systems and planning of construction work, in their processes for fabrication and building, and in their processes for operating and maintaining, as well as decommissioning their facilities [5, 6].

Hence with the implementation of BIM, the issue of clashes can be effectively overcome. BIM process comprises many applications, one of them is 3D Clash Detection, wherein the clashes are identified from the 3D information models in no

S. Malsane (✉) · M. J. Merin · G. J. Gokul · P. Nithin Dev · S. T. Babu
National Institute of Construction Management and Research, Pune, India
e-mail: smalsane@nicmar.ac.in

P. Srivastava et al. (eds.), *Mathematical, Computational Intelligence and Engineering Approaches for Tourism, Agriculture and Healthcare*, Lecture Notes in Networks and Systems 214, https://doi.org/10.1007/978-981-16-3807-7_23

time. The clash detection tool facilitates the identification of clashes between various trade components as mentioned earlier. Clash management includes two procedures: detecting clashes and resolving clashes [7, 8]. A clash detection test usually results in a large number of relevant and non-relevant clashes. The irrelevant clashes have no substantial influence on the construction works and can be solved in the subsequent design or construction phase.

The relevant clashes need to be resolved prior to construction to avoid rework and idle time during construction. The identified relevant clashes should be communicated to the structural, architectural, or the services team (Mechanical, Electrical, and Plumbing) to make necessary changes [9]. This process of detection of the clash, subsequent grouping into relevant and irrelevant clashes, communication of the identified relevant clashes to the concerned team, decision-making as to how to deal with the clash, and final testing if the clashes have been resolved, is a time-consuming and cumbersome process that takes several months.

The occurrence of a large volume of clashes through the CD and its effective communication has posed a challenge to the construction professionals. So a method is devised to handle this army of clashes efficiently. To improve the quality of clash detection results, this study devises a method on how to group, handle, and communicate a high volume of relevant clashes.

2 Literature Review

Contemporary design practices and construction delivery have historically suffered from terrible coordination and irregularities because of the way different department teams manage, operate, and communicate project lifecycle data amongst themselves. After each of the disciplines (Structural, Architectural, MEPF, Environmental Engineering, etc.) have finished their work, clash detection is done. Detecting these inconsistencies is important, as they would badly affect the construction process in terms of enforcing delays, design revisions, materials price increment, and a cascade of problems and cost overruns [3, 2].

The complication of modern-day construction projects has increased drastically and there's no great improvement in productivity. Traditionally, the productivity of the construction industry has been much lower than that of other industries due to a lack of ability to adopt new changes in technology. The productivity of other industries such as the automobile industry has increased steadily over the years by adopting new modified methods, techniques, and technology [10]. Indian construction is not yet up to world standards. It is not making use of the true potential of BIM tools. Most of the architectural and engineering firms in India still use 2D Computer-Aided Design, i.e., CAD drawings. Due to the failure of traditional difficult method of detecting clashes using 2D methods, it is important to bring about a new manner of running and thinking in the creation within the construction industry so that it calls for implanting modern technologies like BIM to enhance the clash detection techniques in preference to the traditional technique [4, 6, 11].

Typically, a clash takes place when building elements of different models occupy the same space [12]. It entails overlaying of drawing to check if there are any conflicts. With BIM though, this technique is improved, as BIM brings automation to clash detection. Clash detection enables effective identification, inspection, and reporting of conflicts in a project model. It is used for checking completed/ongoing projects and decreases the risk of human error during model inspections. Clash detection is essential because different models (architectural, structural, MEP, etc.) are integrated into one master BIM model [13, 1, 14]. Clashes are usually classified according to the tolerance level between the colliding components [15].

- Soft Clash (Clearance): It refers to components that are nearer than a particular distance (a minimum clearance) from each other or in other words it refers to an item that occupies an area that may affect the movement of another item or person.
- Hard Clash: It refers to a building component that is physically yet accidentally penetrating some other building component; that is, two (or more) components that occupy the same physical space in the model.
- 4D/Workflow Clash: Clashes resulting from scheduling clashes of interdisciplinary activities that affect the scheduled duration and the cost of the project.

The complication of modern-day construction projects has increased drastically and there's no great improvement in productivity. Traditionally, the productivity of the construction industry has been much lower than that of other industries due to a lack of ability to adopt new changes in technology.

3 Methodology

Identifying and resolving clashes has been crucial to improving coordination and thus overall efficiency in a project in terms of cost, time, and human effort. To understand clashes, their influence, their effect on the productivity of the project, an extensive literature study was undertaken. Further various published cases were studied as part of data collection to identify the challenges associated with clash detection. The details about how these challenges were overcome were identified through the cases. The authors applied the developed method on a sample project to validate its effectiveness in making the process of clash detection and resolution easier (Fig. 1).

3.1 Issues Associated with Clash Detection

The authors studied various BIM cases to understand BIM implementation and specifically CD-related issues. The authors not only established the issues but also briefed about them. Further, the authors provided a possible BIM-based solution.

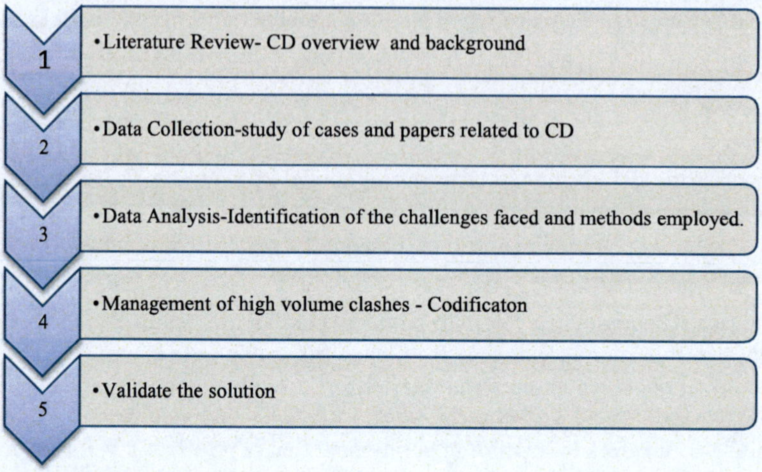

Fig. 1 Research methodology in sequential steps

The need for effectively and efficiently handling large volume clashes was one of the significant issues that came out through the cases (Table 1).

3.2 *Clash Management Method*

From the analysis of various cases, it is understood that there are many kinds of challenges associated with the clash detection and resolution process at different stages of a project. Many of these challenges could be overcome by managing the clashes in an effective and organized manner. The authors found out that one of the most commonly occurring challenges was to handle and convey a high volume of relevant and non-relevant clashes. Therefore, here in this study, the following method is identified for the efficient management of clashes to make this process effective and easier for the participants.

- To Raise a Request for Information (RFI) to obtain the missing information from the design drawings.
- To run the clash detection tests between suitable trade-specific models like structural versus mechanical piping, structural versus plumbing, architectural versus plumbing, etc.
- To classify the clashes as "approved clashes" and "clashes to be reviewed". Approved clashes are those that do not lead to any change orders or reworks, for example, the ducts or fixtures passing perpendicular to any floors and walls. Clashes to be reviewed are those which need to be resolved by the concerned team (architectural, structural, MEP, etc.).

Table 1 Cases to understand CD related issues faced by industry stakeholders

No.	Case title	Clash/issue	Brief description	Solution	Source
01	Shanghai West Railway Station	Distinguish design error from modeling error	Design error takes place due to the overlap of various disciplines and modeling error occurs due to inaccurate modeling by modelers	Clashes need sorting and separation with a simplified method. Codification, in which each clash is assigned a code based upon zone; floor etc. could be of use	Guangbin et al. [16]
02	Diyala University Campus	Inadequate documentation and coordination	Without the use of BIM and CD applications, the project team faced challenges such as too many change orders, lack of communication, and coordination among different subcontractors, professionals, etc.	An Integrated BIM system can result in 3D building models from an earlier stage of the project for better clarity, efficient RFI system, as well as the clashes management system	Abd and Khamees [17]
03	Data Center in Gujarat, India				Hitech-CADD Services [18]
04	Residential Project, Coimbatore, Tamil Nadu	Client's unawareness about the importance of CD	The project team conveyed the importance of resolving clashes before the construction to avoid potential snags, material waste, etc., during construction	The client was notified about the wastage in materials including concrete, bricks, and cement mortar. The saving of resources before and after clash detection was noted and presented	Sidiq and Thamilselvi, [14]
05	Laboratory project in Paris	Handle large number of clashes	Filter out all the relevant clashes from the total army of clashes particularly those that cause change orders	Clash Detection Matrix: It is a double-entry table describing every planned clash test	Moreau [15]

- The issues of a certain category (for example, structural versus plumbing ducts) are grouped to make groups of similar clashes. Hence, different groups are made.

  The groups of clashes are suitably named as, for example,

$$0.01\_01\, \boldsymbol{STR\ vs.\ MD\_(Level)\_(Zone).}$$

(*In which structural components are of highest priority and are represented by 0.01, followed by architectural component represented by 1.01, followed by MEP components represented according to the priorities given based on the project.*)

4 Validation

The authors implemented the developed method on a sample project which has architectural, structural, and MEP components. The detailed steps in which the method was implemented and the final groups of clashes obtained from the analysis are explained below. The procedure followed for the notification of clashes to concerned parties is also explained below.

4.1 Obtaining the Design

A 3D design model was obtained from the architectural, structural, and MEP (Mechanical, Electrical, and Plumbing) teams well ahead of the start of the clash detection process. In this case, the model procured was an integrated model instead of trade-specific, so individual sets were created separating the structural, architectural, and MEP components of the model (Fig. 2).

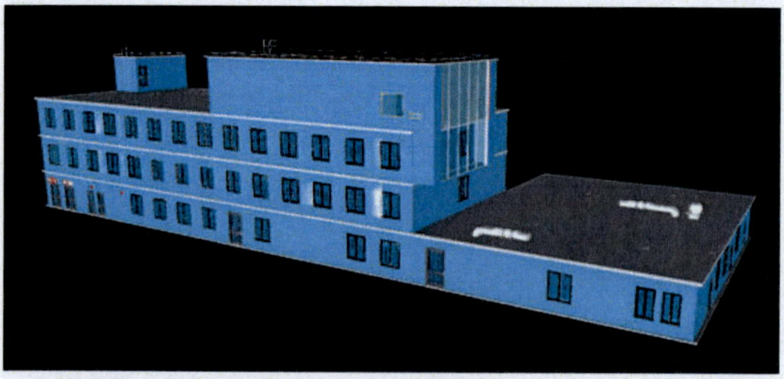

Fig. 2 A 3D integrated sample model procured for validating the developed methodology

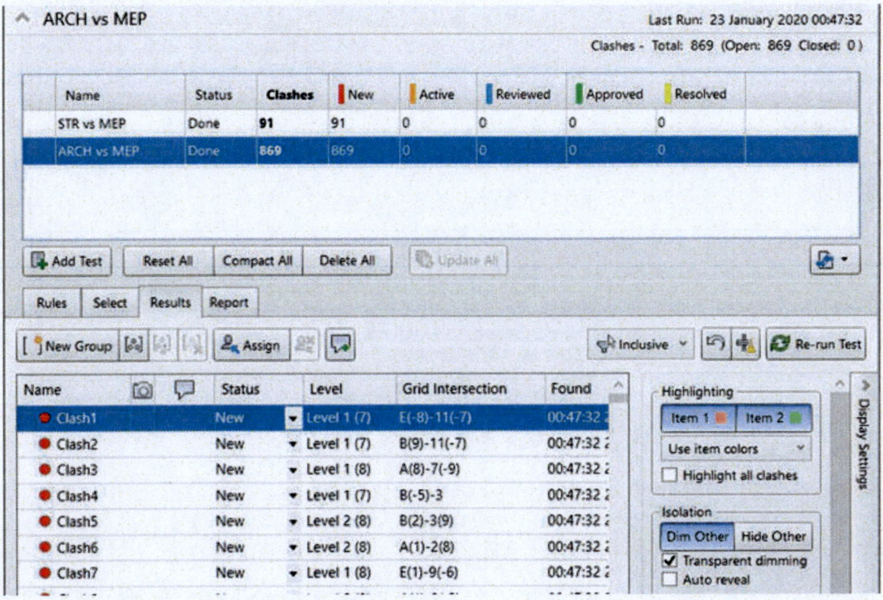

Fig. 3 Clash Detection tests results between various trade models

4.2 *Perform Clash Detection Test*

After the creation of separate sets, clash detection tests were conducted between the structural components versus MEP components and architectural components versus MEP components which resulted in 91 and 869 clashes, respectively. Compared to the thousands of clashes usually arising, this amounts only to a very few clashes. The tests that were conducted and the results obtained are shown in Fig. 3.

4.3 *Grouping of Clashes*

In this step, the large number of clashes obtained were grouped into the main two groups, viz. clashes to be reviewed or 'issues' and clashes that are approved or 'approvable' (see Fig. 4).

Approvable refers to clashes that do not require immediate attention or resolution, i.e., they are resolved directly at construction sites. Issues refer to the clashes which require immediate attention and they must be solved before the start of construction activities.

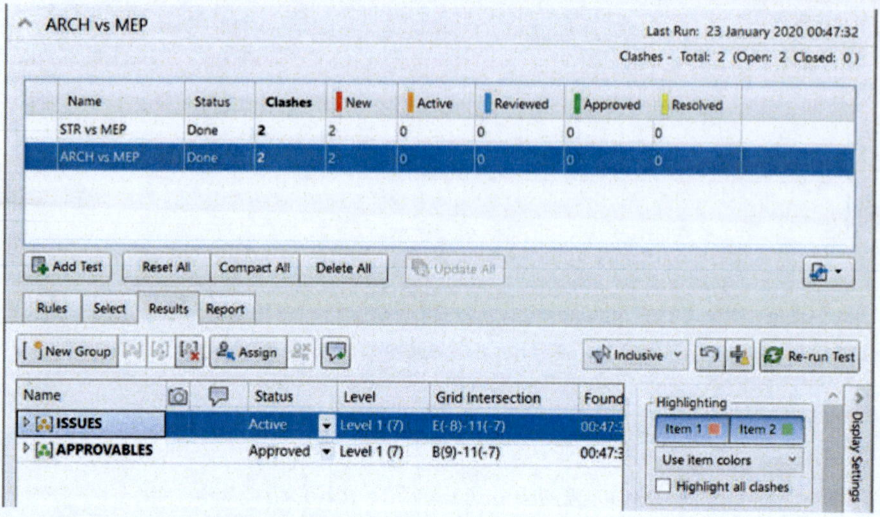

Fig. 4 Clash classification into issues and approvable

4.4 *Codification*

Coding of clashes is the main and integral part of handling the clashes and effectively communicating to different contractors, subcontractors, and other parties involved. To resolve the clashes which are grouped as issues, contractors and subcontractors must understand the nature and area of a clash where it has occurred. Grouping of clashes helps to organize the large amount of clashes formed, based on the required criteria. Codification refers to the grouping of individual clashes of similar components occurring at similar area under a standard code. Here, for codification, the entire model was divided into four zones, viz. A, B, C, and D. It was done for easy identification of clashes of any particular region. Figure 5 shows the division of zones in the model.

Consider an example of the code that is developed and used in this model.

0.01_001.FSvsMD_L2_B.

- 0.01 refers to the involvement of structural components. Similarly, 1.01 refers to architectural components.
- 001 provides the count of clash groups that are present after codification.
- FSvsMD—Components that are clashing are represented by the abbreviated names. In this case, the floor slab clashes with mechanical ducts.
- L2 refers to level 2 of the building model.
- B refers to zone B.

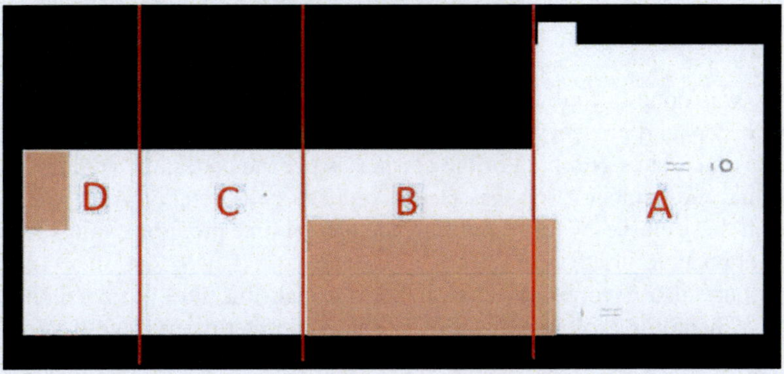

Fig. 5 Division of zones

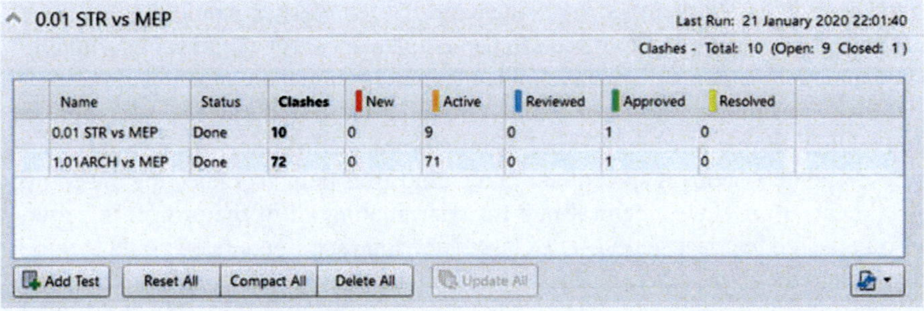

Fig. 6 Reduced clashes after codification

The total number was reduced from 81 to just 10 groups and from 869 to 72 groups including the approvable group. Figure 6 shows the reduced number of clashes and the codes given for similar groups of clashes.

4.5 Report Generation

3D Clash Detection allows report generation in various formats like XML, HTML, and as viewpoints. The authors selected viewpoints as the report format. The report could be easily shared and viewed. The report generated could be then uploaded to the BIM repository so that the designers or contractors of the concerned discipline can access it for correction.

5 Conclusion

This study was done to find a method to manage clashes effectively to save time in change orders and prevent reworks on site. Based on the developed codification method, this facilitates grouping of the clashes as per the codes, the authors like to conclude that the number of clashes can be reduced considerably. As a result, the resolving of these clashes would consume much lesser time providing valuable input to the site at an appropriate time.

The authors also have proposed a method through this study to overcome the challenge of handling the large army of clashes. The method involves effectively segregating the army of clashes as relevant and non-relevant clashes. The relevant clashes were later codified to effectively notify the clashes to the concerned departments. The identified method was applied successfully on a sample project for its validation.

Further, it is concluded that the conveying of such clashes would be very easy and efficient. Conventionally, after clash detection, each clash would be individually studied and sent to concerned stakeholders for resolving. This process is cumbersome, time-consuming, and error-prone. Grouping of similar clashes using codification eradicates this challenge. Applications such as BIM track or BIM 360 docs or web-based management and collaboration platforms help in connecting the entire project team. It provides a central hub for coordinating all of the project information from design to construction. The clash detection report generated could then be uploaded and the status of the clashes that need correction can be viewed using such platforms.

6 Recommendations

Through this study, a manual method of grouping of clashes using the codification technique is proposed to overcome the challenge of handling a large army of clashes. However, a plugin could be incorporated into exiting clash detection related applications which facilitates grouping and coding automatically without depending on manual input.

References

1. Eastman, C., Teicholz, P., Sacks, R., Liston, K.: BIM Handbook—A guide to Building Information Modeling, 2nd edn. John Wiley & Sons. Inc., London (2011)
2. Kensek, K.: Building Information Modeling. Routledge, New York (2014)
3. Akponeware, A., Adamu, Z.: Clash detection or clash avoidance? An investigation into coordination problems in 3D BIM. MDPI, Buildings **75**(7), 1–28 (2017)
4. NITI Task Force: A Report on Project and Program Management. NITI Ayog, India (2019)

5. Azhar, S.: Building information modeling (BIM): trends, benefits, risks, and challenges for the AEC industry. Leadersh. Manage. Eng. **11**(3), 241–252 (2011)
6. Roy, D., Malsane, S., Samanta, P.: Identification of critical challenges for adoption of integrated project delivery. Lean Constr. J. **01**(2018), 01–15 (2018)
7. Hjelseth, E.: Classification of BIM-based model checking concepts. J. Inf. Technol. Constr. **21**, 354–370 (2016)
8. Sidiq, A., Thamilselvi, P.: Study on implementation of clash detection tool in housing project using BIM. Int. J. Adv. Res. Sci. Eng. **7**(6), 215–225 (2018)
9. Bockstael, D., Issa, M.: A methodology for contractor clash detection using building information modelling on commercial construction projects. J. Inf. Technol. Constr. **21**, 233–249 (2015)
10. Gijezen, S., Hartmann, T., Buursema, N. and Hendriks, H.: Organizing 3d building information models with the help of work breakdown structures to improve the clash detection process. Visco-Centre for Visualization and Simulation in Construction, Working paper, pp. 1–18 (2009)
11. Raut, S., Valunjkar, S.: Improve the productivity of building project using building information modelling (BIM) based 4d simulation model. Int. J. Res. Appl. Sci. Eng. Technol. **5**(4), 53–61 (2017)
12. Hu, Y., Lacouture, D.: Clash relevance prediction in BIM-based design coordination using Bayesin statistics. In: Construction Research Congress, ASCE, pp. 649–658 (2018)
13. Bhagwat, P., Shinde, R.: Clash detection-a new tool in project management. Int. J. Sci. Res. Sci. Eng. Technol. **2**(4), 193–197 (2016)
14. Shyamkant, B., Patil, A., Pataskar, S.: Cost and time optimization for construction of residential building by clash detection in building information model (BIM). Int. Res. J. Eng. Technol. **4**(1), 550–553 (2017)
15. Moreau, S.: Navisworks manage—empowering clash detection. Autodesk University **2015**, 1–20 (2015)
16. Guangbin, W., Wei, L., Xuru, D.: Exploring the High-efficiency clash detection between architecture and structure. In: International Conference on Information Management and Engineering 2011, vol. 52(85) (2011)
17. Abd, A., Khamees, A.: As built case studies for BIM as conflicts detection and documentation tool. Cogent Eng. **4**, 1–11 (2017)
18. Hitech-CADD Services, https://www.hitechcaddservices.com. Accessed 21 Sep 2019

Design of a Circular Ring Slot Patch Antenna at 2.45 GHz for Hyperthermia Applications

Abhijeet Singh, Pushpendra Singh, and S. P. Singh

1 Introduction

Hyperthermia is a useful method for cancer treatment. In this, cancer cells of the body are heated up to 113 °F for a definite time period. Research has shown that chemotherapy is more successful with hyperthermia for cancer treatment, as it supports the circulation of drugs to infected body cells.

For hyperthermia applications, we need an applicator that can heat a specific region of the body. Antennas have the beauty of transmitting high-power signals with specific bandwidth so that we can use an antenna for this purpose. Antennas are already used for a lot of medical applications including Microwave CT (computed tomography), hyperthermia treatments, coagulation, and wireless capsule endoscopy [1–4]. A lot of antennas with different sizes and shapes are already designed for hypothermia applications. Researchers are working on reducing the weight, complexity, and size of hyperthermia applicator as well as heating of specific body areas.

Rectangular waveguide and horn antennas are used in the hyperthermia applicator for heating of infected body cells which are bulky and heavy in size, aperture is also fixed for these antennas. These defects make clinical treatment difficult sometimes. We can replace Rectangular waveguide and horn antennas with a patch antenna, as we are aware of the size and weight of the patch antenna [5, 6]. In patch antenna,

A. Singh (✉) · P. Singh
Rajkiya Engineering College Banda, Atarra, Banda 210201, India

P. Singh
e-mail: erpsingh@rediffmail.com

S. P. Singh
IIT- BHU, Varanasi, India
e-mail: spsingh.ece@itbhu.ac.in

P. Srivastava et al. (eds.), *Mathematical, Computational Intelligence and Engineering Approaches for Tourism, Agriculture and Healthcare*, Lecture Notes in Networks and Systems 214, https://doi.org/10.1007/978-981-16-3807-7_24

substrate can be flexible, so it can help in designing flexible hyperthermia applicators for different non-uniform parts of the human body.

Munson was the first person which introduced printed circuit antenna [7] for antenna arrays on the missile, aircraft, and satellites. Arrays of printed circuit dipoles are used for designing a novel "bean-bag" applicator by Mendecki [8] for hyperthermia heating. In spite of printed circuit dipole, a lot of other geometric structures have been used as microstrip antenna for hyperthermia applications. Bahl designed a ring-type structure as microstrip radiators [9]. Stanford University Medical Center developed a clinical test using an annular phased array with a microstrip spiral antenna [10]. Curto [11] proposed a smaller novel compact patch antenna for hyperthermia application at 434 MHz.

A simulation study of a microwave circular ring slot patch antenna at 2.45 GHz in very close proximity to a phantom muscle medium is presented in this paper. The proposed microstrip antenna has a circular patch, a concentric circular ring with a cross-slot at the backside on FR4 epoxy dielectric substrate ($\varepsilon_r = 4.4$, loss tangent = 0.02). Simulation studies of return loss with different positions concerning a phantom muscle model, penetration depth, and SAR distribution pattern are evaluated for various antenna orientations concerning the phantom muscle model. The simulation study is done by a Finite Element Method (FEM) based on Ansoft's HFSS software.

2 Antenna Design

2.1 *Design of Circular Patch Antenna*

The circular patch antenna is used as a basic element for the antenna design. Circular antenna is designed for the dominant TMz_{110} mode using the cavity model formulation. The design procedure is given below:

(a) Specify, f_r(in Hz), and h (in cm).
(b) Determine the actual radius 'a' of the patch as per the following first-order approximation to the solution:

$$a = \frac{F}{\left\{1 + \frac{2h}{\pi \varepsilon_r F}\left[\ln\left(\frac{\pi F}{2h}\right) + 1.7726\right]\right\}^{1/2}}$$

Where

ε_r: Dielectric constant of the substrate,
f_r: Antenna resonant frequency,
h: height of the substrate

$$F = \frac{8.791 \times 10^9}{f_r\sqrt{\varepsilon_r}}$$

2.2 Design of Annular Ring Patch Antenna

The annular ring patch antenna is used as a second element for the antenna design as its effective resonant length is larger in comparison to other patch antennas resonant at the fundamental TM_{11} mode.

The fundamental resonance frequency for the annular ring patch antenna can be calculated with

$$f = \frac{c_0 k}{2\pi\sqrt{\varepsilon_{reff\_ann\_ring}}}$$

Where

$$k = \frac{2}{a_{ARP}+b_{ARP}}, \varepsilon_{reff\_ann\_ring} = \frac{\varepsilon_r+1}{2} + \frac{\varepsilon_r-1}{2}\left[1 + 12\frac{h}{b_{ARP}-a_{ARP}}\right]^{1/2},$$

where $\varepsilon_{reff\_ann\_ring}$ is the effective dielectric constant, and a_{ARP} [cm] and b_{ARP} [cm] are, respectively, the inner and outer radii of the annular ring patch.

2.3 Design of Circular Ring Slot Patch Antenna

As shown in Fig. 1, the proposed circular ring slot patch antenna is made on a 34.6 mm × 34.6 mm FR4 epoxy dielectric substrate ($\varepsilon_r = 4.4$, loss tangent = 0.02). The proposed circular ring slot patch antenna has a circular patch, which is surrounded by a concentric circular ring. The bottom plane contains a circular slot at the center. The bottom plane also contains two rectangular slots, perpendicular to each other, of length 'L1' and width 'w'. Slots on the ground plane are responsible for impedance matching. The 50 Ω feed is placed at the ground plane side at Cartesian coordinate x = **y = 8.6 mm, z = 0 mm** with respect to the center of the proposed circular ring slot patch antenna at the origin (Table 1).

A phantom muscle model was modeled with electrical properties for the frequency of **2.45 GHz (ε_r = 52.729, conductivity = 1.7388 S/m, loss tangent = 0.24194) according** to the Federal Communication Commission (FCC) guidelines.

The dimensions of a phantom muscle model were 83.655 mm × 83.655 mm and 50 mm thick. This phantom muscle slab was used in order to see the proposed patch antenna effect on SAR distribution at various distances.

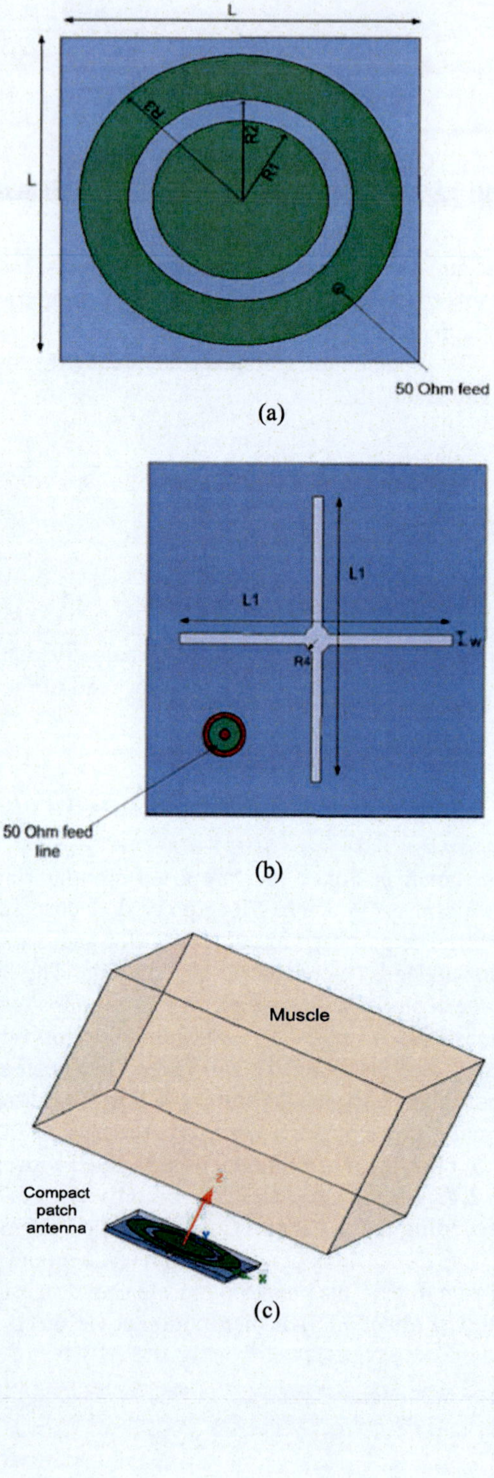

Fig. 1 Proposed circular ring slot patch antenna: **a** the top circular-ring/ patch surface, **b** the bottom ground plane/slots surface, **c** close proximity with phantom muscle

Table 1 Optimized shape parameters of the proposed circular ring slot patch antenna

Parameter	Value (mm)	Parameter	Value (mm)
R1	8.4	R4	1.3
R2	10.73	L1	28
R3	15.48	w	1

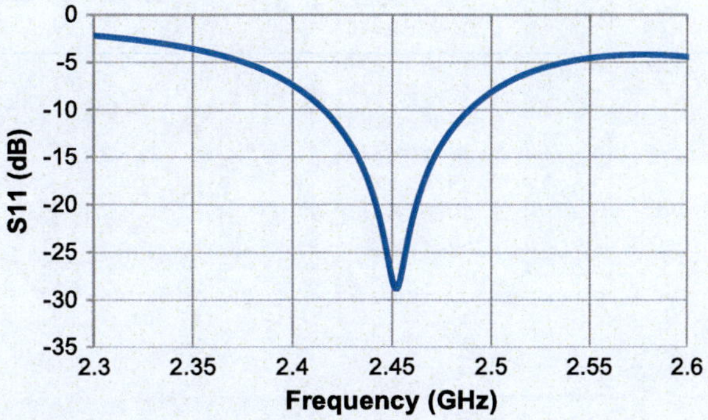

Fig. 2 Return loss for the proposed circular ring slot patch antenna in free space

3 Results and Discussion

3.1 Analysis of the Proposed Circular Ring Slot Patch Antenna in Free Space

The return loss of the proposed antenna is shown in Fig. 2, which is operating at the frequency of 2.45 GHz. Simulation results show that the proposed antenna is resonant at the frequency of 2.45 GHz in free space. The return loss of the proposed antenna at 2.45 GHz is −27.9 dB in free space.

The surface current distribution of the antenna is shown in Fig. 3 at 2.45 GHz.

3.2 Matched Impedance for Different Separation Distances Between the Proposed Circular Ring Slot Patch Antenna and a Phantom Muscle Medium

The S11 simulation in Figs. 4 and 5 is used to compare the antenna matched impedance for different separation distances with phantom muscle medium. In Fig. 4, results show that return loss remains sufficiently stable when the proposed antenna is placed on top of the phantom body muscle when compared to free space at a frequency of 2.45 GHz.

On reducing further, the distance between antenna and muscle medium-small mismatch of impedance occurred, as shown in Fig. 5. Generally, it is found that

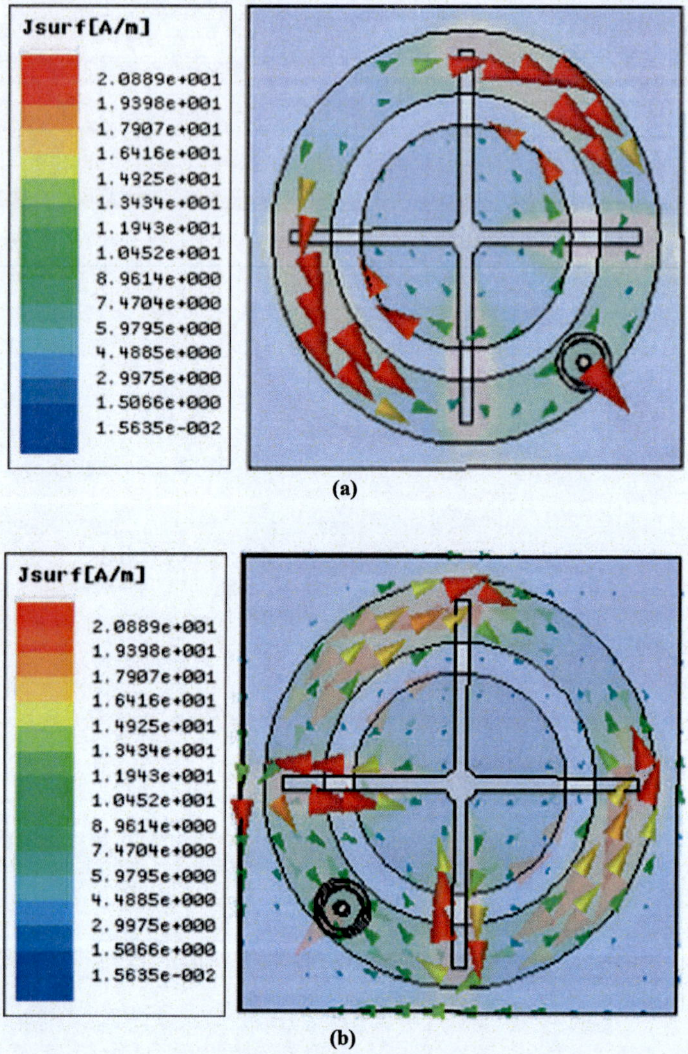

Fig. 3 Surface Current distribution (at 2.45 GHz) of the circular ring slot patch antenna **a** the top circular-ring/ patch surface, **b** the bottom ground plane/slots surface

impedance mismatching has occurred in the ordinary antennas when it is placed on top of the bio-medium. To resolve the problem of impedance mismatching, matching medium is placed between antenna and bio-medium. Because of matching, medium hyperthermia applicator became more complex and heavy in weight. From the results, it is clearly shown that return loss remains stable for different positions with phantom muscle so there is no requirement of a matching medium in the case of the proposed antenna.

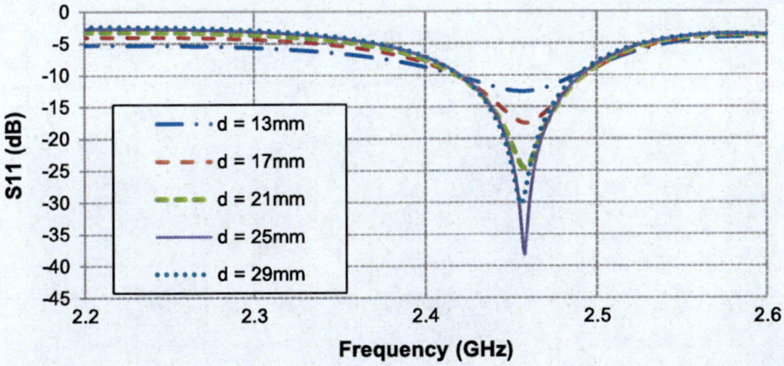

Fig. 4 Return loss for various distances (d = 29 mm to 13 mm) between the proposed circular ring slot patch antenna and muscle medium

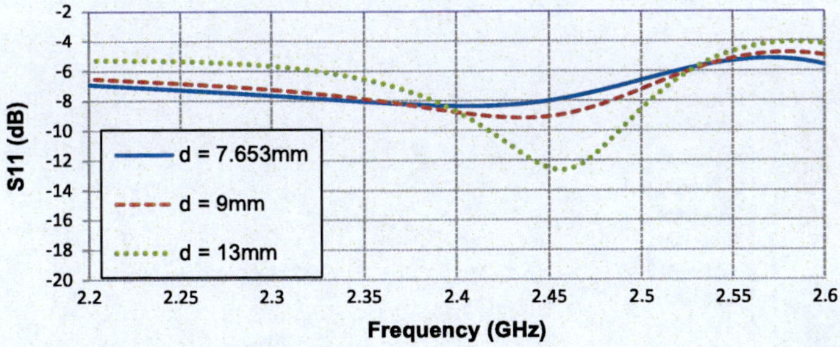

Fig. 5 Return loss for various distances (d = 13 mm to 7.653 mm) between the proposed circular ring slot patch antenna and muscle medium

3.3 Specific Absorption Rate Patterns

SAR shows the exposure of electromagnetic waves on the human body. SAR is defined as a measure of the rate at which energy is absorbed per unit mass of the muscle. Mathematically, the SAR is related to the internal electric field [12] as

$$SAR = \frac{\sigma E^2}{2\rho}$$

where σ (S/m) is the conductivity of muscle in Siemens per meter, E is the magnitude of internal electric field strength in volts per meter, and ρ is the mass density in Kg/m^3.

Figure 6 shows the distribution of electromagnetic waves in the phantom muscle model when the proposed antenna is placed 25 mm away from the phantom muscle. From the figure, it is clearly shown that electromagnetic waves are uniformly distributed in muscle due to the antenna. The peak value of local SAR is 5.2424 W/kg for 25 mm distance. It can be seen that the SAR near to antenna surface is quite high.

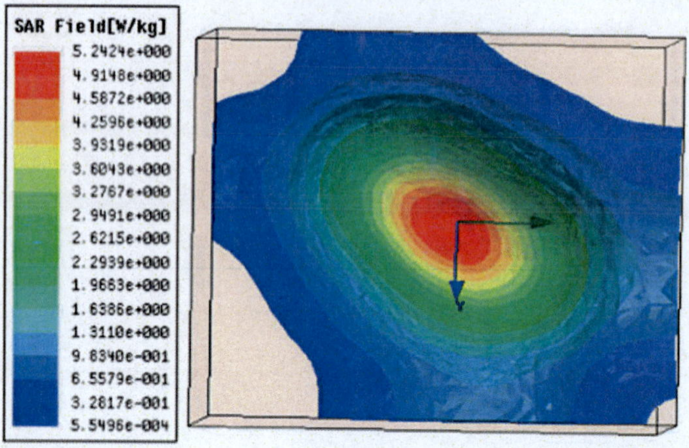

(a)

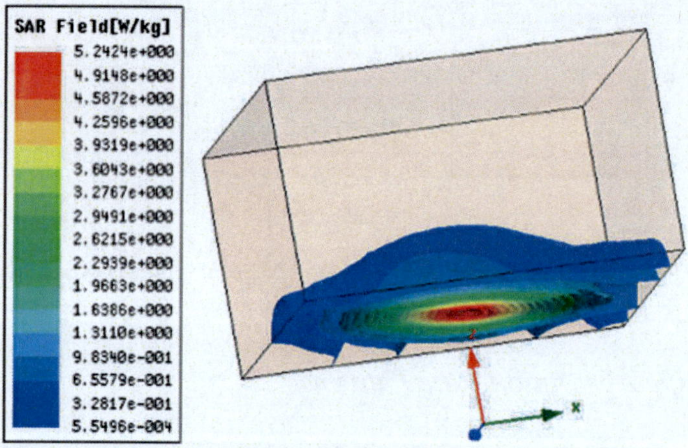

(b)

Fig. 6 SAR at phantom muscle surface for proposed circular ring slot patch antenna at 25 mm distance: **a** bottom view **b** side view

3.4 Penetration Depth

Penetration depth can be explained as the depth inside the human body where SAR value is reduced to 13.5% (i.e., −8.7 dB) [12]. The local SAR at 2.45 GHz of phantom medium is shown in Fig. 7. From Fig. 7, it is observed that on reducing the distance between the circular ring slot patch antenna and phantom muscle, a small increase in penetration depth is achieved.

Figure 8 shows the average SAR (1 g), local SAR, and average SAR (10 g) of the phantom medium at 2.45 GHz when the separation between phantom muscle

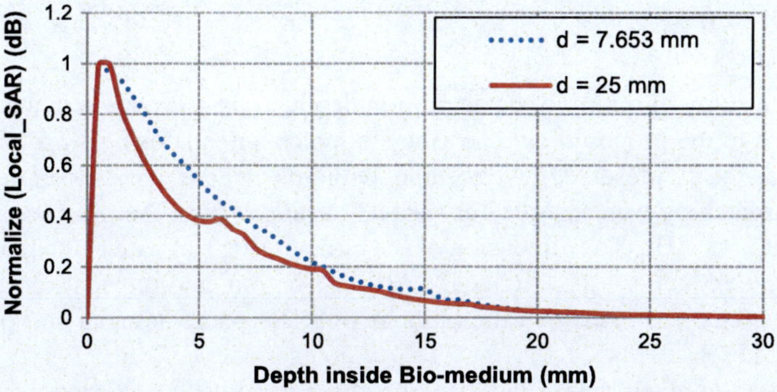

Fig. 7 Local SAR distribution (at 2.45 GHz) in phantom for various separations between proposed circular ring slot patch antenna and phantom muscle along the z-axis

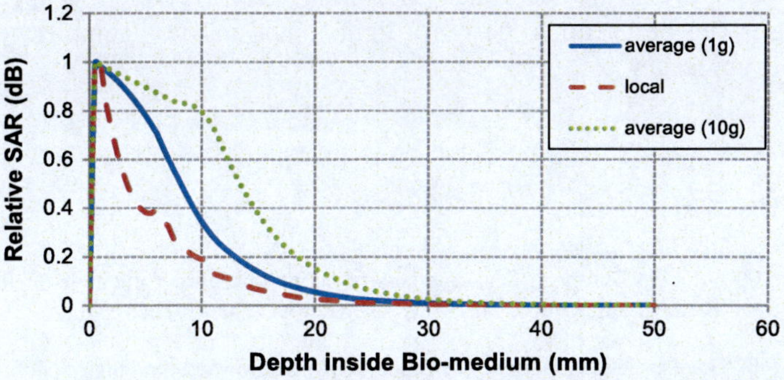

Fig. 8 Relative SAR distribution (at 2.45 GHz) in the phantom when the separation between proposed circular ring slot patch antenna and phantom muscle is 25 mm along the z-axis

and the proposed compact patch antenna is 25 mm. Penetration depths in phantom muscle medium considering local SAR are found 12 mm. However, penetration depth increases to 15.35 mm and 20 mm, respectively, when average SAR (1 g) and SAR (10 g) distributions are considered.

From Table 2, it can be observed that the proposed antenna is compact in size and penetration depth is also improved [5].

Table 2 Characteristics and penetration depth of conventional patch and proposed antenna

Type of applicator	Antenna size (mm^2)	Bandwidth (MHz)	Penetration depth (mm)
Conventional patch antenna	54 × 54	50	18
Proposed	34.6 × 34.6	76	20

4 Conclusions

A simulation study of return loss, penetration depth, and SAR pattern in a proposed phantom due to the proposed circular ring slot patch antenna using Ansoft's HFSS software has been presented. Noticeable detuning of the antenna in terms of impedance matching was observed at various distances from the muscle layer and free space at 2.45 GHz. The dimension of the circular ring slot patch antenna is optimized for size reduction, specific local SAR distribution, and impedance matching for different separation between the circular ring slot patch antenna and phantom muscle at 2.45 GHz.

From the result of return loss, it is clear that there is no need for a matching medium for the proposed antenna as compared to rectangular waveguide and horn antenna. SAR distributions of the proposed antenna are also uniform in a specific region so the proposed antenna can be used for local hyperthermia applicator at 2.45 GHz. The penetration depth of the antenna is also good enough for hyperthermia application. All the results can be useful to design a lightweight and less complex, efficient hyperthermia applicator.

References

1. Bolomey, J.C.: Recent European developments in active microwave imaging for industrial, scientific, and medical applications. IEEE Trans. Microw. Theory Tech. **37**(12), 2109–2117 (1989)
2. Meaney, P.M., Paulsen, K.D., Hartov, A., Crane, R.K.: Microwave imaging for tissue assessment: initial evaluation in multi target tissue-equivalent phantoms. IEEE Trans. Biomed. Eng. **43**(9), 878–890 (1996)
3. Singh, S., Sahu, B., Singh, S.P.: Hyperthermia performance of conformal applicator for limb tumor in presence of water bolus. In: International Symposium on Antennas and Propagation (ISAP), pp. 1–2 (2017)
4. Planche, K., Vinnicombe, S.: Breast imaging in the new era. Cancer Imaging **4**, 39–50 (2004)
5. Singh, S., Sahu, B., Singh, S.P.: Bi-layered/tri-layered bio-media in direct contact with metal diagonal horn for hyperthermia. Int. J. Microw. Wirel. Technol. **10**(8), (2018)
6. Singh, S., Singh, S.P.: Microstrip slot antenna for hyperthermia applications. In: 2015 IEEE Applied Electromagnetics Conference (AEMC), pp. 1–2 (2015)
7. Munson, R.E.: Conformal microstrip antennas and microstrip phased arrays. IEEE Trans. Antennas Propag. AP-**22**, 74–78 (1974)
8. Mendecki, J., Friedenthal, E., Botstein, C., Sterzer, F., Paglione, R.: Therapeutic potential of conformal applicators for induction of hyperthermia. J. Microw. Power **14**, 139–144 (1979)
9. Bahl, I.J., Stuchly, S.S., Lagendijk, J.J., Stuchly, M.A.: Microstrip loop radiators for medical applications. IEEE Trans. Microw. Theory Tech. MTT-**30**, 1090–1093 (1982)
10. Kapp, D.S., Fessenden, P., Samulski, T.V., Bagshaw, M.A., Cox, R.S., Lee, E.R., Lohrbach, A.W., Meyer, J.L., Prionas, S.D.: Stanford university institutional report, phase 1 evaluation of equipment for hyperthermia treatment of cancer. Int. J. Hyperth. **4**(1), 75–115 (1988)
11. Curto, S., McEvoy, P., Bao, X., Ammann, M.J.: Compact patch antenna for electromagnetic interaction with human tissue at 434 MHz. IEEE Trans. Antennas Propag. **57**(9) (2009)
12. Gupta, R.C., Singh, S.P.: Analysis of the SAR distributions in three layered bio-media in direct contact with a water-loaded modified box-horn applicator. IEEE Trans. Microw. Theory Tech. **53**(9), 2665–2671 (2005)

An Exploratory Study of the Spirituality-Oriented Experiences of Tourists

Chai Ching Tan, Phrapalad Somchai Damnoen, Yannakorn Toprayoon, Nittaya Dabjan, and Kanthita Damkam

1 Introduction

Tourism serves many important functions benefiting tourists in terms of the values they receive such as functional and spiritual values, manifested in terms of the meaning the tourists found from their tour experiences [1]. Even the tourism process itself can be spirituality-oriented, for instance, Sears [2] regards tourist attraction sites as functional equivalents to "sacred spaces", and as such, the tourists can be considered, metaphorically, as "pilgrims". The daily tour activities to the tourist attraction sites are thus ritualized in a pilgrimage sense [2].

This study aims to explore how tourists think and act that would influence their revisit intention and recommending their experiences to others. The significance of the tourist experience is emphasized in this study. As argued in Moxley [3], ultimately humans tend to look for spiritual experience when opportunities arise as "spirit" is the very essence of human existence. Having absorbed in a spiritual experience, tourists can find a connection to the universe or with oneself [4], such as in the domains of searching for meaningfulness in life, values for what they are doing or in an attempt of pursuing [5], or something larger that has to do with soul consciousness [6].

C. C. Tan (✉)
Rattanakosin International College of Creative Entrepreneurship (RICE), Rajamangala University of Technology Rattanakosin, Nakhon Pathom, Thailand
e-mail: chaiching.tan@rmutr.ac.th

P. S. Damnoen
Mahachulalongkornrajavidyalaya University, Bangkok, Thailand

Y. Toprayoon
Thai Research Association, Bangkok, Thailand

N. Dabjan · K. Damkam
School of Management, Mae Fah Luang University, Chiang Rai, Thailand

P. Srivastava et al. (eds.), *Mathematical, Computational Intelligence and Engineering Approaches for Tourism, Agriculture and Healthcare*, Lecture Notes in Networks and Systems 214, https://doi.org/10.1007/978-981-16-3807-7_25

Nevertheless, the existing literature in the discipline of tourism touches very little on the experiential domain of spiritual and pilgrimage tourism. This study aims to explore for validation to support the role of tourist experience nature in influencing their attitudes and behaviors in the spirituality-oriented tourism journey, and serves as the basis for further research.

2 Literature Review

Spiritual effort or pilgrimage is differentiated by the intention of the spiritual pursuers in search of meaningfulness in life or for building identity [7]. Sears [2] regards tourism as a pilgrimage effort where the visitors, metaphorically as pilgrims, undergo ritualized tour-like pilgrimage experiences to tourist attraction sites as the functional equivalents to sacred spaces. Nevertheless, the study of tourist behaviors on pilgrimage or spirituality-oriented tourism is not straightforwardly simple, as there is a complex web of factors influencing the so-called tourism phenomenon. From a stimulus-organism-response (S-O-R) paradigmatic angle [8], tourist attitudes, behaviors, experiences, and their trip pattern decisions can be influenced by the stimulus offered by the attraction sites and the journey, as well as the personal motives, expectation and religious beliefs, and faiths. As such, the S-O-R phenomenon shares the socio-cognitive theory of learning [9] in the aspects that the tourists subject themselves consciously [6] to information processing on personal matters and about their encounters. Thus, the study of spirituality-oriented tourism is worthy of a systematic and cross-disciplinary effort. The effort of this study is purported at the exploratory level, by putting aside the personal domain of stimulus which certainly should be variables of interest in further research. Market-size wise, pilgrimage, and spirituality-oriented tourism is now experiencing a resurgence in the tourism marketplace [10] across the globe, culture, and religious traditions [11].

Through interacting with the tour guides [12] who are subject-experts of the spiritual or pilgrimage sites, or with the spiritual gurus, tourists generate the so-called value-in-use. The value-in-use is a result of tourists subjecting themselves to reflexive interpretation and meaning formation [13] of their tour experiences, being manifested in a sense of self that can inform tourist life goals and narratives [14], or lead to non-self-transformation [6]. The values the tourists perceived to receive, such as life satisfaction and significance of experience, essentially, give spiritual tourism the currency of relative advantage in the tourism sector—which is a value proposition to differentiate from the alternative tourism choices such as academic tourism, adventure tourism, agritourism, and ecotourism.

Two important tourist experience nature are often cited in the extant literature, namely cognitive and affective aspects, which are attributable to tourist's mindful cognitive judgments and emotions derived from their tour consumptions [15]. Mindful cognition and emotion are the points of differentiation from other modes of tourism as spiritual effort is consciousness induced, even at consumer behavioral domains [6]. The concept of mindfulness denotes a mental state that

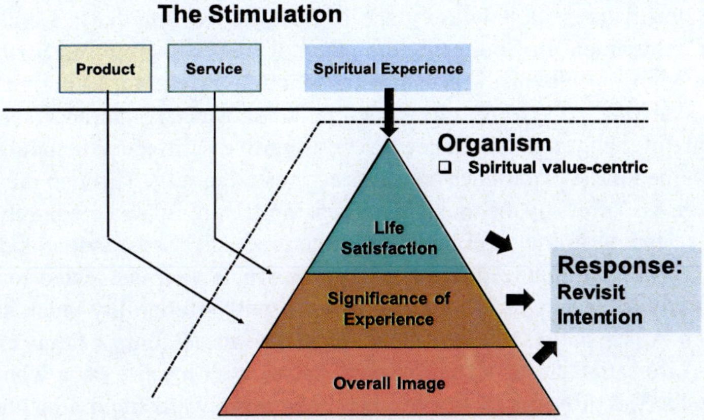

Fig. 1 The S-O-R conceptual model adapted for the spirituality-oriented tourism study

can enable the tourists to learn consciously, and the tourist experiences also tend to be more positive when compared to other modes of tourism [16]. In other words, through spiritual tour stimulation, tourists evaluate their experiences cognitively and affectively, and form favorable images of the tour in view of some value end-states [17], which ultimately influence their revisit intention [18]. Towards this end, the following two hypotheses are assumed for this study:

H1: Both the cognitive and affective tourist experiences should lead to the formation of tourist value, i.e., life satisfaction and significance of tourist experience, and positive global image.

H2: The organistic perceptions in terms of life satisfaction, significance of tourist experience, and global image of the tour will result in revisit intention of the tourists, leading ultimately to recommending others of the tours worthwhile of going.

Global image is a general summative impression of the tour, represented in terms of measurement scale spanning from "very negative to very positive" and "very unworthy of the time spent to very worthy of the time spent", which were adapted and modified from [19]. The stated hypotheses can be structured by using the SOR (Stimulus-Organism-Response) theory of consumer behavior [1] shown in Fig. 1.

3 Method

To operationalize the context describing the spirituality-oriented destination, this study differentiates three destination types. It starts with a tourism-oriented destination, which serves mainly the tourists for esthetic purposes, and involves more spirituality-oriented destinations, namely spiritual, where tourists visit for spiritual guidance, and pilgrimage purposes. Pilgrimage is a journey of faith, associated, for instance, with the life traces of important figures like the Buddha. For Buddhists, pilgrimage destinations include the place of birth of the Buddha in Lumbini, the place of Enlightenment in Bodhgaya, the place of the first sermon in Sarnath, and the place of Mahaparinibanna in Kushinagar. Though differentiated in motives, the tours often share the same travel infrastructure, i.e., hostels, restaurants, souvenirs.

The nature of the tourist experience can be cognitive or affective in nature, which takes root in the tourist experience concepts advocated in [20]. Through the production of experience offers by the tours, tourists gain meaningful and memorable experiences [21], leading to the significance of experience, life satisfaction, and overall image [1]. The operationalized items of the constructs were subjected to validity and reliability pilot tests with the subject-experts, with a reliability index indicated by Cronbach Alpha more than 0.80 thresholds. The measurement items are listed as follows: Life satisfaction—"I am more satisfied with my life, as a whole, after the tour", and "the progress of my life has been on upward trend after the tour". Global image involves measurement items, "In general, and taking into account the experience of the tour, the image I have to the destination is very negative (1) … to (5) very positive, and very unworthy of the time spent (1) … to (5) very worthy of the time spent". Cognitive tourist experience states, "giving the tourist an opportunity to engage in activities that really influence the state of mind and heart" and "indicating a place of religious assets that is capable to influence the tourist at deep level". The affective tourist experience examines the extent the tour "makes the tourist able to calm the mind" and "makes the tourist connecting to soul at deep-feeling level". The significance of tourist experience is one-item measurement: "The experience has had a positive influence in my life". Recommendation is measured by "I told my positive impression to friends and others proactively", and "I persuaded my friends and others to make a visit to the destination". The revisit intention is measured by "I will revisit the destination if future opportunity arises", and "I would visit the destination once again despite of additional travel cost in my budget". The questionnaire-based survey was conducted, using Google Forms, through Internet-based circulation to those who had participated in any of the three types of spirituality-related tourism, namely: tourism-oriented destination, spirituality oriented, and pilgrimage. To obtain more homogeneous replies, the pilgrimage destinations were restricted to only the Indian Buddhism sites, being associated with the life traces of important figures like the Buddha in Lumbini, Bodhgaya, Sarnath, and Kushinagar, India.

4 Data Analysis

The data were collected from the tourists to the tourism-oriented sites, at 67, spirituality-oriented destinations at 56, and pilgrimage destinations at 47, leading to a total of 170 valid responses for structural equation model (SEM) analysis. Table 1 indicates the supporting evidence of convergent and divergent validity, and reliability assessments, with Cronbach Alpha surpassing the 0.80 threshold for well-constructed questionnaires, and the diagonal square root of total variance explained (TVE) more

Table 1 Evidences of convergent and divergent validity of constructs

	CTE	ATE	GI	LS	STE	RI	RE
CTE	0.95						
ATE	0.84	0.95					
GI	0.80	0.77	0.96				
LS	0.83	0.79	0.75	0.99			
STE	0.79	0.79	0.74	0.75	0.811		
RI	0.83	0.69	0.77	0.84	0.64	0.98	
RE	0.79	0.92	0.64	0.74		0.78	0.96
TVE	0.92	0.92	0.93	0.98		0.97	0.93
Alpha	0.0.91	0.91	0.92	0.97		0.92	0.97

Note CTE = Cognitive tourist experience. ATE = Affective tourist experience. GI = Global image. LS = Life satisfaction. STE = Significance of tourist experience. RI = Revisit intention. RE = Recommendation

than the cross-correlations coefficients, signifying divergent validity of the considered constructs within the requirement. With TVE surpassing 0.5 threshold, and the measurement item-factor loading above 0.90, as also shown in the SEM structure in Fig. 2, and Cronbach Alpha all above 0.9, except a single-item construct "Significance of Tourist Experience (STE)", it can be concluded that construct's convergent validity is well established.

The statistical test of variances (ANOVA) results in Fig. 2, which show that tourists to both spirituality-oriented and pilgrimage destinations demonstrate a higher level of "life satisfaction" and "significance of the tourist experiences", at the post-experience

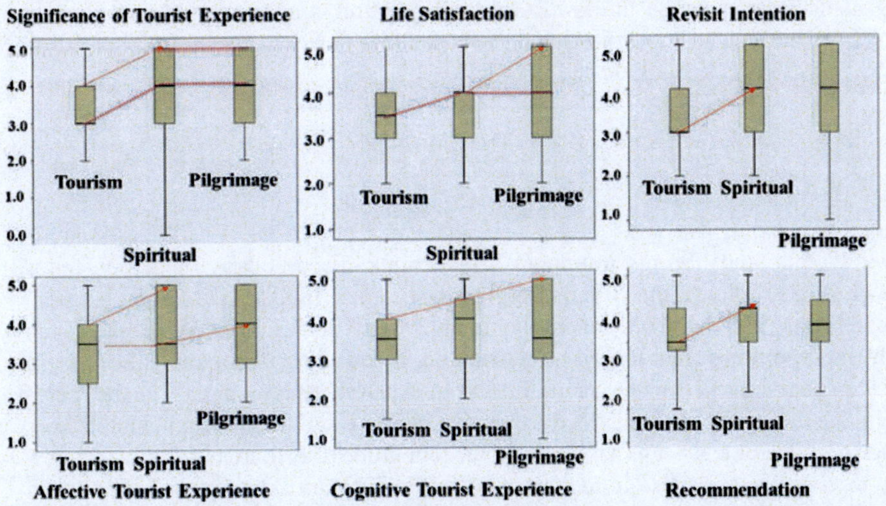

Fig. 2 The categorical differences of the studied constructs

stages. Feedbacks were also made available to a few survey respondents, and they suspect the large variations of the pilgrimage tours to India could be due to the nature of tour packages, and the inconveniences and obstacles faced during tours, such as transportation and tourism amenities like supporting services and infrastructure. Nevertheless, in terms of recommendation, the responses are more in alignment, shown by the narrower standard deviation depicted in Fig. 2.

From the ANOVA test result, the three tourism destinations can be considered to exhibit different degrees of spiritual experience and values to tourists, and they provide the tourists with the opportunity to co-create value with the help of the following value proposition: "To provide a spiritual experience that can lead to life satisfaction". For example, in the value co-creation concept, a monk can be considered as a tour guide on storytelling. Storytelling is important because it makes human values salient and brings the desired message to life, and guides tourists' perceptions of what they are experiencing, helps them determine the value of the experience, and facilitates immersion [22]. In short, a spirituality-oriented tour can make use of the co-creation of resources for creating memorable tourist experiences, such as tour guide's storytelling and guidance to let the tourists feel the events that happened in the past that are of spiritual significance and praying together. Thus, resources are either operand (i.e., the destination artifacts) and operant (i.e., knowledge and skill of tour guides). The Structured Equation Modeling (SEM) statistics analysis delivers good incremental fits such as NFI at 0.931, RFI at 0.903, IFI at 0.946, TLI at 0.90, and CFI at 0.945. However, the absolute fit indexes such as the livelihood maximization $\chi^2/\text{df} = 4.391$, $P = 0.000$ (not significant), and RMSEA at 0.132. The tested model is given in Fig. 3, which provides the supports for the two hypotheses H1 and H2. Life satisfaction and significance of tourist experience are shown also playing a mediator role bridging the relationship between revisit intention and recommendation, which can be explained using the theory of planned behavior (TPB) structure: in that the organistic states provide the level of confidence and confirmation to the tourists, which, in turn, positively influence their intentional state and thus recommendation behaviors. TPB is first advocated in [23].

5 Conclusion

Overall, this study brings important implications. Theoretically, it demonstrates the generic validity of the stimulus-organism-response (S-O-R) paradigm, but both academicians and the practitioners should apply the S-O-R in creative manner, such as by incorporating other theories, for example, TPB and socio-cognitive theories. In spiritual tour context, in which mindfulness of experiences is essential [1], the socio-cognitive theory should consider the dynamic interplays of the environmental factors (such as monk or teacher as a tour guide, the tour guide role, tourists-to-tourists interaction, living experiences in monasteries, and cultural artifacts and heritages), and personal factors in terms of motivation and aspiration. The behavioral domains could be extended in further research to include tourist flow experiences. Specifically, the

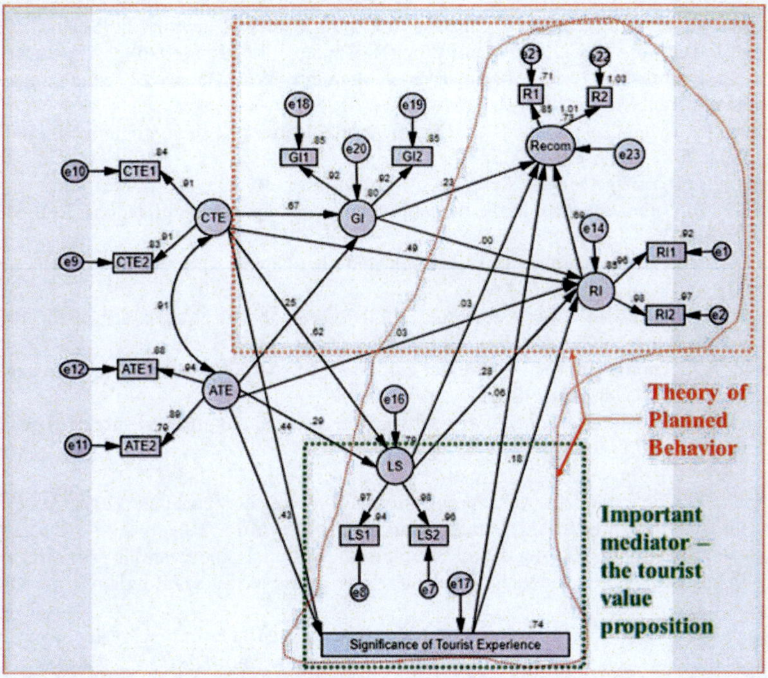

Fig. 3 The tested structural equation model

findings draw attention to two categories of the organistic states a spirituality-oriented tour should target and exploit, namely tour image and tour values. Tour values are represented by the life satisfaction and the significance of tourist experiences. These two are the loyalty drivers which form the customer capital to tour operators.

References

1. Tan, C.C., Sitikarn, B.: Balanced scorecard (BSC) structured business model for tribes-based community-based tourism: cases with Chiang Rai. J. MCU Peace Stud. **7**(4), 899–922 (2019)
2. Sears, J.F.: Sacred Places: American Tourist Attractions in Nineteenth Century. Oxford University Press, New York (1998)
3. Moxley, R.S.: Leadership & Spirit: Breathing New Vitality and Energy into Individuals and Organizations. Jossey-Bass, San Francisco (2000)
4. Young-Eisendrath, P., Miller, M.E.: Beyond enlightened self-interest: the psychology of mature spirituality in the twenty-first century. In: Young-Eisendrath, P., Miller, M.E. (eds.) The Psychology of Mature Spirituality: Integrity, Wisdom, Transcendence, pp. 1–7. Routledge, London (2000)
5. Roof, W.C.: Spiritual Marketplace: Body Boomers and the Remaking of American Religion. Princeton University Press, Princeton (1999)
6. Tan, C.C., Damnoen, P.S.: Buddhist noble eightfold path approach in the study of consumer and organizational behaviors. J. MCU Peace Stud. **8**(1), 1–20 (2020)

7. Baumann, Z.: From pilgrimage to tourist – or a short history of identity. In: Stuart, H., Gay, P.D. (eds.) Questions of Cultural Identity, pp. 18–36. Sage, London (1996)
8. Tan, C.C., Patthracholakorn, A.I.: Towards a community-based theory of brand community engagement. Adv. Sci. Lett. **24**(7), 5167–5170 (2018)
9. Akgun, A.E., Lynn, G.S., Byrne, J.C.: Organizational learning: a socio-cognitive framework. Hum. Relat. **56**(7), 839–868 (2003)
10. Digance, J.: Pilgrimage at contested sites. Ann. Tour. Res. **30**, 143–159 (2003)
11. Reader, I.: Pilgrimage growth in the modern world: meanings and implications. Religion **37**, 210–229 (2007)
12. Tan, C.C.: A fieldtrip approach to study tour guide in shaping student tourist experience. J. Mekong Soc. **13**(2), 23–44 (2017)
13. Klein III, R.E., Kernan, J.B.: Contextual influences on the meaning ascribed to ordinary consumption object. J. Consum. Res. **18**, 311–324 (1991)
14. Arnould, E.J., Thompson, C.J.: Consumer culture theory (CCT): twenty years of research. J. Consum. Res. **31**(4), 868–882 (2005)
15. del Bosque, I.R.., Martin, H.S.: Tourist satisfaction: a cognitive-affective model. Ann. Tour. Res. **35**(2), 551–573 (2008)
16. Langer, E.J.: Mindful learning. Curr. Dir. Psychol. Sci. **9**(6), 220–223 (2000)
17. Holttinen, H.: Contextualizing value propositions: examining how consumers experience value propositions in their practices. Australas. Mark. J. **22**, 103–110 (2014)
18. Sharma, P., Nayak, J.K.: Testing the role of tourists' emotional experiences in predicting destination image, satisfaction, and behavioral intentions: a case of wellness tourism. Tour. Manag. Perspect. **28**, 41–52 (2019)
19. Cossio-Silva, F.J., Revilla-Camacho, M.A., Vega-Vazquez, M.: The tourist loyalty index: a new indicator for measuring tourist destination loyalty. J. Innov. Knowl. **4**, 71–77 (2019)
20. Pine, B.J., Gilmore, J.H.: The Experience Economy: Work Is Theatre and Every Business a Stage. Harvard Business School Press, Boston (1999)
21. Zatori, A.: Experience-centric approach and innovation. In: The Handbook of Managing and Marketing Tourism Experiences, pp. 21–44 (2016). https://doi.org/10.1108/978-1-78635-290-320161002
22. Mathisen, L.: The exploration of the memorable tourist experience. Adv. Hosp. Leis. **8**, 21–41 (2012)
23. Ajzen, I.: The theory of planned behavior. Organ. Behav. Hum. Decis. Process. **50**, 179–211 (1991)

A Correlational Analysis of the Factors Influencing the Transfer of Research and Development to Entrepreneurs

Charmaine Savia Siqueira Lobo

1 Introduction

Entrepreneurship has been a driving force, right from the time of using specialized labour and its importance continues to grow over the years [8]. It is compared to a revolution across the globe, with an intensity comparable to that of the Industrial Revolution [21]. The trajectory of economies and markets has been given a new direction by the active presence of entrepreneurs [1]. The world economy would come to a halt without the never-ending creation of social and business modifications [38]. The field of entrepreneurial research comprises a plethora of diverse perspectives, research ideas, methodologies and traditions, since entrepreneurial activity has touched upon many disciplines. This domain of research involves creativity, having a pro-active outlook and not being risk-averse. Entrepreneurial research can be undertaken at different tiers: by researchers, universities, firms, entrepreneurs and at the macroeconomic level by trained institutions. The entrepreneurial environment should be conducive to undertake research in this field and to optimally transfer the same to the entrepreneurs. Several stimulatory and inhibitory factors have a major role to play [9]. Entrepreneurial research is one of the fastest growing fields of academic research [8].

C. S. Siqueira Lobo (✉)
Fr. Agnel College of Arts and Commerce, Pilar, Goa, India

P. Srivastava et al. (eds.), *Mathematical, Computational Intelligence and Engineering Approaches for Tourism, Agriculture and Healthcare*, Lecture Notes in Networks and Systems 214, https://doi.org/10.1007/978-981-16-3807-7_26

2 Review of Literature

Developing an entrepreneurial culture in institutions of learning strengthens the link between centres of learning and business. This can be done by providing incentives, improvising intellectual property policies and developing managerial skills to facilitate the transfer of new innovations to enterprises. At the same time, universities can maintain their independence in R&D findings. Research networks can facilitate sharing of information across sectors without necessitating a bulk investment by individual parties. Commercialization of research can be facilitated with the establishment of incubators, technology parks and offices. Enterprises are a major source of demand for R&D specialists. In this way, students get encouraged to enrol themselves in such fields. This generates a self-reinforcing virtuous circle of R&D and entrepreneurial activity [35].

In capitalist economies, firms engaged in research are a major determinant of economic growth [4]. Small enterprises under-invest in R&D due to information asymmetries, limited resources, financial problems and uncertainty involved [32]. Studies have found a high correlation between the spurt of nascent enterprises engaged in high technology and the availability of R&D. Investments in R&D increase alliances within firms and promote innovations as well as the use of external sources of knowledge. Firms need to undertake research to be able to know how to innovate and where to do so [30]. Entrepreneurial R&D serves two major functions: it creates knowledge within the firm to stimulate inventions [30], and it enhances the ability of the enterprise to comprehend and absorb knowledge spillovers generated by other institutions' R&D [11].

2.1 Influencial Factors in the Field of Entrepreneurship

Entrepreneurial Finance: Enterprises in the field of science and technology can receive funds from various sources such as crowdfunding platforms, accelerators and seed funds. Studies show that new enterprises continue to depend on bank financing [29]. Others are of the opinion that entrepreneurs can run their enterprises without resorting to external financing through financial bootstrapping and bricolage [3, 36]. Entrepreneurs procure funds from a range of sources. Evidence from Chinese companies suggests that the corporate debt ratio bears an inverse correlation with R&D investment. It has also been found that strict corporate financial control and redundant finances also hinder the R&D investment of a company [23].

Entrepreneurial Education: Imparting education on the subject of entrepreneurship does not mean only teaching people how to manage an enterprise or business of their own. It stimulates the ability to think creatively and encourages a feeling of empowerment and self-worth [12]. Many researchers believe that this educational process should start early in life. Many are of the view that children possess entrepreneurial traits from the time they are born [22]. Hence, education related to entrepreneurship should be imparted right from the tender years. Successful entrepreneurs can be engaged as mentors, advisors and coaches to students. The curriculum can be so designed in a way that it includes experiential methods such as learning by doing [35]. Another ideology is that even though a person may not have the inborn trait of becoming an entrepreneur, he or she can be trained how to become one [15].

Government Entrepreneurship Policies and Programmes: Government intervention can be grouped into two types, directive and facilitative. The former approach tries to identify opportunities and suggests the direction that enterprises should follow. The latter creates a conducive environment for entrepreneurs by providing law and order and social stability that is difficult to be provided by private players. Great things can be achieved through an interaction of entrepreneurship with the government [37]. The regulatory environment should aim at reducing the administrative burdens encountered dealing with setting up of an enterprise [35]. Some studies support government policies in stimulating entrepreneurial activity, while others have found contradictory results. However, too much intervention by the government in the form of public funding lowers the incentives to undertake entrepreneurial activities [6].

Motivation: In order to realize the process of entrepreneurship, the transformation of ideas into action is important [5, 20]. Motivation acts as a link between intention and action. It is also a factor that explains different responses of people to the same stimuli [10]. Depending on their motivations, individuals have different perceptions of decision-making processes, opportunity considerations and risk aversion [31]. Studies have proved that people who are achievement-oriented very often make their career choice in the field of entrepreneurship as compared to other job avenues [24]. Research has shown that in order to have independence and financial freedom, individuals start their own business, thereby controlling their own career even at the cost of taking risks [19].

Socio-Cultural Context: Among the environmental or external factors that affect entrepreneurship, cultural and social stimuli have a major role to play [16, 26]. These factors either facilitate or inhibit the initiative of starting an enterprise and becoming an entrepreneur. Research shows that national culture influences three broad areas of entrepreneurship: (1) overall entrepreneurial indices, for example, the creation of novel business initiatives, (2) characteristics of entrepreneurs and (3) corporate entrepreneurship. If a nation's culture is more individualistic in nature, less risk-aversive, male-dominated and has a low power-distance, all these facilitate entrepreneurship [18]. Distinct institutional orders such as the family, religion, market and government. affect entrepreneurial behaviour in a big way [33, 34]. These orders bring along with them values, laws and norms often conflicting with each other, in turn affecting entrepreneurship.

Physical Infrastructure: Investing in physical infrastructure creates an environment conducive to the emergence of entrepreneurs and enterprises. It facilitates the promotion of novel ideas and initiatives [13], supporting the development of centres of prosperity and progress and bringing together such centres in areas where there is adequate provision of quality infrastructure [25, 27]. The important role played by infrastructure in stimulating entrepreneurship was emphasized in the Global Entrepreneurship Monitor (GEM) model [28]. Physical infrastructure is positively and significantly associated with regional start-up activity in Germany. Investments in infrastructure can stimulate the flow of goods, ideas, capital and people [2].

Internal Market Dynamics: The supply side link, among markets, arises since production processes that supply to a current market are able to supply to a future market, even though there may be changes in the nature of the product over time, as sunk costs have already been incurred [17]. Game theoretical models (GTMs) highlight two reasons motivating firms to engage in research: (i) chance to get higher profits in future and (ii) the fear of being overtaken by potential rival firms who are engaged in innovations [14].

3 Research Gap

A lot of studies have been conducted showing a relation between entrepreneurship and factors such as availability of entrepreneurial finance, entrepreneurship education, government entrepreneurship programmes, motivation, socio-cultural context and physical infrastructure. In most of the cases, a positive and significant relationship has been found. However, no study has been conducted to show whether any of these factors bear a significant relationship to the availability and transfer of entrepreneurial research and development to entrepreneurs. The present study seeks to bridge this gap and contribute to the field of entrepreneurial research and development.

4 Objectives

This paper has two main objectives .

4.1 To Find Out Whether There Exists Any Correlation Between the Transfer of R&D to Entrepreneurs and Factors Such as

i. Entrepreneurial finance,
ii. Entrepreneurial education and
iii. Government entrepreneurship programmes.

4.2 To Find Out Whether There Exists Any Correlation Between the Transfer of R&D to Entrepreneurs and Factors Such as

iv. Motivation,
v. Social and cultural context,
vi. Physical infrastructure and
vii. Internal market dynamics.

5 Limitations

The major limitation of this study is that it relies on data collected by GEM experts from a survey conducted in 2018, thus making it a static analysis as data was not available for previous years on these parameters. Secondly, only 49 countries have been considered for analysis as complete data was available for only these countries.

6 Research Methodology

As the present study lies in the domain of international entrepreneurial research and development, it makes use of global-level data. It has used secondary data collated from the Global Entrepreneurship Monitor (GEM) Report 2018–19. Data for the following countries was analysed: Argentina, Madagascar, Guatemala, the Netherlands, Brazil, India, Chile, Thailand, Peru, Israel, Colombia, Luxembourg, Qatar, Croatia, Slovenia, France, Japan, Cyprus, Germany, the United Kingdom, Switzerland, Greece, Indonesia, Republic of Korea, Iran, Uruguay, Ireland, Spain, Italy, Lebanon, Sweden, Morocco, the United Arab Emirates, Panama, Poland, Canada, Puerto Rico, Bulgaria Taiwan, Russian Federation, Angola, Saudi Arabia, China, Slovak Republic, Egypt, Sudan, Turkey, Austria and the United States.

GEM conducts surveys among a minimum of 36 entrepreneurs in each country, besides assessing the opinion of experts in areas related to entrepreneurship such as finance, government and education. The questions deal with factors that are vital for entrepreneurship to thrive. A nine-point Likert scale has been used where 1 signifies that facilities are highly insufficient and 9 signifies that facilities are highly sufficient. An average figure is then obtained for every main criterion [7]. The present study analyses data collected by GEM on 9 areas, namely R&D transfer, entrepreneurial finance, entrepreneurship education at school stage, imparting education on entrepreneurship in schools and in institutions of higher learning, entrepreneurial programmes by the government, motivation of entrepreneurs, norms of society and culture, physical infrastructure and internal dynamics of the market [7]. The data is analysed using

Table 1 Correlation between R&D transfer and various factors

	R&D transfer	
	Coefficient of correlation	Significance level (2-tailed)
Entrepreneurial Finance	0.619[a]	0.000
Education on entrepreneurship in school	0.612[a]	0.000
Education on entrepreneurship post-school	0.522[a]	0.000
Government Entrepreneurship Programmes	0.822[a]	0.000
Motivation	0.430[a]	0.002
Physical infrastructure	0.512[a]	0.000
Social and Cultural Norms	0.358[b]	0.012
Internal Market Dynamics	0.118	0.419

[a]Significant correlation at 0.01 level
[b]Significant correlation at 0.05 level

SPSS statistical software. Spearman's bi-variate correlation tool was used to test the hypothesis. The level of significance was also assessed.

7 Results

Upon conducting Spearman's correlational analysis, the following results were obtained (Table 1).

8 Discussion

8.1 Transfer of R&D

The transfer of R&D according to the GEM Report refers to the transfer of research from research centres and universities to entrepreneurs. It emphasizes the degree of investment in entrepreneurial R&D in that particular nation and to what extent the research findings can be commercialized.

8.2 *Objective 1*

To find out whether there exists any correlation between transferring R&D to enterprises or entrepreneurs and factors such as Entrepreneurial finance, Entrepreneurial education and Government entrepreneurship programmes.

The Entrepreneurial Finance: This parameter measures the perceptions of experts on the state of finances available for the present and future entrepreneurs. The ease of availability of finance from informal investment avenues, venture capitalists, professional business angels, loans by the government, banks, grants and subsidies and crowdfunding is evaluated. To what extent entrepreneurs have access to equity markets is also analysed [7].

H0: No correlation exists between entrepreneurial finance and the transfer of R&D.
H1: There is a correlation between entrepreneurial finance and the transfer of R&D.

A positive correlation is found between entrepreneurial finance and R&D transfer in the present study. The correlation coefficient is 0.619 and is significant at a 0.01 significance level. Thus, the null hypothesis stands rejected. It means that in countries where entrepreneurs have access to adequate finance, there exists a higher transfer of R&D to enterprises and vice versa. Entrepreneurs are able to pay for and thus access the latest research conducted in their fields. Higher R&D transfer stimulates innovative production and processes among entrepreneurs, thus increasing their revenue.

Entrepreneurial Education: Entrepreneurial education can be divided into two categories:

*Education on Entrepreneurship in Schools***:** To what extent subjects related to entrepreneurship are taught to school students as part of their curriculum is assessed by this criterion. It may also be included as part of various school programmes. It also assesses whether students are being trained in skills and values required to become an entrepreneur.

H0: No correlation exists between imparting education on entrepreneurship in schools and the transfer of R&D.
H2: There is a correlation between imparting education on entrepreneurship in schools and the transfer of R&D.

The results indicate that a significant and positive correlation exists between providing education on entrepreneurship in schools and the transfer of R&D. The coefficient of correlation is 0.612. Thus, the null hypothesis stands rejected and the alternate hypothesis is accepted. Thus, countries which have incorporated entrepreneurship subjects and skills in the curriculum of school-goers tend to have a higher rate of transfer of research and development to entrepreneurs. This may be because interest is generated in young minds at a very early age. As part of their assignments and projects, school students may also be able to generate valuable findings in the field of entrepreneurial research and vice versa. Higher rates of R&D transfer will stimulate higher demand for entrepreneurial skills and knowledge about the field.

Education on entrepreneurship post-school: The stage post-school refers to institutions such as colleges, vocational centres and business schools. It evaluates whether these institutes include subjects related to entrepreneurship in their syllabi. It also assesses whether such institutes are imbibing skills and values essential for entrepreneurship in their students.

H0: There is no correlation between imparting education on entrepreneurship beyond school and the transfer of R&D.
H3: There is a correlation between imparting education on entrepreneurship beyond school and the transfer of R&D.

The analysis shows that a positive and significant correlation exists between imparting education on entrepreneurship post-school and the transfer of R&D. The coefficient of correlation is 0.522. Thus the alternate hypothesis is accepted. It indicates that nations that have entrepreneurial education courses as part of the syllabus of the degree courses or as part of the curriculum of institutions of higher learning have a higher tendency of being able to transfer these research findings to businesses. The reverse also holds true, the higher the rate of transfer of research findings to enterprises, the higher will be the demand for quality research and trained professionals in the field.

Government Entrepreneurship Programmes: Government entrepreneurship programmes take into account whether specific programmes are provided to entrepreneurs by public agencies such as incubators and various subsidies. It also assesses whether specialized agencies are available to guide entrepreneurs in their course of action.

H0: No correlation exists between entrepreneurship programmes supported by the government and the transfer of R&D.
H4: There is a correlation between entrepreneurship programmes supported by the government and the transfer of R&D.

The present study found a positive and strong correlation between government entrepreneurship programmes and the transfer of R&D. The coefficient of correlation is 0.822, indicating that favourable provisions by the government help enterprises in adopting the latest innovations discovered through R&D. Thus, the alternate hypothesis stands accepted. The reverse also holds true, nations with higher rates of R&D transfer will have improved production processes and final products, thus generating higher revenues to firms, who will in turn contribute to the government coffers by way of taxes. This revenue to the government can be used to further stimulate the entrepreneurial environment through various schemes and programmes.

8.3 *Objective 2*

To find out if there is any correlation between the transfer of R&D to entrepreneurs and factors such as Motivation, Social and cultural context, Physical infrastructure and Internal market dynamics.

Motivation: According to the GEM Report, the Motivational Index measures the reasons driving people to become entrepreneurs. It finds out whether entrepreneurs are necessity-driven or opportunity-driven or whether they are improvement opportunity-driven.

H0: No correlation exists between the motivation of entrepreneurs and the transfer of R&D.

H5: There is a correlation between the motivation of entrepreneurs and the transfer of R&D.

Upon analysing the data of 49 nations, it reveals a positive and significant correlation between motivation and R&D transfer. Thus, the alternate hypothesis is accepted. The coefficient of correlation is 0.430. It implies that the higher the rank a country secured on the motivational index, the higher will be the transfer of R&D to the entrepreneur. Entrepreneurs who are high on motivation will try to affiliate themselves to centres of research, thus benefitting both parties. Similarly, higher rates of R&D transfer will boost the motivation level of existing and potential entrepreneurs to adopt and adapt to new changes.

Cultural and Social norms: These factors indicate whether a nation displays a focus on entrepreneurship focus through its culture, behaviour, values and practices. In turn, this provides a source of encouragement to entrepreneurs by providing them support and recognition towards their activities.

H0: No correlation exists between cultural and social factors and the transfer of R&D.

H6: There is a correlation between cultural and social factors and the transfer of R&D.

An analysis of the data reveals that there exists a relatively low but positive and significant correlation between the cultural and social context and transfer of R&D at the 0.05 level of significance. Thus, the alternate hypothesis is accepted. The coefficient of correlation is 0.358. It implies that nations having social and cultural norms favouring entrepreneurship also tend to have higher rates of research and development transfer to the entrepreneurs and vice versa. Higher rates of R&D transfer will bring in innovative, customer-friendly products which will change the societal outlook and make people embrace newer varieties.

Physical Infrastructure: The presence of infrastructure facilities such as business operations, transportation and communication at the country level as well as internationally has been evaluated. The presence and ease of accessibility to facilities such as high-speed Internet, calling services, availability of land and buildings, highways, airports, railways and ports have also been assessed.

H0: No correlation exists between the availability of physical infrastructure and the transfer of R&D.
H7: There is a correlation between the availability of physical infrastructure and the transfer of R&D.

The above table reveals a significant and positive correlation between the availability of physical infrastructure and R&D transfer. Thus, the alternate hypothesis is accepted. 0.512 is the coefficient of correlation. It indicates that nations with better provisions of infrastructure that is accessible to entrepreneurs tend to have a higher transfer of research and development and vice versa. Higher rates of R&D transfer will stimulate the demand for better infrastructure facilities especially better global connectivity.

Internal Dynamics of the Market: This factor measures whether a freely functioning market exists where no single party has the influence to set prices. In such a market, any alterations in demand will result in simultaneous alterations in supply.

H0: No correlation exists between the internal dynamics of the market and the transfer of R&D.
H8: There is a correlation between the internal dynamics of the market and the transfer of R&D.

An analysis of the above table indicates a very weak but positive correlation among the above-mentioned variables. Thus, the alternate hypothesis is accepted. 0.118 is the coefficient of correlation. The results are not significant as the p-value is 0.419. It indicates that a nation having a free and open market economy has a low tendency of transferring research and development to entrepreneurs and vice versa.

9 Conclusion

The first objective sought to find ought whether there exists any correlation between factors such as entrepreneurial finance, entrepreneurial education, government entrepreneurship programmes and R&D transfer. It was found that all these factors are positively and significantly correlated with R&D transfer. The analysis of objective two reveals a significant and positive correlation between the transfer of R&D and corresponding factors including entrepreneurial motivation, provision of physical infrastructure and social and cultural norms. Nations that were ranked high on these factors had higher rates of transfer of R&D from institutions of research and learning to entrepreneurs. However, internal dynamics of the market showed a weak, positive, but not significant correlation with R&D transfer.

Government entrepreneurship programmes were the most highly correlated factor with R&D transfer, indicating the major role that the government has to play to facilitate the transfer of R&D. The next highly correlated factor was entrepreneurial finance. With the onset of the commercialization of R&D, the ability to acquire and use finance by entrepreneurs is an important factor to help transfer R&D to firms.

The third factor that obtained a relatively high correlation was having entrepreneurial education as part of the curriculum of schools. The only factor that obtained a very weak correlation that was not significant was internal market dynamics. It indicates that a freely functioning open market, left to the forces of demand and supply to determine prices, does not often result in the transfer of R&D to entrepreneurs.

10 Policy Implications

In order to ensure that R&D can be optimally transferred and utilized by entrepreneurs, certain factors are significantly correlated with this transfer. To stimulate the environment for the free flow of R&D transfer, the following policy implications may be considered:

- Affiliating centres of research and learning to the industry.
- Transferring R&D at concessional rates to entrepreneurs who keep abreast of the latest research in their field.
- The government can give subsidies and loans at lower rates of interest to entrepreneurs engaging in R&D and those purchasing it from research centres.
- Encouraging post-school educational institutions such as colleges and universities to conduct interactive sessions with successful entrepreneurs, giving students projects related to entrepreneurship and entrepreneurial research and having a minimum of two subjects related to entrepreneurship in each course.
- Incorporating entrepreneurial skills, values and subjects in the curriculum of school students in order to stimulate their thought process in the field of entrepreneurial research right from an early age.
- Providing adequate physical infrastructure facilitates the easy flow of R&D. It is helpful to both sides, those engaging in research and those making use of the findings.
- Building up the entrepreneurial spirit and encouraging research and development among family members, friends and the society at large will help in stimulating an entrepreneurial culture in society.

11 Suggestions for Future Research

- There is an opportunity to conduct an econometric analysis, e.g. a regression analysis using data of more countries.
- Causality test can also be undertaken to find whether there exists a causal relation and its direction between R&D transfer and the various factors influencing it.

References

1. Ahmad, H.: Personality traits among entrepreneurial and professional CEOs in SMEs. Int. J. Bus. Manag. **5**(9), 203–213 (2010)
2. Audretsch, D.B., Heger, D., Veith, T.: Infrastructure and entrepreneurship . Small Bus. Econ. **44**(2), 219–230 (2015)
3. Baker, T., Nelson, R.E.: Creating something from nothing: resource construction through entrepreneurial bricolage. Adm. Sci. Q. **50**(3), 329–366 (2005)
4. Baumol, W.J.: The Free-Market Innovation Machine: Analyzing The Growth Miracle of Capitalism. Princeton University Press, Princeton (2002)
5. Bird, B.J.: Entrepreneurial Behavior. Scott, Foresman & Company, Glenview (1989)
6. Bjornskov, C., Foss, N.: Economic freedom and entrepreneurial activity: some cross-country evidence. Public Choice **134**(3), 307–328 (2006)
7. Bosma, N., Kelley, D.: Global Entrepreneurship Monitor, 2018/2019 Global Report. Grafica Andes, Chile (2018)
8. Carland, J., Carland, J.: Entrepreneurship: an American dream. J. Bus. Entrep. **9**(1), 33–45 (1997). ABI/INFORM
9. Carlsson, B., Braunerhjelm, P., McKelvey, M., Olofsson, C., Persson, L., Ylinenpaa, H.: The evolving domain of entrepreneurship research. Working Paper, Swedish Entrepreneurship Forum (2012)
10. Carsrud, A., Brannback, M.: Entrepreneurial motivations: what do we still need to know? J. Small Bus. Manag. **49**(1), 9–26 (2011)
11. Cohen, W.M., Levinthal, D.A.: Innovation and learning: the two faces of R&D. Econ. J. **99**, 569–596 (1989)
12. Consortium for Entrepreneurship Education Entrepreneurship Everywhere: The Case for Entrepreneurship Education. Columby, USA (2008)
13. Davidsson, P.: Entrepreneurial opportunities and the entrepreneurship nexus: are-conceptualization. J. Bus. Ventur. **30**(5), 674–695 (2015)
14. Easley, D., Kleinberg, J.: Networks, Crowds and Markets. Cambridge University Press, Cambridge (2010)
15. Erikson, T.: Towards a taxonomy of entrepreneurial learning experiences among potential entrepreneurs. J. Small Bus. Enterp. Dev. **10**, 106–112 (2003)
16. European Commission Entrepreneurship Education in Europe: Fostering Entrepreneurial Mindsets Through Education and Learning. Commission of the European Communities, Oslo (2006)
17. Gottinger, H.W.: Innovation, Dynamics of Competition and Market Dynamics. Arch. Bus. Res. **4**(1), 1–16 (2016)
18. Hayton, J.C., George, G., Zahra, S.A.: National culture and entrepreneurship: a review of behavioural research. Entrep. Theory Pract. **26**(4), 33–52 (2002)
19. Kiyosaki, R.: Rich Dad Poor Dad: What the Rich Teach Their Kids About Money - That the Poor and Middle Class Do Not! Rich Dad on Brilliance Audio (2009)
20. Krueger, N.F., Carsrud, A.L.: Entrepreneurial intentions: applying the theory of planned behaviour. Entrep. Reg. Dev. **5**(4), 315–330 (1993)
21. Kuratko, D., Welsch, H.: Strategic Entrepreneurial Growth. South-Western College, Bloomington (2003)
22. Landstrom, H., Sexton, D.: Introduction. In: Landstrom, H., Sexton, D.L. (eds.) Handbook of Entrepreneurship, pp. xix–xxiv. Blackwell Publishers, Oxford (2000)
23. Liwei, C., Xiaoyong, D.: Analysis of the distribution characteristics of R&D investment and the influencing factors of R&D investment intensity based on panel data of 300 thousand industrial enterprises in China. Chin. J. Soft Sci. (2012)
24. McClelland, D.: The Achieving Society. The Free Press, New York (1962)
25. Moseley, M.J.: Growth Centres in Spatial Planning: Pergamon Urban and Regional Planning. Elsevier, Amsterdam (2013)

26. OECD: OECD Employment Outlook. OECD, Paris (2000)
27. Perroux, F.: Economic space: theory and applications. Q. J. Econ. **64**(1), 89–104 (1950)
28. Reynolds, P.D., Hay, M., Bygrave, W.D., Camp, S.M., Autio, E.: Global Entrepreneurship Monitor: 2000 Executive Report, p. 6. Kauffman Center for Entrepreneurial Leadership, Kansas City (2000)
29. Robb, A.M., Robinson, D.T.: The Capital Structure Decisions of New Firms. Rev. Financ. Stud. **27**(1), 153–179 (2014)
30. Rosenberg, N.: Why do firms do basic research (with their own money)? Res. Policy **19**(2), 165–174 (1990)
31. Shane, S., Locke, E.A., Collins, C.J.: Entrepreneurial motivation. Hum. Resour. Manag. Rev. **13**(2), 257–279 (2012)
32. Teece, D.J.: Profiting from technological innovation: implications for integration, collaboration, licensing and public policy. Res. Policy **15**(6), 285–305 (1986)
33. Thornton, P.H.: Markets from Culture: Institutional Logics and Organizational Decisions. Stanford University Press, Stanford (2004)
34. Thornton, P.H., Ocasio, W.: Institutional logics. In: Greenwood, R., Oliver, C., Andersen, S.K., Suddaby, R. (eds.) Handbook of Organizational Institutionalism, pp. 99–129. SAGE, Thousand Oaks (2008)
35. Trade and Development Board Investment, Enterprise and Development Commission (UNCTAD).: Key aspects of entrepreneurship and innovation policy frameworks for enhancing local productive capacities, Geneva. (2011)
36. Winborg, J., Landstrom, H.: Financial bootstrapping in small businesses: examining small business managers' resource acquisition behaviors. J. Bus. Ventur. **16**(3), 235–254 (2001)
37. Yu, T.F.L.: Economic development in latecomer economies: an entrepreneurial perspective. Development Policy Review **16**(4), 353–372 (1998)
38. Zimmerer, T., Scarborough, N.M.: Entrepreneurship and New Venture Formation. East Tennessee State University; Presbyterian College (1996)

Some Common Fixed Point Theorems in Generalized Fuzzy Metric Spaces for Weakly Compatible with CLR_{ST} Property

V. Vinoba, V. Pazhani, and M. Jeyaraman

1 Introduction

In 1965, the idea of a fuzzy set was presented by Zadeh [1]. At that point in 1975, Kramosil and Michalek [2] presented the fuzzy metric space as a generalization of a metric space. In 1994, George and Veeramani [3] adjusted the idea of fuzzy metric spaces with the assistance of continuous t-norms. In 1986, Jungck [4] presented the idea of compatible mappings in metric spaces. In 2002, Aamri and El Moutawakil [5] characterized a property (E. A) for self-mappings that contained the class of non-compatible mappings in metric spaces. In a paper of Ali and Imdad [6], it was called attention to that property (E. A) permits supplanting the completeness necessity of the space with a progressively normal state of the closedness of the range. In 2011, Sintunavarat and Kummam [7] authored the possibility of common limit range property which loosens up the prerequisite on closedness of the subspaces. A few common fixed point theorems were demonstrated by numerous researchers in the system of fuzzy metric spaces through certain relations.

In this paper, we prove some common fixed point theorems for weakly compatible mappings with CLR_{ST} property in generalized fuzzy metric spaces.

V. Vinoba · V. Pazhani
Department of Mathematics, Kunthavai Naacchiyaar Government Arts College for Women (Autonomous), Affiliated to Bharathidasan University, Thanjavur, Tiruchirappalli, Tamilnadu, India

V. Pazhani · M. Jeyaraman (✉)
Department of Mathematics, Raja Doraisingam Goverment Arts College, Affiliated to Alagappa University, Sivagangai, Karaikudi, Tamilnadu, India

P. Srivastava et al. (eds.), *Mathematical, Computational Intelligence and Engineering Approaches for Tourism, Agriculture and Healthcare*, Lecture Notes in Networks and Systems 214, https://doi.org/10.1007/978-981-16-3807-7_27

2 Preliminaries

Definition 2.1 [8] A 3-tuple (X, $\mathcal{M}$, *) is called Generalized fuzzy metric space if X is an arbitrary nonempty set, * is a continuous t-norm and $\mathcal{M}$ is a fuzzy set on $X^3 \times (0, \infty)$, satisfying the following conditions: for each $x, y, z, a \in X$ and $t, s > 0$.

($\mathcal{M}$1) $\mathcal{M}(x, y, z, t) > 0$,
($\mathcal{M}$ 2) $\mathcal{M}(x, y, z, t) = 1$ if and only if $x = y = z$,
($\mathcal{M}$3) $\mathcal{M}(x, y, z, t) = \mathcal{M}(p\{x, y, z\}, t)$, when p is the permutation function,
($\mathcal{M}$4) $\mathcal{M}(x, y, a, t) * \mathcal{M}(a, z, z, s) \le \mathcal{M}(x, y, z, t + s)$,
($\mathcal{M}$5) $\mathcal{M}(x, y, z,.)$: $(0, \infty) \to [0, 1]$ is continuous.

Definition: 2.2 [8] Let (X, $\mathcal{M}$, *) be a generalized fuzzy metric space and $\{x_n\}$ be a sequence in X.

(i) $\{x_n\}$ is said to be converges to a point $x \in X$ if $\lim_{n\to\infty} \mathcal{M}(x, x, x_n, t) = 1$ for all $t > 0$.
(ii) $\{x_n\}$ is called a Cauchy sequence if $\lim_{n\to\infty} \mathcal{M}(x_{n+p}, x_{n+p}, x_n, t) = 1$ for all $t > 0$ and $p > 0$.
(iii) A generalized fuzzy metric space in which every Cauchy sequence is convergent is said to be complete.

Definition: 2.3 [8] A pair ($\mathcal{A}$, $\mathcal{S}$) of self-mappings of a nonempty set X is said to be weakly compatible if they commute at their coincidence points, that is, if $\mathcal{A}p = \mathcal{S}p$ for some $p \in X$, then $\mathcal{A}\mathcal{S}p = \mathcal{S}\mathcal{A}_p$.

Definition: 2.4 [9] Two pairs ($\mathcal{A}$, $\mathcal{S}$) and ($\mathcal{B}$, $\mathcal{T}$) of self-mappings of a generalized fuzzy metric spaces (X, $\mathcal{M}$, *) are said to satisfy the $CLR_{\mathcal{S}\mathcal{T}}$ property if there exists a sequence $\{x_n\}$, $\{y_n\}$ in X such that for all $t > 0$, $\lim_{n\to\infty} \mathcal{A}x_n = \lim_{n\to\infty} \mathcal{S}x_n = \lim_{n\to\infty} \mathcal{B}y_n = \lim_{n\to\infty} \mathcal{T}y_n = p$ for some $p \in \mathcal{S}(X) \cap \mathcal{T}(X)$.

Lemma 2.4 Let (X, $\mathcal{M}$, *) be a generalized fuzzy metric space. If there exists a constant $k \in (0,1)$ such that $M(x, y, z, kt) \ge M(x, y, z, t)$, for all $x, y, z \in X$ and $t > 0$, then $x = y = z$.

3 Main Results

Our results include a class Φ of all mappings ϕ: $[0, 1] \to [0, 1]$ satisfying the following properties:

(ϕ1): ϕ is continuous and non decreasing on [0, 1];
(ϕ2): $\phi(s) > s$ for all $s \in (0, 1)$.

We note that if $\phi \in \Phi$, then $\phi(1) = 1$, and that $\phi(s) \ge s$ for all $s \in [0, 1]$.

Lemma 3.1 Let $\mathcal{A}$, $\mathcal{B}$, $\mathcal{S}$ and $\mathcal{T}$ be four self-mappings of a generalized fuzzy metric space
(X, $\mathcal{M}$, *). Suppose that

(1) the pair ($\mathcal{A}$, $\mathcal{S}$) (or ($\mathcal{B}$, $\mathcal{T}$) satisfies the CLR$_{\mathcal{S}}$) (or CLR$_{\mathcal{T}}$)) property;
(2) $\mathcal{A}$(X) ⊂ $\mathcal{T}$(X)(or $\mathcal{B}$(X) ⊂ $\mathcal{S}$(X));
(3) $\mathcal{T}$(X) (or $\mathcal{S}$(X)) is a closed subset of X;
(4) $\mathcal{B}(y_n)$ converges for every sequence $\{y_n\}$ in X whenever $T(y_n)$ converges (or $\mathcal{A}(x_n)$ converges for every sequence $\{x_n\}$ in X whenever $S(x_n)$ converges);
(5) for all $x, y, z \in X$, $x \neq y \neq z$ there exists $t > 0$: $0 < \mathcal{M}(x, y, z, t) < 1$, for some $\phi \in \Phi$,

$$\begin{aligned}\mathcal{M}(\mathcal{A}x, \mathcal{B}y, \mathcal{B}z, t,) \geq \phi(\min\{&\mathcal{M}(\mathcal{S}x, \mathcal{T}y, \mathcal{T}z, t), \mathcal{M}(\mathcal{A}x, \mathcal{T}y, \mathcal{T}z, t)\\ &\mathcal{M}(\mathcal{S}x, \mathcal{B}y, \mathcal{B}z, t), \mathcal{M}(\mathcal{B}y, \mathcal{B}y, \mathcal{T}z, t)\\ &\mathcal{M}(\mathcal{A}x, \mathcal{T}y, \mathcal{T}y, t)\})\end{aligned} \tag{3.1.1}$$

Then the pairs ($\mathcal{A}$, $\mathcal{S}$) and ($\mathcal{B}$, $\mathcal{T}$) satisfy the CLR$_{ST}$ property.

Proof If the pair ($\mathcal{A}$, $\mathcal{S}$) enjoys the CLR$_{\mathcal{S}}$ property, then there exists a sequence $\{x_n\}$ in X such that,

$$\lim_{n\to\infty} \mathcal{A}x_n = \lim_{n\to\infty} \mathcal{A}x_n = p, \tag{3.1.2}$$

where p ∈ $\mathcal{S}$(X). By Lemma 3.1. (2), $\mathcal{A}$(X) ⊂ $\mathcal{T}$(X), and for each sequence $\{x_n\}$, there exists a sequence $\{y_n\}$ in X such that $Ax_n = Ty_n$.

Therefore, due to the closedness of $\mathcal{T}$(X),

$$\lim_{n\to\infty} \mathcal{T}y_n = \lim_{n\to\infty} \mathcal{A}x_n = p, \tag{3.1.3}$$

So, that p ∈ $\mathcal{T}$(X) and in all p ∈ $\mathcal{S}$(X) ∩ $\mathcal{T}$(X). Thus, we have $Ax_n \to p$, $Sx_n \to p$ and $Ty_n \to p$ as $n \to \infty$. By (3.1.4), sequence $\{By_n\}$ converges and in all we need to show that $By_n \to p$ as $n \to \infty$. Suppose that $\mathcal{B}y_n \to p'$ ($\neq$p) as $n \to \infty$, and then using inequality (3.1.1) with $x = x_n$,$y = y_n$, we have

$$\begin{aligned}\mathcal{M}(\mathcal{A}x_n, \mathcal{B}y_n, \mathcal{B}y_{n+1}, t) \geq \phi(\min\{&\mathcal{M}(\mathcal{S}x_n, \mathcal{T}y_n, \mathcal{T}y_{n+1}, t), \mathcal{M}((\mathcal{A}x_n, \mathcal{T}y_n, \mathcal{T}y_{n+1}, t),\\ &\mathcal{M}(\mathcal{S}x_n, \mathcal{B}y_n, \mathcal{B}y_{n+1}, t), \mathcal{M}(\mathcal{B}y_n, \mathcal{B}y_n, \mathcal{T}_{y_{n+1}}, t),\\ &\mathcal{M}(\mathcal{A}x_n, \mathcal{T}y_n, \mathcal{T}y_n, t)\}.\end{aligned} \tag{3.1.4.}$$

$$\text{As } n\to\infty, \mathcal{M}(p, p, p', t) \geq \phi(\min\{\mathcal{M}(p, p, p, t), \mathcal{M}(p, p, p, t), \mathcal{M}(p, p', p', t),$$

$$\mathcal{M}(p', p', p, t), \mathcal{M}(p, p, p, t)\}) \tag{3.1.5}$$

$$\mathcal{M}(p, p', p', t) \geq \phi(\mathcal{M}(p, p', p', t)) \tag{3.1.6}$$

As $p \neq p'$, we have $0 < \mathcal{M}(p, p', p', t_o) < 1$ for some $t_0 > 0$. Then, in view of the condition (ϕ_2), we get $\phi(\mathcal{M}(p, p', p', t_0)) \geq \mathcal{M}(p, p', p', t_o)$, which is a contradiction, thereby implying $p = p'$, which shows that the pairs $(\mathcal{A}, \mathcal{S})$ and $(\mathcal{B}, \mathcal{T})$ enjoy the CLR_{ST} property.

Theorem 3.2 Let $\mathcal{A}, \mathcal{B}, \mathcal{S}$ and $\mathcal{T}$ be four self- mappings of a generalized fuzzy metric space.

(X, $\mathcal{M}$, *) satisfying an inequality (3.1.1). Suppose that the pairs $(\mathcal{A}, \mathcal{S})$ and $(\mathcal{B}, \mathcal{T})$ enjoy the CLR_{ST} property. Then the pairs $(\mathcal{A}, \mathcal{S})$ and $(\mathcal{B}, \mathcal{T})$ have a coincidence point each. Moreover, $\mathcal{A}, \mathcal{B}, \mathcal{S}$ and $\mathcal{T}$ have a unique common fixed point provided both pairs $(\mathcal{A}, \mathcal{S})$ and $(\mathcal{B}, \mathcal{T})$ are weakly compatible.

Proof Since the pairs $(\mathcal{A}, \mathcal{S})$ and $(\mathcal{B}, \mathcal{T})$ satisfy the CLR_{ST} property, there exist two sequences $\{x_n\}$ and $\{y_n\}$ in X such that

$$\lim_{n\to\infty} \mathcal{A}x_n = \lim_{n\to\infty} S\chi_n = \lim_{n\to\infty} \mathcal{B}y_n = \lim_{n\to\infty} \mathcal{T}_{y_n} = p \tag{3.2.1}$$

where $p \in \mathcal{S}(X) \cap \mathcal{T}(X)$. Since $p \in \mathcal{S}(X)$, there exists a point $u \in X$ such that $\mathcal{S}u = p$.

We show that $\mathcal{M}(Au, p, p, t) = 1$.

If not, then using inequality (3.1.1) with $x = u$, $y = y_n$, we get

$$\begin{aligned}\mathcal{M}(\mathcal{A}u, \mathcal{B}y_n, \mathcal{B}y_{n+1}, t) \geq \phi(\min\{&\mathcal{M}(Su, \mathcal{T}y_n, \mathcal{T}y_{n+1}, t), \mathcal{M}(\mathcal{A}u, \mathcal{T}y_n, \mathcal{T}y_{n+1}, t),\\ &\mathcal{M}(Su, \mathcal{B}y_n, \mathcal{B}y_{n+1}, t), \mathcal{M}(\mathcal{B}y_n, \mathcal{B}y_n, \mathcal{T}y_{n+1}, t),\\ &\mathcal{M}(\mathcal{A}u, \mathcal{T}y_n, \mathcal{T}y_n, t)\}\end{aligned} \tag{3.2.2}$$

As $n \to \infty, \mathcal{M}(\mathcal{A}u, p, p, t) \geq \phi(\min\{\mathcal{M}(\mathcal{S}u, p, p, t), \mathcal{M}(\mathcal{A}u, p, p, t)$, $\mathcal{M}(Su, p, p, t)$,

$$\mathcal{M}(p, p, p, t), \mathcal{M}(\mathcal{A}u, p, p, t)\}) \tag{3.2.3}$$

and so,

$$\mathcal{M}(\mathcal{A}u, p, p, t) \geq \phi(\mathcal{M}(\mathcal{A}u, p, p, t)) \tag{3.2.4}$$

If $\mathcal{A}u \neq p$, then $0 < \mathcal{M}(Au, p, p, t_0) < 1$ for some $t_0 > 0$. Then in view of the condition (ϕ_2) we get $\phi(\mathcal{M}(Au, p, p, t_0)) > \mathcal{M}(Au, p, p, t_0)$ which is a contradiction. Therefore, $\mathcal{A}u = p$ so that $\mathcal{A}u = p = \mathcal{S}u$ which shows that u is a coincidence point of the pair $(\mathcal{A}, \mathcal{S})$.

Also $p \in \mathcal{T}(X)$; there exists a point $v \in X$ such that $\mathcal{T}v = p$.

Now we assert that $\mathcal{M}(p, Bv, Bv, t) = 1$. Assume the contrary, and then using inequality (3.1.1) with $x = u$, $y = v$, $z = v$, we have

$$\mathcal{M}(\mathcal{A}u, \mathcal{B}v, \mathcal{B}v, t) \geq \phi(\min\{\mathcal{M}(Su, \mathcal{T}v, \mathcal{T}v, t), \mathcal{M}(\mathcal{A}u, \mathcal{T}v, \mathcal{T}v, t), \mathcal{M}(Su, \mathcal{B}v, \mathcal{B}v, t),$$

$$\mathcal{M}(\mathcal{B}v, \mathcal{B}v, \mathcal{T}v, t), \mathcal{M}(\mathcal{A}u, \mathcal{T}v, \mathcal{T}v, t)\}) \tag{3.2.5.}$$

which reduces to

$$\mathcal{M}(p, \mathcal{B}v, \mathcal{B}v, t) \geq \phi(\min\{\mathcal{M}(p, p, p, t), \mathcal{M}(p, p, p, t), \mathcal{M}(p, \mathcal{B}v, \mathcal{B}v, t), \mathcal{M}(\mathcal{B}v, \mathcal{B}v, p, t), \mathcal{M}(p, p, p, t)\}) \tag{3.2.6}$$

$$\mathcal{M}(p, \mathcal{B}v, \mathcal{B}v, t) \geq \phi(\mathcal{M}(p, \mathcal{B}v, \mathcal{B}v, t)) \tag{3.2.7}$$

As, Bv $\neq$ p implies $0 < \mathcal{M}(p, Bv, Bv, t_0) < 1$ for some $t_0 > 0$, then in view of condition (ϕ_2), we get $\phi(\mathcal{M}(p, Bv, Bv, t_0)) > \mathcal{M}(p, Bv, Bv, t_0)$, which is a contradiction. Therefore, $\mathcal{B}v = p$ so, that $\mathcal{B}v = p = Tv$ which shows that v is a coincidence point of the pair $(\mathcal{B}, \mathcal{T})$.

Since the pair $(\mathcal{A}, \mathcal{S})$ is weakly compatible and $\mathcal{A}u = \mathcal{S}u$, hence $\mathcal{A}p = \mathcal{A}\mathcal{S}u = \mathcal{S}\mathcal{A}u = \mathcal{S}p$. Now, we show that p is a common fixed point of the pair $(\mathcal{A}, \mathcal{S})$.

To prove this, we show that $\mathcal{M}(Ap, p, p, t) = 1$.

If not, then using inequality (3.1.1) with $x = p$, $y = v$, $z = v$, we have

$$\mathcal{M}(\mathcal{A}p, \mathcal{B}v, \mathcal{B}v, t) \geq \phi(\min\{\mathcal{M}(\mathcal{S}p, \mathcal{T}v, \mathcal{T}v, t), \mathcal{M}(\mathcal{A}p, \mathcal{T}v, \mathcal{T}v, t), \mathcal{M}(\mathcal{S}p, \mathcal{B}v, \mathcal{B}v, t) \; \mathcal{M}(\mathcal{B}v, \mathcal{B}v, \mathcal{T}v, t), \mathcal{M}(\mathcal{A}p, \mathcal{T}v, \mathcal{T}v, t)\}) \tag{3.2.8}$$

and so

$$\mathcal{M}(\mathcal{A}p, p, p, t) \geq \phi(\min\{\mathcal{M}(\mathcal{S}p, p, p, t), \mathcal{M}(\mathcal{A}p, p, p, t), \mathcal{M}(\mathcal{S}p, p, p, t), \mathcal{M}(p, p, p, t), \mathcal{M}(\mathcal{A}p, p, p, t)\}). \tag{3.2.9}$$

Then on simplification, we obtain

$$\mathcal{M}(\mathcal{A}p, p, p, t) \geq \phi(\mathcal{M}(\mathcal{A}p, p, p, t)) \tag{3.2.10}$$

Since, $\mathcal{A}p \neq p$, therefore $0 < \mathcal{M}(Ap, p, p, t_0) < 1$ for some $t_0 > 0$. Then in view of the condition (ϕ_2), we get $\phi(\mathcal{M}(Ap, p, p, t_0)) > \mathcal{M}(Ap, p, p, t_0)$, which is a contradiction. Hence, $\mathcal{A}p = p = \mathcal{S}p$. Therefore, p is a common fixed point of the pair $(\mathcal{A}, \mathcal{S})$.

Also the pair $(\mathcal{B}, \mathcal{T})$ is weakly compatible and $\mathcal{B}v = \mathcal{T}v$; then $\mathcal{B}p = \mathcal{B}\mathcal{T}v = \mathcal{T}\mathcal{B}v = \mathcal{T}p$. To accomplish this, we assert that $\mathcal{M}(p, Bp, Bp, t) = 1$. If not, then using inequality (3.1.1) with $x = u$, $y = p$,$z = p$, we have

$$\mathcal{M}(\mathcal{A}u, \mathcal{B}p, \mathcal{B}p, t) \geq \phi(\min\{\mathcal{M}(\mathcal{S}u, \mathcal{T}p, \mathcal{T}p, t), \mathcal{M}(\mathcal{A}u, \mathcal{T}p, \mathcal{T}p, t), \mathcal{M}(\mathcal{S}u, \mathcal{B}p, \mathcal{B}p, t), \mathcal{M}(\mathcal{B}p, \mathcal{B}p, \mathcal{T}p, t), \mathcal{M}(\mathcal{A}u, \mathcal{T}p, \mathcal{T}p, t)\}) \tag{3.2.11}$$

which reduces to

$$\mathcal{M}(\mathrm{p}, \mathcal{B}\mathrm{p}, \mathcal{B}\mathrm{p}, t) \geq \phi(\min\{\mathcal{M}(\mathrm{p}, \mathcal{B}\mathrm{p}, \mathcal{B}\mathrm{p}, t), \mathcal{M}(\mathrm{p}, \mathcal{B}\mathrm{p}, \mathcal{B}\mathrm{p}, t), \mathcal{M}(\mathrm{p}, \mathcal{B}\mathrm{p}, \mathcal{B}\mathrm{p}, t), \mathcal{M}(\mathcal{B}\mathrm{p}, \mathcal{B}\mathrm{p}, \mathcal{B}\mathrm{p}, t), \mathcal{M}(\mathrm{p}, \mathcal{B}\mathrm{p}, \mathcal{B}\mathrm{p}, t)\}) \quad (3.2.12)$$

and so

$$\mathcal{M}(\mathrm{p}, \mathcal{B}\mathrm{p}, \mathcal{B}\mathrm{p}, t) \geq \phi(\mathcal{M}(\mathrm{p}, \mathcal{B}\mathrm{p}, \mathcal{B}\mathrm{p}, t)) \quad (3.2.13)$$

If $\mathcal{B}\mathrm{p} \neq \mathrm{p}$, then $0 < \mathcal{M}(\mathrm{p}, \mathrm{Bp}, \mathrm{Bp}, t_0) < 1$ for some $t_0 > 0$. Then,the condition (ϕ_2) it follows that $\phi(\mathcal{M}(\mathrm{p}, \mathrm{Bp}, \mathrm{Bp}, t_0)) > \mathcal{M}(\mathrm{p}, \mathrm{Bp}, \mathrm{Bp}, \mathrm{t})$, which is a contradiction. Therefore, $\mathcal{B}\mathrm{p} = \mathrm{p} = \mathcal{T}\mathrm{p}$ which show that p is a common fixed point of the pair $(\mathcal{B}, \mathcal{T})$.

The uniqueness of common fixed point is an easy consequence of inequality (3.1.1)

Example 3.3 Let $(X, \mathcal{M}, *)$ be a generalized fuzzy metric space, where $X = [4, 16]$, with product t-norm defined as $a * b = ab$ for all $a, b \in [0, 1]$ and $\mathcal{M}(x, y, z, t) = \frac{t}{t+|x-y|+|y-z|+|z-x|}$ for all $t > 0$ and $x, y, z \in X$. Define the self-mappings $\mathcal{A}$, $\mathcal{B}$, $\mathcal{S}$ and $\mathcal{T}$ by.

$$\mathcal{A}(x) = \begin{cases} 4 & \text{if } x \in \{4\} \cup (6, 16) \\ 4.5 & \text{if } x \in (4, 6] \end{cases}$$

$$\mathcal{B}(x) = \begin{cases} 4 & \text{if } x \in \{4\} \cup (6, 16) \\ 5 & \text{if } x \in (4, 6] \end{cases}$$

$$\mathcal{S}(x) = \begin{cases} 4 & \text{if } x = 4 \\ 11 & \text{if } x \in (4, 6) \\ 12 & \text{if } x = 6 \\ \frac{x+140}{36} & \text{if } x \in (6, 16) \end{cases}$$

$$\mathcal{T}(x) = \begin{cases} 4 & \text{if } x = 4 \\ 15 & \text{if } x \in (4, 6) \\ 16 & \text{if } x = 6 \\ \frac{x+140}{36} & \text{if } x \in (6, 16) \end{cases}$$

We obtain $\mathcal{A}(X) = \{4, 4.5\} \not\subseteq [4, 4.3) \cup \{15, 16\} = \mathcal{T}(X)$,

$$\mathcal{B}(X) = \{4, 5\} \not\subseteq [4, 4.3) \cup \{11, 12\} = \mathcal{S}(X).$$

Hence, $\mathcal{S}(X)$ and $\mathcal{T}(X)$ are not closed of X and so Theorem 2.1 of Imdad et al. [10] cannot be applied to this example. Next, we choose two sequences $\{x_n\} = \{6 + (1/n)\}$, $\{y_n\} = \{4\}$ and then clearly $\lim_{n\to\infty} \mathcal{A}x_n = \lim_{n\to\infty} \mathcal{S}x_n = \lim_{n\to\infty} \mathcal{B}y_n = \lim_{n\to\infty} \mathcal{T}y_n = 4 \in$

$\mathcal{S}(X) \cap \mathcal{T}(X)$, which shows that both pairs $(\mathcal{A}, \mathcal{S})$ and $(\mathcal{B}, \mathcal{T})$ enjoy the $CLR_{\mathcal{ST}}$ Property. By a routine calculation, one can verify inequality (3.1.1) where ϕ is defined by $\phi(s) = \sqrt{s}$. Furthermore, we obtain that the pairs $(\mathcal{A}, \mathcal{S})$ and $(\mathcal{B}, \mathcal{T})$ are weakly compatible.

Therefore, all the conditions of Theorem 3.2 are satisfied and 4 is a unique common fixed point of $\mathcal{A}, \mathcal{B}, S$ and $\mathcal{T}$, which also remains a coincidence point as well.

Theorem 3.4 Let $\mathcal{A}, \mathcal{B}, S$ and $\mathcal{T}$ be four self-mappings of a generalized fuzzy metric space.

(X, $\mathcal{M}$, *) satisfying inequality (3.1.1). Suppose that the following hypotheses hold:

(1) The pairs $(\mathcal{A}, \mathcal{S})$ and $(\mathcal{B}, \mathcal{T})$ satisfy the common property (E. A)
(2) $\mathcal{S}(X)$ and $\mathcal{T}(X)$ are closed subsets of X.

Then the pairs $(\mathcal{A}, \mathcal{S})$ and $(\mathcal{B}, \mathcal{T})$ have a coincidence point each. Moreover, $\mathcal{A}, \mathcal{B}, S$ and $\mathcal{T}$ have a unique common fixed point provided both pairs $(\mathcal{A}, \mathcal{S})$ and $(\mathcal{B}, \mathcal{T})$ are weakly compatible.

Proof Since the pairs $(\mathcal{A}, \mathcal{S})$ and $(\mathcal{B}, \mathcal{T})$ enjoy the common property (E. A), there exists two sequences $\{x_n\}$ and $\{y_n\}$ in X such that $\lim_{n\to\infty} \mathcal{A}x_n = \lim_{n\to\infty} \mathcal{S}x_n = \lim_{n\to\infty} \mathcal{B}y_n = \lim_{n\to\infty} \mathcal{T}y_n = p$ for some $p \in X$. Since $\mathcal{S}(X)$ and $\mathcal{T}(X)$ are closed subsets of X, hence $\lim_{n\to\infty} \mathcal{S}y_n = p \in \mathcal{S}(X)$. Therefore, there exists a point $u \in X$ such that $\mathcal{S}u = p$. Similarly, $\lim_{n\to\infty} \mathcal{T}y_n = p \in \mathcal{T}(X)$. Therefore, there exists a point $v \in X$ such that $\mathcal{T}v = p$. The rest of the proof runs on the lines of the proof of Theorem 3.2.

Theorem 3.5 Let $\mathcal{A}, \mathcal{B}, S$ and $\mathcal{T}$ be four self-mappings of a generalized fuzzy metric space.

(X, $\mathcal{M}$, *) satisfying all the hypotheses of Lemma 3.1. Then $\mathcal{A}, \mathcal{B}, S$ and $\mathcal{T}$ have a unique common fixed point provided both pairs $(\mathcal{A}, \mathcal{S})$ and $(\mathcal{B}, \mathcal{T})$ are weakly compatible.

Proof In view of Lemma 3.1, the pairs $(\mathcal{A}, \mathcal{S})$ and $(\mathcal{B}, \mathcal{T})$ enjoy the $CLR_{\mathcal{ST}}$ property, there exists two sequences $\{x_n\}$ and $\{y_n\}$ in X such that $\lim_{n\to\infty} \mathcal{A}x_n = \lim_{n\to\infty} \mathcal{S}x_n = \lim_{n\to\infty} \mathcal{B}y_n = \lim_{n\to\infty} \mathcal{T}y_n = p$ for some $p \in \mathcal{S}(X) \cap \mathcal{T}(X)$. The rest of the proof can be completed on the lines of the proof of Theorem 3.2.

Example 3.6 Let (X, $\mathcal{M}$, *) be a generalized fuzzy metric space, where X = [4,16), with product t-norm defined as a *ib = min{a, b} for all a, b ∈ [0, 1] and $\mathcal{M}(x, y, z, t) = \frac{t}{t+D(x,y,z)}$, $t > 0$, where $D(x, y, z) = |x - y| + |y - z| + |z - x|$.

Define the self-mappings $\mathcal{A}, \mathcal{B}, S$ and $\mathcal{T}$ by,

$$\mathcal{A}(x) = \begin{cases} 4 \text{ if } x \in \{4\} \cup (6, 16) \\ \frac{9}{2} \text{ if } x \in (4, 6] \end{cases}$$

$$\mathcal{B}(\mathrm{x}) = \begin{cases} 4 \text{ if } \mathrm{x} \in \{4\} \cup (6, 16) \\ 5 \text{ if } \mathrm{x} \in (4, 6] \end{cases}$$

$$\mathcal{S}(x) = \begin{cases} 4 & \text{if } \mathrm{x} = 4 \\ 14 & \text{if } \mathrm{x} \in (4, 6] \\ \mathrm{x} - 2 & \text{if } \mathrm{x} \in (6, 16). \end{cases}$$

$$\mathcal{T}(\mathrm{x}) = \begin{cases} 4 & \text{if } x = 4 \\ 15 & \text{if } x \in (4, 6] \\ \frac{11x-36}{10} & \text{if } \mathrm{x} \in (6, 16]. \end{cases}$$

Then we have $\mathcal{A}(\mathrm{X}) = \{4, 4.5\} \subseteq [4, 15] = \mathcal{T}(\mathrm{X})$ and $\mathcal{B}(\mathrm{X}) = \{4, 5\} \subseteq [4, 14] = \mathcal{S}(\mathrm{X})$, whereas $\mathcal{S}(\mathrm{X})$ and $\mathcal{T}(\mathrm{X})$ are closed subsets of X.

We choose two sequences $\{x_n\} = \left\{6 + \frac{1}{n}\right\}, \{y_n\} = \{4\}$ then clearly,

$$\lim_{\mathrm{n}\to\infty} \mathcal{A}\mathrm{x_n} = \lim_{\mathrm{n}\to\infty} \mathcal{S}\mathrm{x_n} = \lim_{\mathrm{n}\to\infty} \mathcal{B}y_n = \lim_{\mathrm{n}\to\infty} \mathcal{T}y_n = 4 \in \mathcal{S}(\mathrm{X}) \cap \mathcal{T}(\mathrm{X})$$

which shows that both pairs $(\mathcal{A}, \mathcal{S})$ satisfy the $\mathrm{CLR}_{\mathcal{S}}$ property, and $(\mathcal{B}, \mathcal{T})$ satisfy the $\mathrm{CLR}_{\mathcal{T}}$ property. By a calculation, one can verify inequality (3.1.1) where in ϕ is defined by $\phi(\mathrm{s}) = \sqrt{s}$. Furthermore, we obtain that the pairs $(\mathcal{A}, \mathcal{S})$ and $(\mathcal{B}, \mathcal{T})$ are weakly compatible.

Therefore, all the conditions of Theorem 3.5 are satisfied and 4 is a unique common fixed point of $\mathcal{A}$, $\mathcal{B}$, S and $\mathcal{T}$.

Theorem 3.7 Let $\mathcal{A}$, $\mathcal{B}$, S and $\mathcal{T}$ be four self-mapping of a generalized fuzzy metric space.

(X, $\mathcal{M}$, *). Suppose that

(1) The pairs $(\mathcal{A}, \mathcal{S})$ and $(\mathcal{B}, \mathcal{T})$ enjoy the $\mathrm{CLR}_{\mathcal{ST}}$ property.
(2) For all $x, y, z \in \mathrm{X}$, $t > 0$ and for some $\mathrm{k} \in (0, 1)$.

$$\begin{aligned} \mathcal{M}(\mathcal{A}x, \mathcal{B}y, \mathcal{B}z, kt) \geq \min\{&\mathcal{M}(\mathcal{S}x, \mathcal{T}y, \mathcal{T}z, t), \mathcal{M}(\mathcal{A}x, \mathcal{T}y, \mathcal{T}z, t), \\ &\mathcal{M}(\mathcal{S}x, \mathcal{B}y, \mathcal{B}z, t), \mathcal{M}(\mathcal{B}y, \mathcal{B}y, \mathcal{T}z, t), \\ &\mathcal{M}(\mathcal{A}x, \mathcal{T}y, \mathcal{T}y, t)\}. \end{aligned} \tag{3.7.1}$$

Then the pairs $(\mathcal{A}, \mathcal{S})$ and $(\mathcal{B}, \mathcal{T})$ have a coincidence point each. Moreover, $\mathcal{A}$, $\mathcal{B}$, S and $\mathcal{T}$ have a unique common fixed point provided both pairs $(\mathcal{A}, \mathcal{S})$ and $(\mathcal{B}, \mathcal{T})$ are weakly compatible.

Proof If the pairs $(\mathcal{A}, \mathcal{S})$ and $(\mathcal{B}, \mathcal{T})$ share the $\mathrm{CLR}_{\mathcal{ST}}$ property, then there exist two sequences $\{x_n\}$ and $\{y_n\}$ in X such that $\lim_{n\to\infty} \mathcal{A}\mathrm{x_n} = \lim_{n\to\infty} \mathcal{S}\mathrm{x_n} = \lim_{n\to\infty} \mathcal{B}\mathrm{y_n} = \lim_{n\to\infty} \mathcal{T}\mathrm{y_n} = \mathrm{p}$, for some $\mathrm{p} \in \mathcal{S}(\mathrm{X}) \cap \mathcal{T}(\mathrm{X})$. Since $\mathrm{p} \in \mathcal{S}(\mathrm{X})$, there exists a point $\mathrm{u} \in \mathrm{X}$ such that $\mathcal{S}_\mathrm{u} = \mathrm{p}$. Now we have to show that $\mathcal{A}\mathrm{u} = \mathrm{p}$. On using inequality (3.7.1), we have

$$\mathcal{M}(\mathcal{A}\text{u}, \mathcal{B}y_n, \mathcal{B}y_n, kt) \geq \min\{\mathcal{M}(\mathcal{S}\text{u}, \mathcal{T}y_n, \mathcal{T}y_n, t), \mathcal{M}(\mathcal{A}\text{u}, \mathcal{T}y_n, \mathcal{T}y_n, t),$$
$$\mathcal{M}(\mathcal{S}\text{u}, \mathcal{B}y_n, \mathcal{B}y_n, t),$$
$$\mathcal{M}(\mathcal{B}y_n, \mathcal{B}y_n, \mathcal{T}y_n, t), \mathcal{M}(\mathcal{A}\text{u}, \mathcal{T}y_n, \mathcal{T}y_n, t)\}. \tag{3.7.2}$$

Letting $n \to \infty$ and using Lemma 2.4.

$$\begin{aligned}\mathcal{M}(\mathcal{A}\text{u}, \text{p}, \text{p}, kt) \geq \min\{&\mathcal{M}(\text{p}, \text{p}, \text{p}, t), \mathcal{M}(\mathcal{A}\text{u}, \text{p}, \text{p}, t), \mathcal{M}(\text{p}, \text{p}, \text{p}, t), \mathcal{M}(\text{p}, \text{p}, \text{p}, t),\\ &\mathcal{M}(\mathcal{A}\text{u}, \text{p}, \text{p}, t)\}\\ &= \min\{1, \mathcal{M}(\mathcal{A}\text{u}, \text{p}, \text{p}, t), 1, 1, \mathcal{M}(\mathcal{A}\text{u}, \text{p}, \text{p}, t)\}\\ &= \mathcal{M}(\mathcal{A}\text{u}, \text{p}, \text{p}, \text{t}).\end{aligned}$$

Appealing to Lemma 2.4, we obtain $\mathcal{A}\text{u} = \text{p}$ and so $\mathcal{A}_\text{u} = \text{p} = \mathcal{S}\text{u}$ which shows that u is a coincidence point of the pair $(\mathcal{A}, \mathcal{S})$.

Also $\text{p} \in \mathcal{T}(\text{X})$, there exists a point $\text{v} \in \text{X}$ such that $\mathcal{T}\text{v} = \text{p}$. Now, we have to assert that $\mathcal{B}\text{v} = \text{p}$. On using inequality (3.7.1), we get

$$\begin{aligned}\mathcal{M}(\mathcal{A}\text{u}, \mathcal{B}\text{v}, \mathcal{B}\text{v}, kt) \geq \min\{&\mathcal{M}(\mathcal{S}\text{u}, \mathcal{T}\text{v}, \mathcal{T}\text{v}, t), \mathcal{M}(\mathcal{A}\text{u}, \mathcal{T}\text{v}, \mathcal{T}\text{v}, t), \mathcal{M}(\mathcal{S}\text{u}, \mathcal{B}\text{v}, \mathcal{B}\text{v}, t),\\ &\mathcal{M}(\mathcal{B}\text{v}, \mathcal{B}\text{v}, \mathcal{T}\text{v}, \text{t}), \mathcal{M}(\mathcal{A}\text{u}, \mathcal{T}\text{v}, \mathcal{T}\text{v}, \text{t})\}\end{aligned}$$

$$\begin{aligned}\mathcal{M}(\text{p}, \mathcal{B}\text{v}, \mathcal{B}\text{v}, kt) \geq \min\{&\mathcal{M}(\text{p}, \text{p}, \text{p}, t), \mathcal{M}(\text{p}, \text{p}, \text{p}, t), \mathcal{M}(\text{p}, \mathcal{B}\text{v}, \mathcal{B}\text{v}, t),\\ &\mathcal{M}(\mathcal{B}\text{v}, \mathcal{B}\text{v}, \text{p}, t), \mathcal{M}(\text{p}, \text{p}, \text{p}, t)\}\\ &= \min\{1, 1, \mathcal{M}(\text{p}, \mathcal{B}\text{v}, \mathcal{B}\text{v}, t), \mathcal{M}(\text{p}, \mathcal{B}\text{v}, \mathcal{B}\text{v}, t), 1\}\\ &= \mathcal{M}(\text{p}, \mathcal{B}\text{v}, \mathcal{B}\text{v}, t).\end{aligned}$$

In view of Lemma 2.4, we have $\mathcal{B}\text{v} = \text{p}$, that is, $\mathcal{B}\text{v} = \text{p} = \mathcal{T}\text{v}$ which shows that v is a coincidence point of the pair $(\mathcal{B}, \mathcal{T})$.

As the pair $(\mathcal{A}, \mathcal{S})$ is weakly compatible and $\mathcal{A}\text{u} = \mathcal{S}\text{u}$, therefore $\mathcal{A}\text{p} = \mathcal{A}\mathcal{S}\text{u} = \mathcal{S}\mathcal{A}\text{u} = \mathcal{S}\text{p}$. Now, we show that p is a common fixed point of the pair $(\mathcal{A}, \mathcal{S})$. To prove this, using inequality (3.7.1), we have

$$\begin{aligned}\mathcal{M}(\mathcal{A}\text{p}, \mathcal{B}\text{v}, \mathcal{B}\text{v}, kt) \geq \min\{&\mathcal{M}(\mathcal{S}\text{p}, \mathcal{T}\text{v}, \mathcal{T}\text{v}, t), \mathcal{M}(\mathcal{A}\text{p}, \mathcal{T}\text{v}, \mathcal{T}\text{v}, t), \mathcal{M}(\mathcal{S}\text{p}, \mathcal{B}\text{v}, \mathcal{B}\text{v}, t),\\ &\mathcal{M}(\mathcal{B}\text{v}, \mathcal{B}\text{v}, \mathcal{T}\text{v}, t), \mathcal{M}(\mathcal{A}\text{p}, \mathcal{T}\text{v}, \mathcal{T}\text{v}, t)\}.\end{aligned}$$

$$\begin{aligned}\mathcal{M}(\mathcal{A}\text{p}, \text{p}, \text{p}, kt) \geq \min\{&\mathcal{M}(\mathcal{A}\text{p}, \text{p}, \text{p}, t), \mathcal{M}(\mathcal{A}\text{p}, \text{p}, \text{p}, t), \mathcal{M}(\mathcal{A}\text{p}, \text{p}, \text{p}, t),\\ &\mathcal{M}(\text{p}, \text{p}, \text{p}, t), \mathcal{M}(\mathcal{A}\text{p}, \text{p}, \text{p}, t)\}\\ &= \min\{\mathcal{M}(\mathcal{A}\text{p}, \text{p}, \text{p}, t), \mathcal{M}(\mathcal{A}\text{p}, \text{p}, \text{p}, t), \mathcal{M}(\mathcal{A}\text{p}, \text{p}, \text{p}, t), 1, \mathcal{M}(\mathcal{A}\text{p}, \text{p}, \text{p}, t)\}\\ &= \mathcal{M}(\mathcal{A}\text{p}, \text{p}, \text{p}, t).\end{aligned}$$

Owing to Lemma 2.4, we get $\mathcal{A}\text{p} = \text{p} = \mathcal{S}\text{p}$. Therefore, p is a common fixed point of the pair $(\mathcal{A}, \mathcal{S})$. Since the pair $(\mathcal{B}, \mathcal{T})$ is weakly compatible and $\mathcal{B}\text{v} = \mathcal{T}\text{v}$, hence $\mathcal{B}\text{p} = \mathcal{B}\mathcal{T}\text{v} = \mathcal{T}\mathcal{B}\text{v} = \mathcal{T}\text{p}$. On using inequality (3.7.1), we get

$$\mathcal{M}(\mathcal{A}\text{u}, \mathcal{B}\text{p}, \mathcal{B}\text{p}, kt) \geq \min\{\mathcal{M}(\mathcal{S}\text{u}, \mathcal{T}\text{p}, \mathcal{T}\text{p}, t), \mathcal{M}(\mathcal{A}\text{u}, \mathcal{T}\text{p}, \mathcal{T}\text{p}, t), \mathcal{M}(\mathcal{S}\text{u}, \mathcal{B}\text{p}, \mathcal{B}\text{p}, t), \mathcal{M}(\mathcal{B}\text{p}, \mathcal{B}\text{p}, \mathcal{T}\text{p}, t), \mathcal{M}(\mathcal{A}\text{u}, \mathcal{T}\text{p}, \mathcal{T}\text{p}, t)\}$$

$$\mathcal{M}(\text{p}, \mathcal{B}\text{p}, \mathcal{B}\text{p}, kt) \geq \min\{\mathcal{M}(\text{p}, \mathcal{B}\text{p}, \mathcal{B}\text{p}, t), \mathcal{M}(\text{p}, \mathcal{B}\text{p}, \mathcal{B}\text{p}, t), \mathcal{M}(\text{p}, \mathcal{B}\text{p}, \mathcal{B}\text{p}, t), \mathcal{M}(\mathcal{B}\text{p}, \mathcal{B}\text{p}, \mathcal{B}\text{p}, t), \mathcal{M}(\text{p}, \mathcal{B}\text{p}, \mathcal{B}\text{p}, t)\}$$

$$\mathcal{M}(\text{p}, \mathcal{B}\text{p}, \mathcal{B}\text{p}, kt) \geq \mathcal{M}(\text{p}, \mathcal{B}\text{p}, \mathcal{B}\text{p}, t).$$

By Lemma 2.4, we obtain $\mathcal{B}\text{p} = \text{p} = \mathcal{T}\text{p}$ which shows that p is a common fixed point of the pair $(\mathcal{B}, \mathcal{T})$. Uniqueness of common fixed point are an easy consequence of the inequality (3.7.1) in view of Lemma 2.4.

References

1. Zadeh, L.A.: Fuzzy sets. Inform. Control **8**, 338–353 (1965)
2. Kramosil, I., Michalek, J.: Fuzzy metric and statistical spaces. Kybernetica **11**, 336–344 (1975)
3. George, A., Veeramani, P.: On some results in fuzzy metric spaces. Fuzzy Sets Syst. **64**, 395–399 (1994)
4. Jungck, G.: Compatible mappings and common fixed points. Internat. Math. J. Maths. Sci. **9**, 771–779 (1986)
5. Aamri, M., El Moutawakil, D.: Some new common fixed point theorems under strict contractive conditions. J. Math. Anal. Appl. **270**(1), 181–188 (2002)
6. Ali, J., Imdad, M.: An implicit function implies several contractions. Sarajevo J. Math. **4**(4), 269–285 (2008)
7. Sintunavarat, W., Kumam, P.: Common fixed point theorems for a pair of weakly compatible mappings in fuzzy metric spaces. J. Appl. Math. **2011**, Article ID 637958
8. Sedghi, S., Shobe, N.: Fixed point theorem in $\mathcal{M}$-fuzzy metric spaces with property (E). Adv. Fuzzy Math. **1**(1), 55–65 (2006)
9. Chauhan, S: Fixed points of weakly compatible mappings in fuzzy metric spaces satisfying common limit in the range property. Ind. J. Math. **2011**, Article ID 637958
10. Ali, I.M.M, Hasan, M.: Common fixed point theorems in fuzzy metric spaces employing common property (E.A.). Math. Comput. Model. **55**(3–4), 770–778 (2012)

Performance Prediction of Technical Artifacts in Heart Disease Diagnosis

Alka Gulati and Kanak Saxena

1 Introduction

The software artifacts analysis includes techniques to develop the quality of the software being produced or acquire despite its size and domain. An analysis of artifacts includes a particular goal, input, output, user resources, and a group of activities to be performed in a specific order. Improving the precision of heart diseases detection has been investigated by many researchers in the literature [1]. Many researchers have investigated the improvement in precision of heart diseases detection in literature.

The physical examination, symptoms, and signs of the patient help in diagnosing heart disease. In heart disease, the issue related to smoking, cholesterol, obesity, history of heart disease in the family, BP, and no physical exercise are treated to be high-risk factors [2]. In the exigent circumstances of health issues in today's era, an automated prediction system predicting a patient's heart condition can be offered to provide accordingly warnings and treatment on time and effective [3, 4].

On the basis of artifacts of heart, specialists can take clinical decisions intelligently. The evaluation of artifacts can help in diagnosing patients properly and administering efficient treatments. To make any system effective, the functionality should be well defined. The importance of identifying artifacts arose at the first phase of system development only [5]. In a healthcare system it becomes necessary to create models for classifying functional and nonfunctional artifacts. In the 3-tier artifact model, functional artifacts are classified as technical artifacts, process artifacts, and device artifacts. Technical artifacts with the functions are objects. Functions are interrelated. They exist by human intervention. They include technological scalability, feasibility, and compatibility agent.

A. Gulati (✉)
Lakshmi Narain College of Technology, Bhopal, Madhya Pradesh, India

K. Saxena
Department of CS and IT, Samrat Ashok Technological Institute, Vidisha, Madhya Pradesh, India

P. Srivastava et al. (eds.), *Mathematical, Computational Intelligence and Engineering Approaches for Tourism, Agriculture and Healthcare*, Lecture Notes in Networks and Systems 214, https://doi.org/10.1007/978-981-16-3807-7_28

Heart disease has become a very common disease in the age group from 30 to 65. The youth is being affected a lot by this disease. Therefore, it has become mandatory to be conscious about health. The analysis of the artifacts like Body Mass Index of the body, Lipid profile, Diabetics, Chest pain while physical exercise, family history of heart disease can give a prediction of heart disease and can reduce the mortality rate [6].

In the following sections, Sect. 2 gives details of the data sets, Sect. 3 gives literature survey, Sect. 4 illustrates the proposed technique in heart disease prediction, Sect. 5 provides the case study of the medical domain in support of the proposed technique, Sect. 6 provides the performance metrics, and Sect. 7 provides the results.

2 Data Sets

Datasets of heart diseases are obtained from UCI datasets (University of California) [7]. For the proposed work UCI datasets are used and data is collected of 298 patients. The artifacts available in the data sets are 52 in number. They contain the symptoms and accordingly their value. In the research, 52 artifacts are reduced to 12 artifacts. The artifacts are all categorized as technical. The technical artifacts used are Age, Chest pain, TrestBP, Cholesterol, Fasting blood sugar, Restecg, Pain on exercise (Exang), Thalach, Oldpeak, Slope, Ca, Thal. [8, 9].

3 Literature Review

In the healthcare domain, the artifacts are the software-oriented scripts that can execute on the medical inputs as treated/untreated data set that can be further ready for the analysis purpose [10]. The results can be refined by the further independent or interdependent replicated analysis through the experiments. The technical artifacts aspire is to promote the perspective experiments rather than design the experiments as a single entity. This is advantageous as the potential of the experiments were clarified and the results can be analyzed and can able to predict the solutions on the basis of the unpredictable associations of the sign and symptoms.

Stefano et al. [11] discussed the technical artifacts in the view of the human intervention and similarity with the natural entities. The authors presented in three different categories of ontological, engineering, and technical artifacts. They focus on the conceptual analysis of the technical artifacts through the plan analysis. The physical object plan of the event introduces the uncertainty aspect whenever the new technical object is introduced in the new category of the technical artifacts [11].

Xiao [12] illustrates the implication of the artifacts in health care in three different areas. They are (i) Technological (ii) Methodological, and (iii) Theoretical. In the technological area, the physical artifacts must be replaced or modified by the induction/deployment of the new technologies, whereas for the methodological the highest priority must be with respect to the user requirement with the impact on the artifacts to have ease in collaborative work. At last, the theoretical aspect of the new and innovative distributed cognition modeling provides the commencing position [12].

Ineke et al. [13] focused on the literacy of the technical artifacts due to their routine usage in the current scenario. They introduced the technical artifacts in the educational domain with the empirical study performed on the mapping of the teachers' perspective and the nature of the artifacts [13].

Allen et al. [14] focus on the easy affordability and accessibility of health care in the future must be provided well in time, place, and more importantly with diagnostic accuracy and medication. The system must be so designed that can assist the experts as well as the patients even in adverse ecological conditions [14].

Allahverdi et al. [15] designed a fuzzy-based expert system to determine the risk of coronary heart disease in patients. The system gave the risk ratio and recommendation for normal life, diet, and medical treatment [15].

Anooj [3], developed a decision support system for the prediction of risk in heart patients using weighted fuzzy rules. The system has two phases. In the first phase, weighted fuzzy rules are generated automatically. In the second phase, decision support system based on fuzzy rules is developed [3].

In the future, the healthcare domain has the emphasis on the detection of dementia well on time, to maintain the patient records for which the technique is used of the block chain. For the early detection of cancer, the automatic visual evaluation—a machine learning algorithms helps in the precise diagnosis. In the present living conditions, loneliness is the major cause of the illness and it is of high risk that it can be the cause of death due to cardiovascular disease. Due to the difficulty in the respiratory system, the future is ready to prepare the devices which can help in breathing. The device "SoeMac" is designed to help in restoring oxygen especially when the person is in sleep mode.

4 Proposed Method

The proposed method uses computational techniques for technically diagnosing the severity of heart diseases. It is focused on technical artifacts. These artifacts will be used for evaluation which will help in developing accurate software which will improve predictions. Figure 1 describes the proposed method.

The following steps are used in pursuing the proposed method:

1. Apply Technical Factors on 52 artifacts.
   (i) Apply weights (w) to the artifacts.
   (ii) Sum $= \sum_{Pi} w \times r$, where r is the rating of factors.
   (iii) $|T - Sum| \geq \varepsilon$.
   Where T is the total of artifacts and ε is the threshold value.
   (This process helps in the reduction of number of artifacts.)

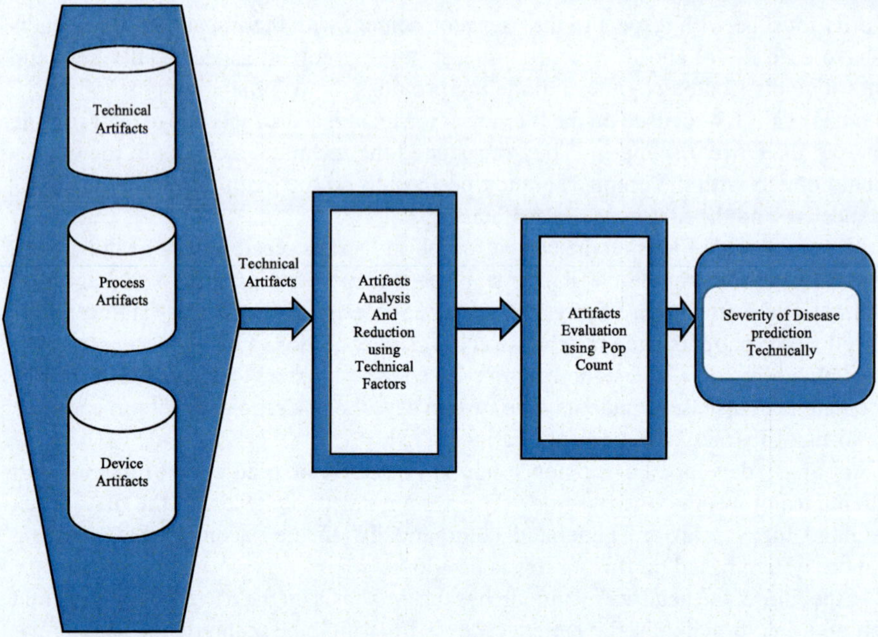

Fig. 1 Proposed technique

2. For the presence of symptoms use binary value 1 else 0.
   The aim is to have a focus on the more severe patients as S will result in the same pop count to more patients, but the occurrence probability of the symptoms may affect in presence of the others.
3. On the basis of the disease's signs and symptoms, the binary weights will generate.
4. Assigning the relative binary weights.
   Generate , for n independent symptoms.$S = \{s_i\}_{i=1}^{n}$
   Estimate the binary weights w(p) using the leading one's and compute the weighted pop count of patient p as

$$\hat{l}(p) = \log_{\epsilon}\left|S_y^*(p_i)\right| \cong int\left(\hat{l}(p)\right)$$

   The weight is estimated as

$$w = E\{i \in int\left(\hat{l}\right)\}|i = l$$

5. By comparing codes of the patients and practitioners get the estimated pop count.

Table 1 Factor values

Technical factor values	Representation
0	No effect
1	Very less effect
2	Moderate effect
3	Intermediary effect
4	Significant effect
5	Very Significant Effect

To consign 0 and 1 to the artifacts of various patients, we have set individual artifact values for different artifacts. If in given data sets value is more than the artifact value then assign1 otherwise assign 0.

Use cases are used for collecting user's requirements [16]. Thirteen standard technical factors as defined by Karner exist to approximate the impact on efficiency on technical issues that affect a project [17]. For the reduction of artifacts, the technical factors of the use case point approach were used. In this approach, the rating is given to the artifacts, the weights are assigned to artifacts and at last, the total sum is found.

Standard 13 technical factors are used to evaluate the artifacts. The technical factors are performance throughput, efficiency of end-user, internal processing complexity, reusability code, installation ease, usability ease, portability, changeability ease, concurrency, security factors, accessibility directly of third parties, facilities for special user training, distributed system. Each factor is weighted according to its relative impact. For every project, the evaluation of technical factors is done by the development team. An apparent complexity value as a rating between 0 and 5 is assigned to technical artifacts. The factors' rating values and their representation are shown in Table 1. The multiplication of apparent complexity factor rating is done with the factor weight to get the calculated factor for technical artifacts.

5 Case Study of Medical Domain

Identifying artifacts in the medical domain is a challenging task. Artifacts of medical domain projects should be well-intended and executed to create accurate, secured, flexible and integrated solutions. The medical domain is preferred for the support of the technique and is applied on heart disease datasets. The functional artifacts are classified as process artifacts, device artifacts, and technical artifacts [18]. Process artifacts are measures and metrics needed from various dimensions of the target domain. Device artifacts are real-world entities, roles, structures, behavior, and functions Technical artifacts are objects generated by humans deliberately in order to accomplish a practical function.

The technical factors are applied on 52 technical artifacts to estimate the impact on artifacts for finding their importance in finding the severity of patients. After applying technical factors, we realized that if we apply technical factors to 12 artifacts only

Table 2 Technical artifacts of the medical domain

S. no.	Artifacts	Ratings	Weights	Ratings * weight
1	AGE	5	1.5	7.5
2	CP	5	2	10
3	TRESTBPS	5	1.5	7.5
4	CHOL	4	1.5	6.0
5	FBS	5	2	10
6	RESTECG	4	2	8
7	THALACH	5	2	10
8	EXANG	5	2	10
9	OLDPEAK	4	2	8
10	SLOPE	4	2	8
11	CA	5	2	10
12	THAL	4	2	8

then the sum of factors is the same as 52 artifacts. Thereby 52 artifacts have been reduced to 12 artifacts. Therefore technically we can use 12 artifacts instead of 52. To find the severity technically, artifacts are prioritized by considering their weights and on the basis of information contained by them. The reduced technical artifacts are shown in Table 2.

In this method, the artifacts are prioritized by taking into account the information they contain. This will reduce the complexity and size of software being developed.

Details of 12 Input Artifacts used in the Proposed Method are as given below.

1. Age: ranges from 30 to 74 years.
2. CP is the representation of Chest Pain with 4 variations
   a. 1: typical angina,
   b. 2: typical angina,
   c. 3: non-angina pain, and
   d. 4: asymptomatic pain.
3. Trestbps is the representation of BP in the resting situation with a range from 95 to200.
4. Chol is the representation of Cholesterol with a range from 150 to 400 mg/dl.
5. BPS is the representation of fasting blood sugar with a range from 110 to300.
6. ECG is the representation of electrocardiographic in the resting situation with 3 variations
   a. 0: normal,
   b. 1: ST-T wave abnormal, and
   c. 2: probable/definite left ventricular hypertrophy.
7. Thal Ach is the representation of maximum heart rate with a range from 72 to 203.

8. Exang is the representation of exercise induce angina with two variations.
    a. 0: No and
    b. 1: Yes
9. Oldpeak is the representation of ST wave depression induced by exercise in a resting situation unremitting.
10. Slope is the representation of the slope of exercise at hit the highest point and ST-segment
    a. 1: un sloping,
    b. 2: flat, and
    c. 3: downsloping.
11. CA is the representation of the number of major vessels blocked.
12. Thal with values as.
    a. 3: normal,
    b. 6: fixed defect, and
    c. 7: reversible defect.

The algorithm uses the pop count method to evaluate the artifacts to find out severity technically in patients of heart diseases on the practitioner's perception. The pop count method is used as follows:

Considering the patients' symptoms as artifacts data set $P = \{P_1, P_2, P_3, \ldots, P_n\}$, where each P_i represents a set of symptoms, as $P_i = \{S_1, S_2, S_3, \ldots, S_m\}$, and $P_i = S_{ij}$ represents the jth symptoms of ith patients. According to the symptoms and the practitioner's value for various artifacts, 0 and 1 are assigned to artifacts. If the value is in the range then 0 is assigned else 1 is assigned as shown in Table 3.

To find out the patients' severity technically, the pop count method is used by counting leading ones. By applying function countbit (n): For c in range (n): n& = n – 1, the count will be obtained. After using the function on set P, the person with maximum count showed the severity technically of Heart disease.

After the application of pop count method, severity technically was found. Thus the artifacts are evaluated using pop count method in the operational tier of the artifact model [19]. The predictions were made on the maximum value of pop counts of 1. The patient with the highest number of 1's has the highest severity. According to specialists, the major artifacts used in predicting heart diseases are BP, Chest Pain, chol, Pain_exer. The values are arranged according to the priority of artifacts, one's at leftmost position in bit representation of patients show higher risk than ones at rightmost position. Table 3 shows a sample of the binary weight codes of patients.

The patient with maximum value is highly prone to heart disease. Here P9 has the highest severity of heart disease. In the proposed system, patients severity level can be predicted technically on the basis of the presence of signs and symptoms of patients.

Table 3 Binary codes of patients

Patient	Trestbp	Chol	Cp	Fbs	Restecg	Thalach	Exang	Age	Oldpeak	Slope	Ca	Thal	Sum
P5	0	1	0	0	0	0	0	1	0	1	0	0	3
P6	0	0	0	0	1	0	0	1	1	1	1	0	5
P7	0	0	0	0	0	0	1	1	0	1	0	0	3
P8	0	0	0	0	1	0	0	1	0	1	1	1	5
P9	0	0	0	1	1	0	1	1	1	1	1	1	8
P20	0	1	0	0	1	0	0	1	0	1	0	0	4
P21	0	1	0	0	1	0	0	1	1	1	0	0	5
P22	0	0	0	0	1	0	0	1	1	1	1	1	6
P23	0	0	0	0	1	0	1	1	1	1	1	1	7
P24	0	0	0	0	0	0	0	1	1	1	0	0	3
P25	0	0	0	0	0	0	0	1	0	1	0	0	2
P26	0	1	0	0	0	0	0	1	1	1	0	0	4
P27	0	0	0	1	1	0	1	1	1	1	0	1	7
P28	0	1	0	0	0	0	0	1	1	1	0	0	4
P29	0	0	0	0	0	0	1	1	0	1	1	1	5
P30	0	0	0	0	0	0	0	1	0	1	1	0	3
P31	0	0	0	0	0	0	0	1	0	1	0	0	2
P32	0	0	0	0	0	0	1	1	0	1	1	0	4
P33	0	0	0	0	0	0	0	1	0	1	1	0	3
P34	0	0	0	0	1	0	1	1	1	1	0	1	6
P35	0	0	0	1	1	0	0	1	0	1	0	1	5

(continued)

Table 3 (continued)

Patient	Trestbp	Chol	Cp	Fbs	Restecg	Thalach	Exang	Age	Oldpeak	Slope	Ca	Thal	Sum
P43	0	0	0	0	0	0	0	1	0	1	1	0	3
P44	0	0	0	0	1	0	0	1	1	1	0	0	4
P45	0	0	0	0	1	0	0	1	0	1	1	0	4
P49	0	0	0	0	1	0	0	1	0	1	1	0	4
P50	0	1	0	0	1	0	0	1	0	1	1	0	5

6 Performance Metrics

In this paper confusion output matrix is used for predicting the performance of pop count method. The confusion matrix uses four variables [20]. They are TP (True Positive), TN (True Negative), FP (False Positive), and FN (False Negative). The performance metrics are:

Accuracy: It is the primary metric used to find the performance. It is computed as the percent of the cases classified correctly, and can be computed as

$$\text{Accuracy} = \frac{TP + TN}{TP + TN + FP + FN}$$

Sensitivity (True Positive Rate): It describes the probability to predict the result as positive when the disease is present. It can be computed as

$$\text{Sensitivity (TPR)} = \frac{TP}{TP + FN}$$

Positive predicted value (PPV): The probability that the disease is present when the test is positive. It can be computed as

$$\text{PPV} = \frac{TP}{TP + FP}$$

7 Result

The pop count method was applied on the Cleveland heart disease datasets having 298 patient details. The results are taken after optimizing artifacts. Out of 298 samples, 137 belonged to the presence of heart disease and 161 belonged to the absence of heart disease. The confusion matrix, as a result, is shown in Table 4. Predictions done by the method for TPs were 120 out of 137. True Negatives given by method are 145 out of 161. The percent values of the proposed method are shown in Table 5. The values are for accuracy, sensitivity (TPR), and precision (PPV).

Table 4 Confusion Matrix

Actual	Predicted		
	Present	Absent	Actual total
Present	120 (TP)	17 (FN)	137
Absent	16 (FP)	145 (TN)	161
Predicted total	136	162	298

Table 5 Performance Accuracy of the proposed method

Performance metrics	Values (%)
Accuracy	88.92
Sensitivity (TPR)	87.59
Precision (PPV)	88.23

8 Conclusions

The paper deals with the selection of software artifacts for heart disease prediction with the impact of software artifacts based on 298 patients. The proposed method performs analysis using reduction of artifacts by applying technical factors and evaluation of artifacts using pop count method. The artifacts are reduced on the basis of the priority of artifacts in predicting heart disease. Finally, the technical level of severity on the basis of pop counts of heart diseases is predicted. Confusion matrix is used to find the accuracy of the proposed method. This will help the specialist to diagnose technical severity in patients with the information of technical artifacts and in artifact evaluation while developing software with the interdependencies of various artifacts. The proposed method focuses on the development of a severity prediction model using minimal artifacts which is reliable and accurate for predicting cardiovascular disease in a patient. The major idea of the proposed method is to diagnose patients on their present condition and avoid any misdiagnosis of patients. At rural medical centers, clinicians can take patient's blood pressure, FBS, and Cholesterol information and their currents conditions and can help in prediction and if the severity is there they can immediately refer to secondary centers, and then further investigations like restecg can be done. Hence, early treatment can prevent the deaths caused by late heart disease diagnosis.

References

1. AHA: American Heart Association (2017). http://www.heart.org
2. Straszecka, E.: Combining uncertainty and imprecision in models of medical diagnosis. Inf. Sci. **176**(20), 3026–3059 (2006)
3. Anooj, P.K.: Clinical decision support system: risk level prediction of heart disease using weighted fuzzy rules. J. King Saud Univ. Comput. Inf. Sci. **24**(1), 27–40 (2012)
4. Ali, S., Chia, P., Ong, K.: Graphical knowledge-based protocols for chest pain management. In: Proceedings of the Conference on Computers in Cardiology, Hannover, Germany, 1999, pp. 309–312
5. Banodha, U., Saxena, K.: Dynamic software architecture for medical domain using pop counts. Int. J. Adv. Comput. Sci. Appl. (IJACSA) **5**(2), 100–104 (2014)
6. France, R., Rumpe, B.: Model-driven development of complex software: a research roadmap. In: Future of Software Engineering, ICSE, Minneapolis, IEEE, May 2007, pp. 37–54
7. Heart Disease. http://archive.ics.uci.edu/ml/datasets/HeartDisease
8. Dave, A., Pandi, G.: Analysis of heart disease prediction system using artificial neural network. J. Emerg. Technol. Innov. Res. (JETIR) **5**(11), 336–342 (November 2018)

9. Pete, I., Balasubramaniam, D.: Handling the differential evolution of software artefacts: a framework for consistency management. In: IEEE 22nd International Conference on Software Analysis, Evolution, and Reengineering (SANER) (2015)
10. Tomasi, A., Meneghetti, A.: Binary Linear Code Weighted Distribution Estimation by Random Bit Stream Compression (2018). arXiv:1806.02099
11. Stefano, B., Maarten, F., Paweł, G., Yoshinobu, K., Riichiro, M., Pieter, E.V.: Technical artifact: an integrated perspective, Formal Ontologies Meet Industry, IOS Press, pp. 1–15 (2011)
12. Xiao, Y.: Artifacts and collaborative work in healthcare: methodological, theoretical, and technological implications of the tangible. J. Biomed. Inform. **38**(1), 26–33 (February 2005)
13. Ineke, F., Wim, S., Marc, J.V.: Teaching and learning the nature of technical artifacts. Int. J. Technol. Des. Educ. **21**, 277–290 (2011)
14. Allen, S. et al.: Global Health Care Outlook Shaping the Future, Deloitte, pp. 1–41 (2019)
15. Allahverdi, N., Torun, S., Saritas, I.: Design of a fuzzy expert system for determination of coronary heart disease risk. In: International Conference on Computer systems and technologies (2007)
16. Xiao, Y.: Artifacts and collaborative work in healthcare: methodological, theoretical, and technological implications of the tangible. J. Biomed. Inform. **38**(1), 26–33 (2005)
17. Karner, G.: Resource estimation for objectory projects. Objective systems (1993)
18. Gulati, A., Saxena, K.: A study of perspective artifact model of health care domain. In: International Conference on Social Networking and Computational Intelligence (SCI-2K19), RGPV Bhopal, 08–09 March 2019
19. Gulati, A., Saxena, K.: Assessment of healthcare with use case point. Int. J. Adv. Res. Comput. Sci. March-April **8**(2), 217–221 (2018)
20. Raza, K.: Improving the prediction accuracy of heart disease with ensemble learning and majority voting rule, U- Healthcare Monitoring Systems, January 2019, pp. 179–196

Nonstandard Analysis of Almost Periodicity

S. Alagu and R. Kala

1 Introduction

This short communication encompasses three branches of Mathematics, namely, Harmonic Analysis, Operator Theory and Nonstandard Analysis. For the preliminaries on the topics covered, the Reference numbers (in the Reference list at the end) are enlisted as below against the three subjects:

1. Harmonic Analysis—[5–7, 9, 10],
2. Operator Theory—[8, 9],
3. Nonstandard Analysis—[1–4].

In the first part, nonstandard analytic generalization is carried out for Almost periodic functions and in the second part, it is carried out for Almost periodic operators.

Now we get into the content.

2 Almost Periodic Functions

Let G be a locally compact abelian group and $C(G)$ be the Banach space of bounded, complex-valued functions on G. For $x \in G$, let R_x denote the translation operator on $C(G)$ defined by $(R_x\varphi)(y) = \varphi(y - x)$ for all $y \in G, \ \varphi \in C(G)$. Throughout, we assume that we work in an enlargement $^*G$ of G [1, 3].

$\mathbb{N}, \mathbb{R}, \mathbb{C}$ will denote the sets of natural numbers, real numbers, complex numbers, respectively. For any relevant set A, *A will denote the $*$ - transform of A. For an appropriate function φ, $^*\varphi$ will denote the $*$ - transform of φ and so on.

S. Alagu (✉) · R. Kala
Department of Mathematics, Manonmaniam Sundaranar University, Tirunelveli, Tamilnadu, India

P. Srivastava et al. (eds.), *Mathematical, Computational Intelligence and Engineering Approaches for Tourism, Agriculture and Healthcare*, Lecture Notes in Networks and Systems 214, https://doi.org/10.1007/978-981-16-3807-7_29

Definition 1.1 [5] $\varphi \in C(G)$ is said to be an Almost periodic function if $G_\varphi = \{R_x\varphi : x \in G\}$ is relatively compact in $C(G)$. G_φ is the orbit of f under the group action $x \to R_x f$.

It is known that the set of almost periodic functions on G is a Banach subalgebra of $C(G)$ and is equal to $C(G)$ if and only if G is compact.

An equivalent characterization of Almost periodicity [5] is as follows.

Proposition 1.1 $\varphi \in C(G)$ is Almost periodic if and only if φ can be approximated uniformly by trigonometric polynomials of the form $P(x) = \sum_{i=1}^{n} a_i\gamma_i(x),\ x \in G,\ a_i \in \mathbb{C}$ and $\gamma_1, \gamma_2, \ldots, \gamma_n$ are elements of the dual group Γ of G. —— **(*).**

By transfer principle, we have the following nonstandard criterion:

Proposition 1.2 Let $AP(G)$ denote the Banach algebra of Almost periodic functions on G. Then, $\varphi \in {}^*AP(G)$ if and only if φ can be approximated uniformly by polynomials of the form $P(x) = \sum_{i=1}^{n} a_i\gamma_i(x),\ x \in {}^*G, n \in {}^*\mathbb{N}, a_i \in {}^*\mathbb{C}, \gamma_i \in {}^*\Gamma$ for I = 1,2,...n. (**).

In view of the importance of G_φ, we record a known property of G_φ:

For an Almost periodic, every $g \in \overline{G_\varphi}$ is Almost periodic and hence can be approximated by trigonometric polynomials of the form (*), where the a_i's can be obtained from integrals of translates of φ [5].

Consequently, we have the following nonstandard criterion, by transfer principle:

Proposition 1.3 For an Almost periodic φ, every $\theta \in {}^*(\overline{G_\varphi})$ can be approximated uniformly by trigonometric polynomials of the form (**), where the a_i's can be obtained from integrals of translates of $^*\varphi$.

The notation ^{S}A, the standard part of any set A, could be found in [2].

In the internal set-theoretic language, we have the following.

Proposition 1.4 For a standard Almost periodic φ, every $g \in^s (\overline{G_\varphi})$ can be approximated by polynomials of the form (*), where in this case, the elements n, a_i, γ_i could be nonstandard for nonstandard g. For standard g, the a_i's can be obtained from integrals of translates of φ.

A similar analysis is carried out for operators.

3 Almost Periodic Operators

Definition 2.1 [9] By a homogenous Banach space B on a locally compact group G, we mean a Banach space of complex-valued functions on G satisfying the following:

(i) B is invariant under translations. That is, $f \in B$ implies $R_x f \in B$ for all $x \in G$.

(ii) The norm in B is translation invariant. That is, $\|R_x f\| = \|f\|$ for all $x \in G$, $f \in B$.

(iii) Functions in B translate continuously. That is, $\lim_{x \to 0} \|R_x f - f\| = 0$ for every $f \in B$.

One may take $C_0(G)$ or $L^p(G),\ 1 \le p < \infty$ for B.

Definition 2.2 [9] Let $\mathcal{L}(B)$ denote the Banach algebra of bounded linear operators on B. $T \in \mathcal{L}(B)$ is said to be

(i) invariant if $TR_x = R_x T$ for every $x \in G$.

The set of invariant operators on B is denoted by $\mathcal{L}_0$.

(ii) Almost periodic if $G_T = \{R_x T R_{-x} : x \in G\}$ is relatively compact in $G_T = \{R_x T R_{-x} : x \in G\}$.

The set of Almost periodic operators on B is denoted by $\mathcal{L}_A$.
G_T is the orbit of T under the group action $x \to R_x T R_{-x}$.
As preliminary observations, we have

Proposition 2.1 [9] $\mathcal{L}_0, \mathcal{L}_A$ are closed subalgebras of $\mathcal{L}(B)$ and hence are Banach algebras.

We observe that if $T \in \mathcal{L}_0$, then $G_T = \{T\}$ and hence $\mathcal{L}_0 \subseteq \mathcal{L}_A$.

We wish to identify a bounded function φ on G with the multiplication operator M_φ on B, where $\big(M_\varphi f\big)(y) = \varphi(y) f(y)$ for $f \in B, y \in G$; of course, with the condition on φ ensuring $s \in B \Rightarrow \varphi f \in B$. This is clearly satisfied for $B = L^p(G),\ 1 \le p < \infty$. For $B = C_0(G)$, we assume that φ is a bounded continuous function on G since φf has to be continuous for $f \in C_0(G)$.

With this perspective, we place in the record the following lemma.

Lemma 2.2 $M_{R_x \varphi} = R_x M_\varphi R_{-x}$ for $x \in G$.

Proof For $x \in G, y \in G$.

$$\begin{aligned}
\big[\big(M_{R_x\varphi}\big) f\big](y) &= (R_x \varphi f)(y) \\
&= (R_x \varphi)(y) f(y) \\
&= \varphi(y - x) f(y) \\
\big(R_x M_\varphi R_{-x} f\big)(y) &= \big[M_\varphi R_{-x} f\big](y - x) \\
&= \varphi(y - x)(R_{-x} f)(y - x) \\
&= \varphi(y - x) f(y)
\end{aligned}$$

Therefore, $M_{R_x \varphi} = R_x M_\varphi R_{-x}$.

This is the basis for the group action $x \to R_x T R_{-x}$ on $\mathcal{L}(B)$ motivated by the group action $\varphi \to R_x \varphi$ on $C(G)$ and hence for the formation of G_T from G_φ.

It is easy to check that $\|M_\varphi\| = \|\varphi\|_\infty$ in our standard examples for B, further assuming that φ is continuous when $B = C_0(G)$ besides being bounded [9].

Now from $\{R_x M_\varphi R_{-x} : x \in G\} = \{M_{R_x\varphi} : x \in G\}$ and from the above-mentioned isometry, we get that, in our standard examples, M_φ is an Almost periodic operator if and only if φ is an Almost periodic function.

With all this ground work, now we are prepared for the nonstandard analysis of Almost periodic operators.

Analogous to Proposition 1.1, we have the following.

Proposition 2.3 [9] Every $T \in \mathcal{L}_A$ is the norm limit of finite sums of the form $\sum_{i=1}^{n} M_{\gamma_i} U_{\gamma_i}$, each U_{γ_i} being invariant. — (***).

By transfer principle, this extends to $*\mathcal{L}_A$ in a natural way.

Proposition 2.4 If $V \in\ * \mathcal{L}_A$, then V is the extended norm limit of sums of the type (***), where $n \in {}^*\mathbb{N}$ and $\gamma_1, \gamma_2, \ldots, \gamma_n$ are elements of ${}^*\Gamma$. — (****).

Similarly if $V \in \overline{G_T}$, then V is the norm limit of sums of the type (***), where additionally we have that U_{γ_i}'s are obtained as vector-valued integrals of translates of T [9].

Again by Transfer,

Proposition 2.5 If $V \in {}^*\left(\overline{G_T}\right)$, then V is the extended norm limit of sums of the type (****), where additionally we have that U_{γ_i}'s are obtained from certain integrals of the translates of T.

In the internal set-theoretic language, we have the following.

Proposition 2.6 For a standard almost periodic T, every $V \in^s \left(\overline{G_T}\right)$ is the norm limit of finite sums of the type (***), where in this case, the elements $n, \gamma_i, U_{\gamma_i}$ could be nonstandard for nonstandard V. For standard V, the U_{γ_i}'s are obtained from integrals of the translates of T.

Acknowledgements The first author thanks the Department of Science and Technology (DST), India, for its financial support through INSPIRE fellowship.

References

1. Abraham Robinson, Nonstandard Analysis, North Holland Publishing Company, 1966.
2. Alain Robert, Nonstandard Analysis, John Wiley and Sons, 1985.
3. Albert, E.: Hurd and Peter A. An Introduction to Nonstandard Real Analysis, Academic Press, Loeb (1985)
4. Nelson, E.: Internal Set theory, A new approach to NSA. Bull. Amer. Math. Soc. **83**, 1165–1198 (1977)
5. Hewitt, E., Ross, K.A.: Abstract Harmonic Analysis, vol. I. Springer-Verlag, Berlin-Heidelberg-NewYork (1963)
6. Katznelson, Y.: An Introduction to Harmonic Analysis. John Wiley and Sons Inc., New York (1968)

7. Loomis, L.H.: An Introduction to Abstract Harmonic Analysis. Von Nostrand, New York (1953)
8. Ronald G. Douglas, Banach Algebra techniques in Operator Theory, Academic Press, New York, 1972.
9. Tewari, U.B., Somasundaram, S.: Almost Periodicity in Operator Algebras. Rendiconti dell' Istituto di Matemtica dell' Universita di Trieste **24**, 39–54 (1992)
10. Walter Rudin, Fourier Analysis on Groups, Interscience Publishers, No.12, New York, 1962.